Lepidoptera: Yponomeutoidea I (Argyresthiidae
Scythropiidae, and Yponomeutidae)

World Catalogue of Insects

VOLUME 12

The titles published in this series are listed at *brill.com/wci*

Lepidoptera: Yponomeutoidea I (Argyresthiidae, Attevidae, Praydidae, Scythropiidae, and Yponomeutidae)

By

Jon A. Lewis and Jae-Cheon Sohn

Edited by

Bernard Landry

BRILL

LEIDEN | BOSTON

Cover illustration: © Mississippi Entomological Museum.

Library of Congress Cataloging in Publication Control Number: 2006356329

This publication has been typeset in the multilingual "Brill" typeface. With over 5,100 characters covering Latin, IPA, Greek, and Cyrillic, this typeface is especially suitable for use in the humanities.
For more information, please see www.brill.com/brill-typeface.

ISSN 1398-8700
ISBN 978-90-04-25474-9 (hardback)
ISBN 978-90-04-26426-7 (e-book)

This book is printed on acid-free paper.

Contents

Acknowledgements

We would like to express our cordial thanks to Donald Davis (United States National Museum of Natural History, Wash. D.C., USA) [USNM], John Brown (USDA Systematic Entomological Laboratory, Wash. D.C., USA) [USDA-SEL], and Jean-François Landry (Agriculture and Agri-Food Canada, Ottawa, Canada), for their help in preparing this catalogue. The first author (JAL) is especially grateful to Ronald W. Hodges (USDA-SEL, retired) for first stimulating his interest in Yponomeutoidea and encouraging him to compile a catalogue. The second author (JCS) especially thanks Charles Mitter (University of Maryland, College Park, Maryland, USA) for his guidance, encouragement and financial support during JCS's thesis work on the phylogeny of Yponomeutoidea. We appreciate several colleague researchers, including David Adamski (USDA-SEL), Edward Baraniak (Adam Mickiewicz University, Poznań, Poland), Zlata Gershenson (Schmalhausen Institute of Zoology, Kyiv, Ukraine), Peter Huemer (Tiroler Landesmuseen-Betriebsgesellschaft, Innsbruck, Austria), Utsugi Jinbo (National Museum of Nature and Science, Tokyo, Japan), Ole Karsholt (Zoologisk Museum, Denmark), Akito Kawahara (University of Florida, Gainesville), Kim Mitter (University of Maryland), Mamoru Owada (National Museum of Nature and Science, Tokyo, Japan), Gregory R. Pohl (Canadian Forest Service, Northern Forestry Centre, Alberta), Margarita G. Ponomarenko (Far Eastern Branch of Russian Academy of Science, Vladivostok, Russia), Marty Schlabach (Cornell University), Chun-Sheng Wu (Chinese Academy of Science, Beijing, China), and Andreas Zwick (Australian National Insect Collection, Canberra, Australia), for providing us references, useful tips or translations. Richard Bray and Reed Watkins (Smithsonian volunteers) and Eric Metzler (Ohio Biological Survey), all of whom retrieved reference materials from the Thomas library in Wittenberg University, Ohio, where C. Dos Passos left his extremely valuable entomological literature containing many rare publications. Cudos also to Suzanne A. Smailes and the rest of the staff at the Thomas library for allowing access to the rare book collection. Thanks also are due to Daniel Burckhardt for supplying information on the Müller-Rutz collection housed in the Naturhistorisches Museum, Switzerland. We are indebted to several museum curators, incluing Axel Hausmann (Zoological State Collection, Munich, Germany), James Hogan (Oxford University Museum of Natural History, United Kingdom), Lauri Kaila (Finnish Museum of Natural History), Martin Lödl (Museum of Natural History, Vienna, Austria), Catriona McPhee (Museum Victoria, Melbourne), Wolfram Mey (Museum für Naturkunde, Berlin, Germany), László Ronkay (Hungarian Museum of Natural

History, Budapest), Joël Minet (Museum National d'Histoire Naturelle, Paris, France), and Kevin Tuck (Natural History Museum, London, United Kingdom), in locating type specimens and verifying old records. Without the encouragement and assistance of all these individuals and others, completion of this work would not be possible. We thank the following colleagues for reviewing the manuscript and making many helpful suggestions: Bernard Landry (Museum d'Histoire Naturelle, Geneva Switzerland), and the anonymous reviewers.

List of Abbreviations

Specimen Depositories

AMNH American Museum of Natural History, New York, USA.
AMNZ Auckland War Memorial Museum, Auckland, New Zealand.
AMUP Department of Systematic Zoology, A. Mickiewicz University, Poznań, Poland.
ANIC Australian National Insect Collection, CSIRO, Division of Entomology, Canberra, Australia.
ANSP Academy of Natural Sciences, Philadelphia, Pennsylvania, USA.
BMNH The Natural History Museum, London, England.
BPBM Bernice P. Bishop Museum, Honolulu, Hawaii, USA.
BSCM Buryatsky Scientific Centre Museum, Siberian Division of Russian Academy of Science, Buryatsky, Russia.
CMNV Civic Museum of Natural History, Verona, Italy.
CMNZ Canterbury Museum, Christchurch, New Zealand.
CNC Canadian National Collection of Insects, Arachnids, and Nematodes, Ottawa, Ontario, Canada.
CUIC Cornell University Insect Collection, Ithaca, New York, USA.
DEIB Deutsches Entomologisches Institut, Müncheberg (near Berlin), Germany.
DEMV Department of Entomology, Museum of Victoria, Melborne, Australia.
DEUU Department of Entomology, University of Uppsala, Sweden.
DZPU Department of Zoology, Punjab University, Patiala, India.
ECMP Entomological Collection, Bureau of Science, Manila, Philippines.
EIHU Entomological Institute, Hokkaido University, Sapporo, Japan.
ELEU Entomological Laboratory, Ehime University, Matsuyama, Japan.
ELKU Entomology Laboratory, Kyushu University, Fukuoka City, Fukuoka, Japan.
EMEC Essig Museum of Entomology, University of California, Berkeley, USA.
IES Instituto de Biología y Sistemática, Havana, Cuba.
IMCK Indian Museum of Calcutta/Kolkata, India.
INBIO Instituto Nacional de Biodiversidad, Santo Domingo de Heredia, Costa Rica.
IPEG Institut für Pflanzenschutzforschung Eberswalde, Germany.
IRSN Institut royal des Sciences naturelles de Belgique, Brussels, Belgium.
ISEZ Institute of Systematics and Evolution of Animals, Zoological Institute, Polish Academy of Sciences, Krakow, Poland.
IZAS Institute of Zoology, Academia Sinica, Beijing, China.
LMVA Leopold Museum, Vienna, Austria.

LNK Staatliches Museum für Naturkunde, Karlsruhe, Germany. (formerly Landessammlungen für Naturkunde, Karlsruhe).

LSL Linnean Society of London, London, England.

MCAS Museum of the California Academy of Sciences, San Francisco, California, USA.

MCGF McGuire Center for Lepidoptera and Biodiversity, Gainesville, Florida, USA (including the Florida State Collection of Arthropods).

MCSN Museo Civico di Storia Naturale, Milan, Italy

MCZ Museum of Comparative Zoology, Harvard University, Cambridge, Massachusetts, USA.

MFSN Museo Friulano di Storia Naturale, Udine, Italy.

MGAB Muzeul Naţional de Istoria Naturala "Grigore Antipa," Bucharest, Romania.

MNCN Museo Nacional de Ciencias Naturales, Madrid, Spain.

MNHN Muséum National d'Histoire Naturelle, Paris, France.

MNZ New Zealand Arthropod Collection, Landcare Research (formerly Dominion Museum and National Museum (DSIR) of New Zealand), Aukland, New Zealand.

MRAC Musée Royal de l'Afrique Centrale, Tervuren, Belgium.

MUIC Massachusetts University Insect Collection, Amherst, Massachusetts, USA.

MZLU Zoologiska Museet vid Lunds Universitet (Zoological Museum, University of Lund), Lund, Sweden.

NAFU Entomology Institute, College of Plant Protection, Northwestern Agricultural & Forestry University of Sci-Tech, Shaanxi, China.

NASU I. I. Schmalhausen Institute of Zoology, National Academy of Sciences of Ukraine, Kyiv, Ukraine.

NHMS Naturhistorisches Museum, Basel, Switzerland.

NHMW Naturhistorisches Museum Wien, Vienna, Austria.

NHNW National Museum of Namibia, Windhoek, Namibia.

NIAS National Institute of Agricultural Sciences, Tokyo, Japan.

NMPC National Museum (Natural History), Prague, Czech Republic.

NMS Senckenberg Naturmuseum, Frankfurt, Germany.

NRM Swedish Museum of Natural History (Naturhistoriska Riksmuseet), Stockholm, Sweden.

NSU National Sun Yat-sen University, Kaohsiung, Taiwan.

NUIC Nankai University Insect Collection, College of Life Sciences, Tianjin, China.

OLML Oberösterreichisches Landesmuseum, Linz, Austria.

OUM Hope Museum, Oxford University Museum of Natural History, Oxford, England.

RMBB Regional Museum, Banská, Bystrica, Slovakia.

RMNH	Naturalis Biodiversity Center, Leiden, Netherlands (formerly Rijksmuseum).
SAMA	South Australian Museum, Adelaide, South Australia, Australia.
SAMC	South African Museum, Capetown, South Africa.
SASZ	Institute of Forest Ecology, Slovak Academy of Sciences, Zvolen, Slovakia.
SFIT	Swiss Federal Institute of Technology, Zurich, Switzerland.
SNMB	Slovak National Museum, Bratislava, Slovakia.
TLMF	Tiroler Landesmuseum Ferdinandeum, Innsbruck, Austria.
TMB	Termeszettudomanyi Muzeum Allattara (Hungarian Natural History Museum), Budapest, Hungary.
TMP	Transvaal Museum, Pretoria, South Africa.
UMCE	University Museum, Cambridge, England.
UNAM	Universidad Nacional Agraria La Molina, Lima, Peru.
UOP	Entomological Laboratory, University of Osaka Prefecture, Sakai, Japan.
USNM	Smithsonian Museum of Natural History, Wash., D.C., USA.
VOBC	Vitor O. Becker collection, Reserva Serra Bonita, Camacan, Bahia, Brazil.
WAMA	West Australia Museum, Albany, Australia.
WPNM	Westfälischen Provinzialmuseum für Naturkunde, Münster, Germany.
ZMAK	Zoologisches Forschungsinstitut und Museum Alexander König, Bonn, Germany.
ZMAN	Naturalis Biodiversity Center, Leiden (former collections of Zoölogisch Museum Amsterdam) Netherlands.
ZMAS	Zoological Institute, Russian Academy of Sciences, St. Petersburg, Russia.
ZMB	Museum für Naturkunde (formerly Zoologisches Museum), Humboldt University, Berlin, Germany.
ZMH	Finnish Museum of Natural History, University of Helsinki, Finland.
ZMKU	Kiev University Zoological Museum, Kiev, Ukraine.
ZMO	Zoologisk Museum, Oslo, Norway.
ZMUC	Zoological Museum, University of Copenhagen, Denmark.
ZSM	Zoologische Staatssammlung, Munich, Germany.

Literature Abbreviations Used

Abb. Ber. Erg. Schmett.	Abbildungen zur Berichtigung und Ergänzung der Schmetterlingskunde, besonders der Microlepidopterologie als Supplement zu Treitschke's und Hübner's europaeischen Schnmetterlingen, mit erläuterndem. Text.
Abh. Senck. Nat. Ges.	Abhandlungen Senckenbergischen Naturforschenden Gesellschaft.

Acta Zootaxon. Sin.	Acta Zootaxonomica Sinica.
Amat. Papill.	L'Amateur de papillons. Revue de Lépidoptérologie. Travaux et Observations d'Amateurs, Paris.
Ann. Ent. Soc. Amer.	Annals of the Entomological Society of America.
Ann. Hist.-Nat. Mus. Nation. Hung.	Annales Historico-Naturales Musei Nationalis Hungarici.
Ann. Mag. Nat. Hist.	Annals and Magazine of Natural History.
Ann. Mus. Nac. Hist. Nat. Buenos Aires	Anales del Museo Nacional de Historia Natural Bernardino Rivadavia, Buenos Aires (Entomologia).
Ann. Nat. Hofmus. Wien	Annalen des K. K. Naturhistorischen Hofmuseums, Wien.
Ann. Naturhist. Mus. Wien	Annalen des Naturhistorischen Museums in Wien.
Ann. S. Afr. Mus.	Annals of the South African Museum.
Ann. Soc. Ent. Fr.	Annales de la Société entomologique de France.
Ann. Soc. Linn. Lyon	Annales de la Société Linnéenne de Lyon.
Ann. Transv. Mus.	Annals of the Transvaal Museum.
An. Soc. Cient. Argent.	Anales de la Sociedad Científica Argentina.
Arb. Morph. Taxon. Ent. Berl.	Arbeiten über morphologische und taxonomische Entomologie aus Berlin-Dahlem.
Archiv Naturgesch. (A)	Archiv für Naturgeschichte (Abteilung A).
Archiv Naturgesch. (N.F.)	Archiv für Naturgeschichte Zeitschrift für Systematische Zoologie (Neue Folge).
Attempt Syst. Cat. Br. Tin. Pter.	An Attempt at a systematic catalogue of the British Tineidae & Pterophoridae.
Atti Accad. Pont. Roma	Atti dell'Accademia Pontificia dei Nuovi Lincei (Roma).
Atti R. Ist. Sci. Nat. Napoli	Atti del Reale Istituto d'Incoraggiamento alle Scienze Naturali di Napoli.
Atti Soc. Ital. Milano	Atti della Società Italiana di Scienze Naturali e del Museo Civico di storia naturale in Milano.
Atti Soc. Ital. Sci. Nat.	Estratto dagli Atti della Società Italiana di Scienze Naturali.
Beitr. Gesch. Scselhmett.	Beiträge zur Geschichte der Schmetterlinge.
Beitr. Landesk. Osterr. Enns.	Beiträge zur Landeskunde Oesterreich's unter der Enns.

Beitr. zur Ent.	Beiträge zur Entomologie.
Ber. Lepidopt. Tausch.	Berichte des Lepidopterologischen Tauschvereins in Jena.
Berlin. Ent. Ztschr.	Berliner Entomologische Zeitschrift.
Biol. Centr. Amer.	Biologia Centrali-Americana, Insecta, Lepidoptera-Heterocera.
Biol. sbornik	Biological sbornik Blagoveshchensk Pedagogical Institute, Blagoveshchensk.
Bol. Serv. Plagas Forestales	Boletin del Servicio de Plagas Forestales.
Boll. Ist. Ent. Univ. Bologna	Bollettino dell' Istituto di Entomologia, Universita degli Studia di Bologna.
Boll. Lab. Ent. R. Ist. sup. agr. Bologna	Bollettino del Laboratorio di Entomologia del R. Istituto Superiore Agrario di Bologna.
Boll. Lab. Zool. Agr. Portici	Bollettino del Laboratorio di Zoologia Generale e Agraria della R. Scuola Superiore d'Agricoltura in Portici.
Bresl. Zeitschr. Ent.	Zeitschrift für Entomologie. Breslau.
Bull. Br. Mus. Nat. Hist. (Ent)	Bulletin of the British Museum of Natural History (Entomology).
Bull. Buff. Soc. N.S.	Bulletin of the Buffalo Society of Natural Sciences.
Bull. Hill Mus.	Bulletin of the Hill Museum.
Bull. Illinois Lab. Nat. Hist.	Bulletin of the Illinois State Laboratory of Natural History.
Bull. Inst. Roy. Sci. nat. Belg.	Bulletin de l'Institut royal des Sciences naturelles de Belgique.
Bull. Soc. Ent. Fr.	Bulletin de la Societé entomologique de France.
Bull. Soc. Ent. Mulh.	Bulletin de la Société entomologique de Mulhouse.
Bull. Soc. Imp. Nat. Mosc.	Bulletin de la Société impériale des naturalistes de Moscou.
Bull. Soc. Sci. Nat. Maroc	Bulletin de la Société des sciences naturelles du Maroc, Rabat.
Bull. Univ. Osaka Pref. (B).	Bulletin of the University of Osaka Prefecture, Series B.
Bull. U.S. Geol. Surv.	Bulletin of the United States Geological Survey.
Bull. U.S. Nat. Mus.	Bulletin of the United States National Museum.

Can. Ent.	The Canadian Entomologist.
Caroli Linn. Ent.	Caroli Linnaei Entomologia, Faunae Suecicae descriptionibus aucta. 2.
Cat. Brit. Tin. & Pteroph.	Catalogue of the British Tineidae and Pterophoridae.
Cat. East. Aust. Lepid. Het. Oxford. Univ. Mus.	Catalogue of Eastern and Australian Lepidoptera Heterocera in the collection of the Oxford University Museum.
Cat. Indian Insects	Catalogue of Indian Insects.
Cat. Lep. Eur. Faun.	Catalog der Lepidopteren des Europaeisc-hen Faunengebietes II Microlepidoptera. Dresden.
Cat. Lep. Ins. E. Ind. Co.	Catalogue of Lepidopterous Insects in the East-India Company Museum.
Cat. Lep. Russ.	Catalogue of the Lepidoptera of Russia.
Cat. Sis. Sin. Lep.	Catalogo Sistematico y Sinonimico de los Lepidopteros de la Peninsula Iberica y Baleares.
Cat. Type Specimens Microlepid. Br. Mus. Nat. Hist. Descr. E. Meyrick	Catalogue of the Type Specimens of Microlepidoptera in the British Museum (Natural History) described by Edward Meyrick.
Chars. Undesc.	Characters of Undescribed Lepidoptera Heterocera.
Check List Brit. Ins.	A Check List of British Insects.
Check List Lep. Jap.	Check List of the Lepidoptera of Japan.
Checklist Microlep. Kor.	A Checklist of the Microlepidoptera in Korea.
Checklist S. Afr. Lepid. Het.	Checklist of the South African Lepidoptera Heterocera. Transvaal Museum, Pretoria.
Cinc. Quart. Journ. Sci.	Cincinnati Quarterly Journal of Science.
Corrisp. Zool.	Corrispondenza Zoologica Destinata a Diffondere nel Regno delle due Sicilie tutto cio che si va Discuoprendo Entro e Fuori Europa (e vice versa) Risguardante la Zoologia in Generale, Azzolino e Compagno, Naples.
Corr.-Blatt zool.-min. Ver. Regensb.	Correspondez-Blatt des Zoologisch-mineralogischen Vereins in Rogenburg.

Descr. Lep. Atk.	Descriptions of New Indian Lepidopterous Insects from the collection of the late Mr. W.S. Atkinson, M.A., F.L.S., &c., Rhopalocera, Heterocera.
Die Schmett. Philipp. Inseln.	Reisen im Archipel der Philippien von Dr C. Semper. Zweiter Theil. Wissenschaftliche Resultate. Sechste Band. Die Nachtfalter. Heterocera.
Diss. Mus. Nat. Acad. Upsal.	D. [Docente] D. [Deo] Museum naturalium Academiae Upsaliensis.
Dokl. Akad. Nauk Arm. SSR	Doklady Akademii Nauk Armyanskoi SSR.
Dt. Ent. Z. Iris	Deutsche Entomologische Zeitschrift "Iris" herausgegeben vom Entomologischen Verein Iris zu Dresden.
Encycl. Hist. Nat. (Papillons nocturnes)	Encyclopédie d'histoire naturelle ou traité complet de cette science d'après les travaux des naturalistes les plus éminents de tous les pays et de toutes les époques, Buffon, Daubenton, Lacépède, G. Cuvier, Geoffroy St. Hilaire, Latreille, de Jussieu, Brogniart etc. Ouvrage résumant les observations des auteurs anciens et comprenant toutes les découvertes modernes jusqu'à nos jours.
Ent. Ann.	Entomologist's Annual.
Ent. Ber. (Amsterdam)	Entomologishe Berichten (Amsterdam).
Ent. Carn.	Entomologia Carniolica exhibens insecta Carnioliae indigena et distributa in ordines, genera, species, varietates, methodo Linneana.
Ent. Fennica	Entomologica Fennica.
Ent. Gaz.	Entomologist's Gazette.
Ent. Mag.	The Entomologist's Magazine.
Ent. M. Mag.	The Entomologist's Monthly Magazine.
Ent. News	Entomological News.
Ent. Obozr.	Entomologicheskoye Obozreniye.
Ent. Rec. J. Var.	Entomologist's Record and Journal of Variation.
Ent. Scand.	Entomologica Scandinavica.
Ent. Syst. Suppl.	Entomologia systematica emendata et aucta, secundum classes, ordines, genera, species, adjectis synonymis, locis, observationibus, descriptionibus. Supplement.

Ent. Tidskr.	Entomologisk Tidskrift.
Ent. Weekly Intel.	The Entomologist's Weekly Intelligencer.
Entom. Beyt. zu des Ritter	Entomologische Beyträge zu des Ritter Linne' zwölften Asgabe des Natursystems.
Entomol. Gall.	Entomologica gallica.
Enum. Lep. Hal.	Enumeratio Lepidopterorum Haliciae Orientalis.
Enumer. insect. Museo G. J. Billberg	Enumeratio Insectorum in Museo G.J. Billberg.
Esperiana Mem.	Esperiana Memoir.
Eur. Microlep. Ind. Meth.	Europaeorum Microlepidopterorum index methodicus, sive Pyrales, Tortrices, Tineae et Alucitae Linnaei, secundum novum naturalemque ordinem dispositae, nominibus genuinis restitutis, synonymia accurate elucidata, locis indicatis, novisque speciebus aut larvis brevidescriptis. Pars 1 sistens Tortrices, Phycidas, Crambidas, Tinearumque initium.
Exot. Micr.	Exotic Microlepidoptera.
Fau. Reg. Napoli. Lepid.	Fauna del Regno de Napoli. – Lepidotteri. Napoli, Luglio.
Fauna Jap. Ypon. s. l.	Fauna Japonica Yponomeutidae s. lat. (Insecta: Lepidoptera), Keigaku Publ. Co., Tokyo.
Fauna Suecica	Fauna Suecica (Sistens Animalia Sueciae Regni: Mammalia, Aves, Amphibia, Pisces, Insecta, Vermes. Distributa per Classes & Ordines, Genera & Species, cum Differentiis Specierum, Synonymis Auctorum, Nominibus Incolarum, Locis Natalium, Descriptionibus Insectorum.). Stockholm.
Flanzenf. Ins.	Die Pflanzenfeinde aus der Klasse der Insekten. Ein nach Pflanzenfamilien geordnetes Handbuch sämmtlicher auf den einheimischen Pflanzen bisher beobachteten Insekten zum Gebrauch für Entomologen, Insektensammler, Botaniker, Land- und Forstwirthe und Gartenfreunde.
Forst.-Ins.	Die Forstinseckten.
Gen. Ins.	Genera Insectorum.
Gen. Spec. Ins.	Genera et species insectorum e generosissimi auctoris scriptis extraxit, digessit, Latine quoad

	partem reddidit, et terminologiam insectorum Linneanam addidit.
Gesch. Europ. Schmett.	Geschichte Europäischer Schmetterlinge. Augsburg.
Gorta. Botan. Zool.	Gortania-Atti del Museo Friulano di Storia Naturale Botanica Zoologia.
Hist. Nat. Crust. Ins.	Histoire naturelle, générale et particulière des Crustacés et des Insectes.
Hist. Nat. Lép. France	Histoire naturelle des lépidoptères ou papillons de France.
Hor. Soc. Ent. Ross.	Horae Societatis Entomologicae Rossicae.
Icon. Ins. Rar.	Icones Insectorum rariorum cum nominibus eorum trivialibus locisque e C. Linnaei systema naturae allegatis.
Icon. Règne Anim. Cuvier	Iconographie du Règne Animal de G. Cuvier.
Ill. Brit. Ent.	Illustrations of British Entomology.
Ill. Brit. Ent. Haust.	Illustrations of British Entomology or, a Synopsis of Indigenous Insects: containing their generic and specific distinctions. Haustellata.
Ill. Het. Coll. Brit. Mus.	Illustrations of Typical Specimens of Lepidoptera Heterocera in the collection of the British Museum.
Ill. Zyg-Bomb. N-Am.	Illustrations of the Zygaenidae and Bombycidae of North America.
Introd. Mod. Classif. Insects 2 (Synopsis Genera Br. Insects)	An Introduction to the Modern Classification of Insects, in Synopsis of the Genera of British Insects.
Isis Oken	Isis von Oken.
J. Ins. Sci.	Journal of Insect Science.
J. Linn. Soc. Lond.	Journal of the Linnean Society of London, Zoology.
J. Nat. Conserv.	Journal of Nature Conservation.
J. Nat. Hist.	Journal of Natural History.
J. Cincinn. Soc.	Journal of the Cincinnati Society of Natural History.
Jahr. Nass. Ver. Natur.	Jahrbücher des Nassauischen Vereins für Naturkunde (Wiesbaden).
Jour. Bomb. Nat. Hist. Soc.	Journal of the Bombay Natural History Society.
Jour. N.Y. Ent. Soc.	Journal of the New York Entomological Society.
K. Svenska. Vetensk. Akad. Handl.	Kongliga Svenska Vetenskapsakademiens Handlingar. Uppsala & Stockholm.
Kon. Ned. Akad. Weten. Verh. Afdel. Nat., Tweede Reeks	Koninklijke Nederlandse Akademie van Wetenschappen Verhandelingel Afdeling Natuurkunde, Tweede Reeks.

Kungl. Vetensk. Akad. Nya Handl.	Kungliga Vetenskaps-Academiens Nya Handlingar.
K. Svenska Fregatten Eugenies Resa	Kongliga Svenska Fregatten Eugenies Resa Omkring Jorden.
K.V. Akad. Handl.	Kongliga Vetenskaps-Akademiens Handlingar 1852.
Lep. Brit.	Lepidoptera Britannica.
Lep. Cat.	Lepidopterorum Catalogus, Berlin.
Lep. Ceyl.	The Lepidoptera of Ceylon.
Lep. Schweiz	Lepidopteren-Fauna der Schweiz.
Lep. Volgo-Ural.	Fauna Lepidopterologica Volgo-Uralensis.
Lepid. Euro. Cat. Method.	Lepidopterorum Europaeorum Catalogus Methodicus
Lepid. N.Y. & Neigh. States	Lepidoptera of New York and Neighboring States.
Lepid. Novae	Lepidoptera Novae.
Lepid. Südarabien u. Insel Sokótra	Lepidopteren aus Südarabien und von der Insel Sokótra.
Linn. Ent.	Linnea Entomologica.
List Chin. Ins.	List of Chinese Insects. Sun Yat-sen University Press, Guangzhou.
List Lepid. Ins. BM	List of the specimens of Lepidopterous Insects in the Collection of the British Museum.
List Lepid. Maine, Pt. 2	A List of the Lepidoptera of Maine, Part 2. Department of Entomology, Orono.
Mant. Ins.	Mantissa insectorum sistens eorum species nuper detectas adjectis characteribus genericis, differentiis specificis, emendationibus, observationibus.
Mat. Microlep.-Fauna Chin. Prov. Kiangsu, Chekiang, Hunan	Materialien zu einer Microlepidopteren Fauna der Chinesischen Provinzen Kiangsu, Chekiang, und Hunan. Berlin.
Mem. Dep. Agric. India (Ent.)	Memoirs of the Department of Agriculture in India. Entomological Series.
Mém. Anim. 1	Mémoires sur les animaux sans vertébres 1.
Mém. Mus. H. n. Belg., hors ser.	Mémoires du Musée Royal d'Histoire naturelle de Belgique, hors série.

Mém. Hist. Nat. Prov.	Mémoires pour servir à l'histoire naturelle de la Provence.
Mém. Inst. Sci. Madag.	Mémoires de l'Institut scientifique de Madagascar. Séries E.
Mém. Soc. Emul. Doubs	Mémoires de la Société d'émulation du Doubs.
Memor. Soc. Fau. Flo. Fenn.	Memoranda Societatis pro Fauna et Flora Fennica.
Microlep. Spec. Nov.	Microlepidopterorum species novae – Cracoviae.
Mitt. Ent. Zurich	Mitteilungen der Entomologia Zürich und Umgebung.
Mitt. Münch. Ent. Ges.	Mitteilungen der Münchner Entomologischen Gesellschaft.
Mitt. Mus. Nat. Berl., Dtsch. entomol. Z.	Mitteilungen aus dem Museum für Naturkunde in Berlin, Deutsche Entomologische Zeitschrift.
Mitt. Schweiz. Ent. Ges.	Mitteilungen der Schweizerischen Entomologischen Gesellschaft.
Mus. Nat. Acad. Upsal	D.[Docente] D.[Deo] Museum Naturalium Academiæ Upsaliensis.
Nach. der Bayer. Ent.	Nachrichtenblatt der Bayerischen Entomologen.
Nat. Hist. Brit. Ins.	The Natural History of British Insects.
Nat. Hist. Tin.	The Natural History of the Tineina.
Naturaliste	Le Naturaliste.
Not. Sällsk. Faun. Fenn. Förh.	Notiser ur Sällskapets pro Fauna et Flora Fennica Förhandlingar.
Neu. Beitr. Schmett.	Neuere Beiträge Schmetterlingskunde mit Abbildungen nach der Natur.
Neue Schmett.	Neue Schmetterlinge aus Europa und den angrenzenden Ländern.
Nota lepid.	Nota lepidopterologica.
Nouv. Revue Ent. (N.S.)	Nouvelle Revue d'Entomologie, Nouvelle Série.
Nyt Mag. Nat.	Nyt Magazin f. Naturvidenskab.
N.Z.J. Zool.	New Zealand Journal of Zoology.
Obs. Nat.	Observer of Nature.
Ochr. Rost.	Ochrana Rostlin (Prague).
Opusc. Ent.	Opuscula Entomologica.

Öst Akad. Wiss. Abt.	Österreichische Akademie der Wissenschaften Mathematisch-Naturwissenschaftliche Klasse Abhandlungen.
Pacif. Sci.	Pacific Science.
Pap. Exot.	Papillons Exotiques des trois parties du monde l'Asie, l'Afrique et l'Amérique rassemblés et décrits par Pierre Cramer.
Pap. Proc. R. Soc. Tasm.	Royal Society of Tasmania: Papers and Proceedings.
Pet. Nouv. Ent.	Petites nouvelles entomologiques.
Pol. J. Ent.	Polish Journal of Entomology (=Polskie Pismo Entomologiczne).
Philipp. Jour. Sci., Ser. A	Philippine Journal of Science, Series A.
Préc. Caract. gén. Ins.	Précis des Caractères génériques des Insectes, disposés dans un ordre naturel par le Citoyen Latreille.
Proc. Acad. Nat. Sci. Phil.	Proceedings of the Academy of Natural Sciences of Philadelphia.
Proc. Ent. Soc. Philad.	Proceedings of the Entomological Society of Philadelphia.
Proc. Ent. Soc. Wash.	Proceedings of the Entomological Society of Washington.
Proc. Integ. Man. Dyn. For. Defol. Ins.	Proceedings: integrated management and dynamics of forest defoliating insects 1999 August 15–19. Victoria, BC.
Proc. K. Ned. Akad. Wet. C.	Proceedings Koninklijke Nederlandse Akademie van Wetenschappen. Series C. Biological and Medical Sciences.
Proc. Linn. Soc. N.S. Wales	Proceedings of the Linnean Society of New South Wales.
Proc. R. Soc. Qd.	Proceedings of the Royal Society of Queensland.
Proc. R. Soc. Tas.	Royal Society of Tasmania: Papers and Proceedings.
Proc. USNM.	Proceedings of the United States National Museum.
Proc. Zool. Soc. Lond.	Proceedings of the Zoological Society of London.
Prod. Lepidopt. Brit.	Prodromus Lepidopterorum Britannicorum.
Rec. Auckland Inst. Mus.	Records of the Auckland Institute Museum.

Rec. Dominion Mus.	Records of the Dominion Museum.
Rec. Ind. Mus.	Records of the Indian Museum.
Reichenbachia	Reichenbachia, Staatliches Museum für Tierkunde Dresden.
Reise öst. Fregatte Novara (Zool.)	Reise der Österreichischen Fregatte Novara um die Erde (Zoologischer Theil).
Rep. Brit. Ornith. Uni. Exped.	Report of the British Ornithologists' Union Expedition.
Rev. Bras. Ent.	Revista Brasileira de Entomologia.
Rev. Franc. Lep.	Revue française de lépidoptérologie.
Rev. russe Ent.	Revue russe d'entomologie.
Revue Fr. Ent. (N.S.)	Revue française d'Entomologie (Nouvelle série).
Rudolstädter Nat. Schr.	Rudolstädter Naturhistorische Schriften.
Ruwenzori Exped. 1952	Ruwenzori Expedition 1952.
Samml. aussereurop. Schmett.	Sammlung neuer oder wenig bekannter aussereuropäischer Schmetterlinge.
Samml. Eur. Schmett. Tin.	Sammlung Europäischer Schmetterlinge. Tineidae.
Samml. Nat. Phys. Aufs.	Sammlung naturhistorischer und physikalischer Aufsätze. Nürnberg, Raspe.
Samml. Vögel u. Schmett.	Sammlung auserlesener Vögel und Schmetterlinge.
Schmett. Deutsch.	Die Schmetterlinge Deutschlands und der Schweiz. Abt. II. Kleinschmetterlinge, Bd. 2, Die Motten und Federmotten.
Schmett. Eur.	Die Schmetterlinge Europas.
Schmett. Europa	Die Schmetterlinge von Europa.
Schmett. Wien	Ankündigung eines Systematischen Werkes von den Schmetterlingen der Wienergegend Wien.
SHILAP Revta. lepid.	SHILAP Revista de Lepidopterologia.
Smith. Contr. Zool.	Smithsonian Contributions to Zoology.
Smith's List Lep. Bor. Am.	List of the Lepidoptera of Boreal America by John B. Smith.
Soc. Scient. Fenn., Comm. Biol.	Societas Scientiárum Fennica. Commentationes Biologicae.
Sond. Abhand. West. Prov. Nat. 1	Sonderabdruck aus den "Abhandlungen aus dem Westfälischen Provinzialmuseum für Naturkunde" 1. Die sogenannten "Kleinschmetterlinge" Westfalens.
Stett. Ent. Zeit.	Stettiner Entomologische Zeitung.
Suppl. Cat. Brit. Tin. & Pteroph.	A Supplementary Catalogue of the British Tineidae & Pterophoridae.
Suppl. Entom. Sist.	Supplementum Entomologiae Sistematicae.

Syn. Cat. Lepid. Het.	A Synonymic Catalogue of Lepidoptera Heterocera.
Sys. Ent.	Systema Entomologicae.
Syst.-alph. Verz. Europ. Schmett.	Systematisch alphabetisches Verzeichniss aller bisher bei den Fürbildungen zur Sammlung Europäischer Schmetterlinge, ausgegebenen Benennungen mit Vormerkung auch Augsburgscher Gattungen. Augsburg, Verfasser, 1822.
Syst. Bearb. Schmett. Eur.	Systematische Bearbeitung der Schmetterlinge von Europa.
Syst. Cat. Brit. Ins.	Systematic Catalogue of British Insects.
Syst. Cat. Brit. Tin. & Pteroph.	Systematic Catalogue of the British Tineidae & Pterophoridae.
Syst. Nat.	Systema Naturae.
Tentamen	Tentamen determinationis digestionis atque denominationis singularum stirpium Lepidopterorum, peritis ad inspiciendum et dijudicandum communicatum a Jacobo Hübner.
Tijdschr. v. Ent.	Tijdschrift voor Entomologie.
Tin. Pter. Schweiz	Die Tineen und Pterophoren der Schweiz.
Trans. Am. Ent. Soc.	Transactions of the American Entomological Society.
Trans. Am. Ent. Soc. Phila.	Transactions of the American Entomological Society, and Proceedings of the Entomological Section of the Academy of Natural Sciences.
Trans. Ent. Soc. Lond.	Transactions of the Entomological Society of London.
Trans. Lep. Soc. Japan	Transactions of the Lepidopterological Society of Japan (Tyō To Ga).
Trans. Linn. Soc. Lond.	Transactions of the Linnaean Society of London.
Trans. N.Y. State Agr. Soc.	Transactions of the New York State Agricultural Society.
Trans. N.Z. Inst.	Transactions of the New Zealand Institute.
Trans. Proc. N.Z. Inst.	Transactions and Proceedings of the New Zealand Institute.
Trans. R. Soc. S. Austr.	Transactions of the Royal Society of South Australia.
Trans. Roy. Ent. Soc. Lond.	Transactions of the Royal Entomological Society of London.
Trav. N.Z.	Travels in New Zealand.
Trop. Lepid.	Tropical Lepidoptera.
Trudy vses. ent. Obshch.	Trudy Vsesoyuznogo Entomologicheskogo Obshchestva (St. Petersburg).
Trudy Burya. Inst. Estest. Nauk	Trudy Buryatskogo Instituta Estestevennykh Nauk.

USNM Bull.	U.S. National Museum Bulletin.
Verh. K. ned. Akad. Wet.	Verhandelingen der Koninklijke Nederlandse Akademie van Wetenschappen, afd. Natuurkunde.
Verh. Natur. Vereins Karlsruhe	Verhandlungen des Naturwissenschaftlichen Vereins in Karlsruhe.
Verh. Zool.-Bot. Ges. Wien	Verhandlungen der kaiserlich-königlichen Zoologisch-Botanischen Gesellschaft in Wien.
Veröff. Tirol. Landes. Ferdin.	Veröffentlichungen des Tiroler Landesmuseum Ferdinandeum Ferdinandeum im Alpenraum.
Verz. bekannt. Schmett.	Verzeichniss bekannter Schmettlinge [sic].
Voll. Natur. der Sch. Forst.	Vollständige Naturgeschichte der Schädlichen Forstinsecten.
Wien. Ent. Monatsschr.	Lepidopterologische Fragmente V.; in Wiener Entomologische Monatsschrift.
Zool. Meded. Leiden	Zoologische Mededeelingen.
Zool. Rec.	Zoological Record.
Zool. Zhur.	Zoologischeskii Zhurnal.
Zoologist	The Zoologist, London.
Ztschr. Wien. Entomol. Ges.	Zeitschrift des Wiener Entomologischen Gesellschaft.
Ztschr. Wien. Entomol. Ver.	Zeitschrift des Wiener Entomologen-Vereines.
Zuträge Samml. Exot. Schmett.	Zuträge zur Sammlung exotischer Schmetterlinge: bestehend in Bekundigung einzelner Fliegmuster neuer oder rarer nichteuropäischer Gattungen. 1825. 3.
1st Ann. Rpt. Mass.	First Annual Report on the Injurious and Beneficial Insects of Massachusetts.
6000 Ill. Ins. Jap.	6000 Illustrated Insects of the Japan-Empire.

Other Abbreviations

auct.	auctorum (of authors)
Co.	County
comb. nov.	combinatio nova (new combination)
CSIRO	Commonwealth Scientific and Industrial Research Organization, Canberra, Australia.
Dept	Department
Dist.	district
f.	forma (form)

fig.	figure
figs.	figures
ICZN	International Code of Zoological Nomenclature.
in litt.	in litteris (in correspondence or "manuscript form")
Isl.	Island
Mt.	Mountain
pers. comm.	personal communication
pl.	plate
pls.	plates
Pref.	Prefecture
Prov.	Province
ssp.	subspecies
Supp.	Supplement
syn. nov.	synonymum novum (new synonym)
Terr.	Territory
USA	United States of America
var.	variety
Vol.	volume
Wash., D.C.	Washington, District of Columbia

Introduction

The superfamily Yponomeutoidea, including Yponomeutidae, are one of the basal groups of ditrysian microlepidoptera. The systematic definition of the superfamily had varied until Kyrki (1984, 1990). Recent molecular studies (e.g. Mutanen et al. 2010; Sohn et al. 2013) have confirmed Kyrki's proposal for the taxonomic boundaries of Yponomeutoidea. Kyrki (1984) established the monophyly for Yponomeutoidea on the basis of a single synapomorphy, viz. the presence of pleural lobes on the 8th abdominal segment in males. Later, he added another possible synapomorphy, a transverse ridge on the 2nd abdominal sternite (Kyrki, 1990). In spite of the recent progress toward resolution of yponomeutoid monophyly, there is yet no consensus for the phylogenetic relationships of the subordinate groups within Yponomeutoidea.

Yponomeutidae is the largest family of Yponomeutoidea although it varies in the estimation of the number of included species, depending on the definition of the family. Kyrki (1990) placed six subfamilies within Yponomeutidae: Yponomeutinae, Saridoscelinae, Scythropiinae, Attevinae, Praydinae, and Argyresthiinae. He proposed five autapomorphies to unite these heterogeneous groups under Yponomeutidae: i.e. ocelli lost; pterostigma present from Sc to R; 8th abdominal sternite strongly sclerotized; claws of larval thoracic legs long and fairly straight; crochets of abdominal prolegs bi- or multiserial. These charactristics were however questioned by Dugdale et al. (1998). Recently, van Nieukerken et al. (2011) revised Kyrki's proposal with the results from molecular studies (e.g. Mutanen et al. 2010; Cho et al. 2011) and included only the first three subfamilies, i.e. Yponomeutinae, Saridoscelinae and Scythropiinae, within Yponomeutidae and separated Praydidae, Attevidae and Argyresthiidae as three separate families. Moreover, Sohn et al. (2013) proposed a separate family status for *Scythropia* Hübner, hence Scythropiidae. In sum, only two subfamilies, Yponomeutinae and Saridoscelinae, remain within Yponomeutidae according to the latest classification. These changes necessitate recalculating the known species diversity of Yponomeutidae and result in 288 described species worldwide assigned to Yponomeutidae; 163 species to Argyresthiidae; 53 species to Attevidae; 51 species to Praydidae; and one to Scythropiidae (updated from van Nieukerken et al. 2011). When starting this catalogue, we compiled the species using Kyrki's 1990 system. Although there is no phylogenetic support to unite the families included in our catalogue, we retained our original compilation of the species for convenience, but rearranged it according to the latest classification system.

 | DOI 10.1163/9789004264267_002

The five yponomeutoid families treated in this catalogue are notable for including a number of pest species. For example, the small ermine moths (*Yponomeuta padella* complex), belonging to Yponomeutidae, often cause complete defoliation of trees in Northern Europe (e.g. Leather, 1986; Alonso et al. 2000) and sometimes become a serious threat to the entire ecosystem (OSU, 2010, *electronic source*). The Apple Ermine Moth (*Yponomeuta malinellus*), an occasional pest to the apple industry, has been accientally introduced into the USA (Unruh et al. 1993). Argyresthiidae includes a few serious forest or orchard pests (Carter, 1984) such as the Apple Fruit Miner (*Argyresthia conjugella* Zeller), the Cherry Fruit Moth (*Argyresthia pruniella* (Clerck)), the Larch Shoot Moth (*Argyresthia laevigatella* (Herrich-Schäffer)) and the Juniper Ermine Moth (*Argyresthia trifasciata* Staudinger), originally from the Alps, and now spreading into other parts of Europe. Two species of Praydidae, *Prays citri* (Citrus Flower Moth) and *Prays oleae* (Olive Kernel Borer), are economically important pests in Mediterranean countries (Carter, 1984). Attevidae includes several showy micromoths that are rarely considered as serious pests, but one species, *Atteva fabriciella* (Swederus) (Ailanthus Webworm Moth), has been rigorously studied to minimize damage on a medicinal plant, *Ailanthus triphysa* (Dennst.) Alston (Simaroubaceae), in India (Verma, 1992).

There have been only a few catalogues to compile the world species of Yponomeutidae. However, most of those are outdated (e.g. Meyrick, 1914c) or the geographic and/or taxonomic coverage was incomplete (e.g. Friese, 1960; Gershenson & Ulenberg, 1998). We provide the first catalogue of all the known species belonging to the five Yponomeutoidea families once conforming to Yponomeutidae sensu Kyrki (1990), i.e. Argyresthiidae, Attevidae, Praydidae, Scythropiidae and Yponomeutidae, in this volume, and we plan to treat all other families in future works.

Classification

Despite its importance in tracing the early evolution of Lepidoptera, and the economic losses caused by its species on crops and trees, superfamily Yponomeutoidea has been neglected by systematists and its biodiversity and phylogeny remain poorly understood. Especially problematic is the lack of a robust phylogeny, including a synapomorphy-based definition for the superfamily itself. Until the early 20th century, the taxa currently placed in Yponomeutoidea comprised scattered suprageneric groups of Tineina or Tineae, two collective microlepidopteran group names no longer in use (e.g. Zeller, 1839; Bruand, 1849 [1850]; Stainton, 1854; Meyrick, 1928e), or

Tineidae (e.g. Staudinger & Rebel, 1901; Handlirsch, 1925). Although Stephens (1829) had already distinguished them from other microlepidopteran groups, it was Fracker (1915) who first erected a superfamily for Yponomeutoidea. Shortly later, Mosher (1916) examined lepidopteran pupae and supported the separate superfamily status of Yponomeutoidea. However, as these propositions lacked unambiguous defining characters, the group remained highly heterogeneous and included many genera that now belong to other superfamilies. A succession of subsequent authors advanced increasingly restrictive re-definitions of Yponomeutoidea (e.g. Meyrick, 1928d; Börner, 1939; Friese, 1960; Common, 1970; Brock, 1971; Heppner, 1977; Kuznetzov & Stekolnikov, 1977), but failed to achieve a stable classification because they lacked explicit analyses of phylogeny. Kyrki (1984, 1990), in the first cladistic study, significantly modernized the Yponomeutoidea, in which he included only eight families: Acrolepiidae, Bedelliidae, Glyphipterigidae, Heliodinidae, Lyonetiidae, Plutellidae, Yponomeutidae, and Ypsolophidae. However, the lack of robustness of Kyrki's phylogeny hindered acceptance of his classification, leaving other hypotheses, such as those of Moriuti (1977b) and Heppner (1998), still in contention. Recent molecular studies (Mutanen et al. 2010; Sohn et al. 2013) have resulted in fairly good supports for the systematic boundaries of Yponomeutoidea as defined by Kyrki, although differing from his hypothesis in the inter-relationships of the included family and subfamily groups. A significant discrepancy can be found from what constitutes Yponomeutidae.

Yponomeutidae have long been confused with Plutellidae and the two groups have been repeatedly separated and merged (see Friese, 1960, for a detailed history). The delimitation of Yponomeutidae continues to be debated, due to the lack of robust synapomorphies to define the group. Disagreement on systematic status is especially pronounced regarding four of the six subfamilies which Kyrki (1990) included under Yponomeutidae, as follows: (i) Kyrki's Argyresthiinae is treated as a separate family by several other researchers, such as Moriuti (1977) and Heppner (1998); (ii) Praydinae have been argued to be affiliated with Plutellidae rather than Yponomeutidae (Pierce and Metcalfe, 1935; Friese, 1960; Heppner, 2005); (iii) Kyrki's (1990) postulate that the monogeneric Scythropiinae are the sister group of all other yponomeutids is undermined by conflicting traits shared instead with the *Prays* group; and (iv) Attevinae as defined by Kyrki (1984, 1990), for which close relatives are hard to discern, were argued by Heppner (1998) to require separate family status. Sohn et al. (2013) revised all previously proposed classification systems for Yponomeutidae and found that Yponomeutidae sensu Kyrki (1990) needs to be divided into five separate families. This new classification has been adopted for this catalogue. The revised definitions of the families are given as follows:

Argyresthiidae: An association of *Argyresthia*-group with Yponomeutidae has been disputed in the recent molecular phylogenetic studies (Mutanen et al. 2010; Sohn et al. 2013). From this, the group now constitutes its own family, Argyresthiidae. Monophyly for Argyresthiidae has been recovered with very strong support (Sohn et al. 2013). The family had been thought to be monobasic, defined by unique features of the male genitalia including a laterally produced vinculum and sensilla ornaments on the socii (Dugdale et al. 1998) but a novel lineage (Sohn et al. 2013), lately found in the Neotropical Region, is possibly related with *Argyresthia* and necessitates reconsideration of the argyresthiid autapomorphies.

Attevidae: This unique moth group has been associated with Yponomeutidae in superficial appearance of adults and also sociality of larvae. Heppner (1998) disputed its association with Yponomeutidae. This hypothesis has been supported by Sohn et al. (2013), based on molecular data. The Attevidae can be defined by four autapomorphies (Kyrki, 1984): the presence of chaetosema; reduction of the hindleg tibia and tarsus, especially in the male; the presence of two subventral setae on the larval meso- and metathorax; and concealment of the labial palps in the pupa.

Praydidae: The members of this group are easily distinguished from other yponomeutoids by an unusually broad male 8th sternum and by female apophyses anteriores lacking a branched costa at the base (Friese, 1960; Moriuti, 1977a). Although the moths had frequently been associated with plutellids (Friese, 1960; Moriuti, 1977a; Heppner, 1998), at least two studies, i.e. Kyrki (1990) and Ulenberg (2009), proposed a close relationship between the *Prays* and *Atteva* groups, based on two synapomorphies, the lack of a pecten on the antennal scape and the presence of a larval cranial seta P1 that lies on or above the line defined by setae Af2–P2. A possible additional synapomorphy is the presence of less than four segments in the maxillary palp.

Scythropiidae: The position of *Scythropia* has likewise been controversial. The predominant view has associated this genus with Plutellidae (Friese, 1960; Moriuti, 1977a; Heppner, 1998). Recently, Sohn et al. (2013) revised its systematic position with molecular data and suggested that it constitutes its own family, Scythropiidae. Sohn et al. (2013) also found that the family is possibly associated with Bedelliidae and Heliodinidae. Since these relationships have never been detected from previous morphological studies, the hypothesis still remains provisional. Scythropiidae is monobasic with a unique European species, *Scythropia crataegella* Linnaeus.

Yponomeutidae: Yponomeutidae sensu *lato* in Moriuti (1977a) and Yponomeutidae sensu Kyrki (1990) appeared clearly polyphyletic according to two recent molecular studies (Mutanen et al. 2010; Sohn et al. 2013). The latest working hypothesis on the monophyly of Yponomeutidae (Sohn et al. 2013) is exactly corresponding to Yponomeutinae sensu Moriuti (1977a) in taxonomic boundaries. From this, only two subfamilies, Yponomeutinae and Saridoscelinae, constitute Yponomeutidae. Moriuti (1977a) proposed two synapomorphies for this group, the presence of spine-like setae on the adult abdominal tergites and a seta V1 on the larval head that is as large as a long tactile seta.

Biogeography

Yponomeutidae have been conventionally considered to be a primarily Palearctic or Holarctic group, supposedly due to their preference to temperate climate. Yponomeutidae as defined by Kyrki (1990) seem to support this convention (fig. 1). According to this definition, 40% of the total number of species occur in the Palearctic Region. The proportion of Holarctic species (Palearctic + Nearctic) reaches 53% of the total number of species (fig. 1). Li & Jin (2012) presented a similar observation, although they recognized that the Oriental species diversity of Yponomeutidae (37% of total) is a very close second after the Palearctic one (38.5% of total). Such difference may be partly due to the exclusion of Argyresthiidae from Li & Jin's estimation. The species diversity of Yponomeutidae is significantly lower in the Southern Hemisphere than in the Northern Hemisphere (fig. 1). Especially, its lowest species diversity can be observed in the Neotropical Region where several other lepidopteran groups are quite diverse (Heppner, 1998). It is possible that this unusual species diversity pattern simply represents different inventory efforts rather than an actual depauperate yponomeutid fauna in the Neotropics.

Figure 2 demonstrates a reinterpretation of the same biogeographic data as Figure 1, tabulated by families according to the most recent classification (Sohn et al. 2013). Scythropiidae, comprising only one Palearctic species, is not shown. Their biogeographic composition show apparent Holarctic predominance from only two groups, Yponomeutidae and Argyresthiidae. Among the Holarctic Yponomeutidae (n = 155), the Palearctic species are predominant (n = 136). Such Palearctic predominance is not as strong in argyresthiids (Palearctic/Nearctic = 76/52). Yponomeutidae includes a relatively high number of Oriental species (n = 84) due to the high species diversity of Niphonymphini in South China and Southeast Asia. Argyresthiidae appears almost depauperate in the Southern Hemisphere, but a recent discovery of a diverse novel argyresthiid lineage (see Sohn et al. 2013) from the Neotropical Region may re-shape their biogeography. Attevidae are almost exclusively pantropical with only one exceptional species expanding to temperate USA.

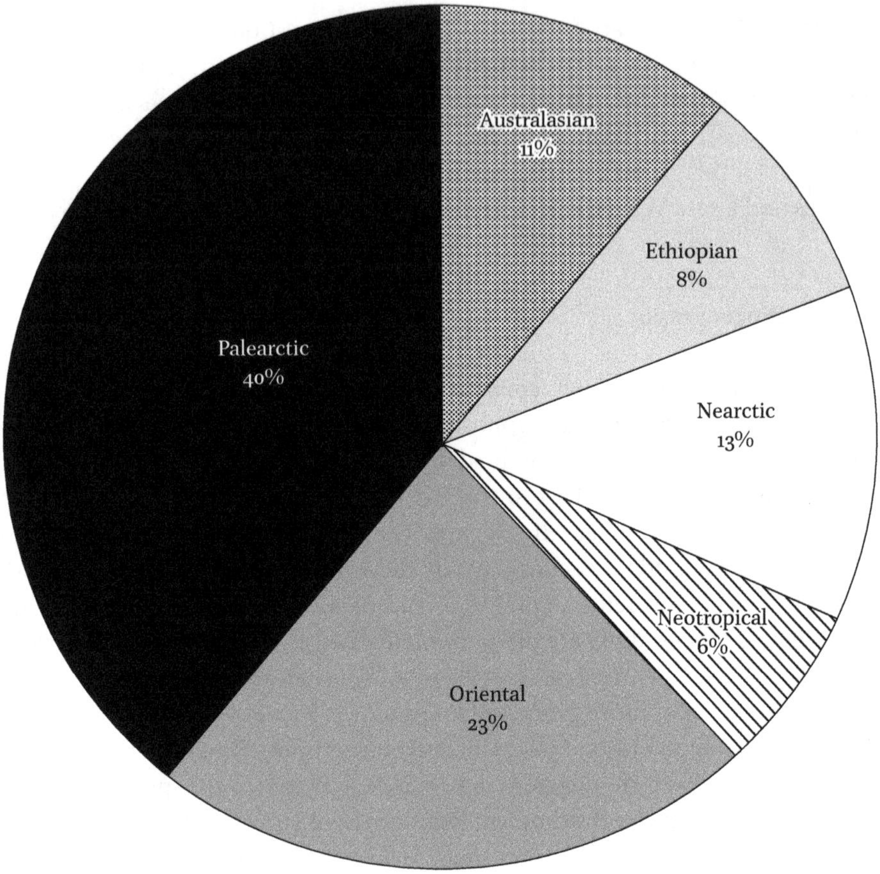

FIGURE 1 *Biogeographic composition of the species belonging to Yponomeutidae sensu Kyrki (1990)*

Praydidae is mostly comprised by Palearctic and Oriental species (20 versus 14) although the discovery of several new species related to *Atemelia* from the Nearctic and Neotropics (J.C. Sohn, unpublished) suggests a comparably high species diversity in the New World.

Host Plants

Yponomeutidae, especially *Yponomeuta*, have provided model systems for the study of evolution and speciation influenced by insect-plant relationships (Menken et al. 1992). Most previous studies have focused on host plant acceptance in a species complex or at the population level within a species (reviewed

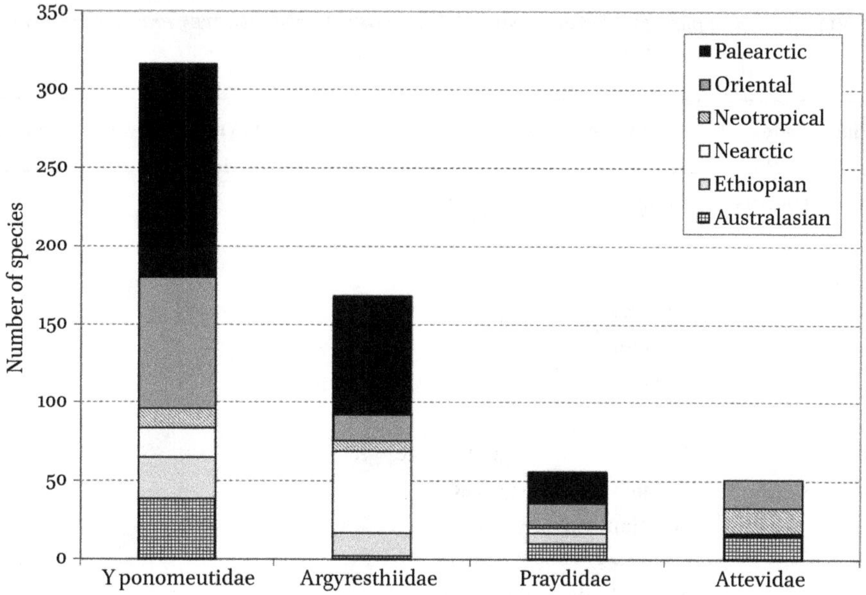

FIGURE 2 *Biogeographic composition of the Yponomeutidae, Argyresthiidae, Praydidae, and Attevidae*

in Menken et al. 1992). At least two studies, Ulenberg (2009) and Turner et al. (2010), scrutinized the evolution of host plant association in Yponomeutidae or *Yponomeuta*, based on phylogenetic approaches. Those analyses, however, are restricted to Holarctic species. Sohn et al. (2013) provided the first robust phylogeny-based, evolutionary hypothesis of the yponomeutoids/host plant families associations. This hypothesis demonstrates three aspects of the yponomeutoid host use patterns: (i) yponomeutoid larval host use shows a high proportion of oligophagy, possibly representing phylogenetic conservatism in diet breadth, (ii) each yponomeutoid family shows a nonrandom useage of host plant families, possibly associated with phytochemicals or other plant characteristics, and (iii) in spite of such nonrandomness, a coevolution between Yponomeutoidea as a whole and their host plant families has not been corroborated.

Table 1 shows the plant families used by the larvae of the five yponomeutoid families included in this catalogue. Since they do not share a common ancestor (polyphyly), the host use patterns need to be interpreted on a family by family basis as follows.

Argyresthiidae: A total of 9 plant orders and 12 families, including two gymnosperm families, are used by argyresthiid larvae. Most argyresthiid species are

TABLE 1 *Host plant families of Argyresthiidae* (ARGY), *Attevidae* (ATTE), *Praydidae* (PRAY), *Scythropiidae* (SCYT), *and Yponomeutidae* (YPON)

Plant Clade	Plant Order	Plant Family	ARGY	ATTE	PRAY	SCYT	YPON
Angiosperms	Apiales	Araliaceae		1			
	Celastrales	Celastraceae					37
	Cornales	Cornaceae	1				
	Dipsacales	Caprifoliaceae			3		
	Ericales	Ericaceae	3				7
	Fagales	Betulaceae	5		1		3
	Fagales	Fagaceae	4				3
	Fagales	Juglandaceae			1		
	Fagales	Myricaceae	1		1		
	Gentianales	Rubiaceae					1
	Lamiales	Oleaceae			7		5
	Malpighiales	Salicaceae	1				3
	Myrtales	Myrtaceae	1				1
	Rosales	Rhamnaceae					2
	Rosales	Rosaceae	14			1	15
	Rosales	Ulmaceae	2		1		
	Santalales	Loranthaceae					3
	Santalales	Santalaceae					3
	Sapindales	Burseraceae			1		
	Sapindales	Rutaceae			7		
	Sapindales	Sapindaceae	1				
	Sapindales	Simaroubaceae		8			
	Saxifragales	Altingiaceae					1
	Saxifragales	Crassulaceae					3
	Saxifragales	Grossulariaceae					3
	Saxifragales	Saxifragaceae					12
Gymnosperms	Cupressales	Cupressaceae	28				
	Pinales	Pinaceae	19				6

needle-leaf miners on two conifer families, Cupressaceae (35%) and Pinaceae (23.8%), and also internal feeders in Rosaceae (17.5%) among angiosperm families.

Attevidae: The host plants are known only for ten species of the family and they almost exclusively belong to Simaroubaceae (88.9% of total).

Praydidae: The larvae feed on at least eight angiosperm families belonging to five orders, most frequently Oleaceae (31.8%) and Rutaceae (31.8%).

Scythropiidae: The only included species, *Scythropia crataegella*, feeds on several genera of Rosaceae, although some, possibly erroneous records from other plant families are also known.

Yponomeutidae: Host records are available for about 106 species of Yponomeutidae. These are associated with 11 orders and 17 families of plants, including the gymnosperm family Pinaceae. The largest proportion of Yponomeutidae larvae feed on Celastraceae (34.3%), followed by Rosaceae (13.9%) and Saxifragaceae (11.1%). Among the yponomeutid subgroups, Niphonymphini are exclusively associated with Fagales and as far as known, and Saridoscelinae feed only on Ericaceae. Several researchers (e.g. Friese, 1960; Moriuti, 1977a; Turner et al. 2010) noticed a repeated usage of Celastraceae in a broad range of yponomeutid genera. From this, Ulenberg (2009) suggested that the ancestor of Yponomeutidae might have been associated with this plant family. This argument, however, is unlikely because none of the basal yponomeutids feed on Celastraceae in their larval stage.

Taxonomic History

The families included in this catalogue have long been treated under Yponomeutidae sensu *lato*, Plutellidae, or even Tineidae. The earliest records of these groups include four species described in the 10th edition of Linnaeus's *Systema Naturae* (1758): i.e. *Phalaena* (*Tinea*) *bonnetella* and *P.* (*T.*) *goedartella*, both now belonging to *Argyresthia* (Argyresthiidae); and *P.* (*T.*) *evonymella* and *P.* (*T.*) *padella*, both now belonging to *Yponomeuta* (Yponomeutidae). *Phalaena Tinea fabriciella* Swederus, 1787, and *Tinea fraxinella* Bjerkander, 1784, were the earliest described species of Attevidae and Praydidae respectively. Since these early taxonomic works, about 620 species have been described by several authors. Ten of these authors described more than ten new species in these families (Table 2). The majority of the species (133 in total) were described by Edward Meyrick (1854–1938), who laid the foundation for modern micro-lepidopteran taxonomy. He described the largest number of valid species in each family catalogued here except Scythropiidae (Table 2). Other important contributors (Table 2) include S. Moriuti, who described 61 new species mostly from Japan; P.C. Zeller, T. de G. Walsingham, and Z.S. Gershenson whose works were based on the world fauna; Huemer for the European species; H. Li and coauthors for the Chinese species; J.-C. Sohn for the Chinese and Costa Rican species; A. Busck for the North American species; A.J. Turner

TABLE 2 *Numbers of valid species-group names introduced by authors who described more than 10 species.* ARGY = *Argyresthiidae;* ATTE = *Attevidae;* PRAY = *Praydidae;* YPON = *Yponomeutidae*

	ARGY	ATTE	PRAY	YPON	Total
Meyrick, E.	25	13	15	82	133
Moriuti, S.	20		9	32	61
Zeller, P.C.	18	3	1	8	30
Li, H.			3	15	18
Gibeaux, C.	12		2	2	16
Gershenson, Z.S.	1			14	15
Busck, A.	12			3	15
Sohn, J.-C.			3	12	15
Huemer, P.	1			13	14
Turner, A.J.	1	2	2	9	14
Walsingham, T. de G.	4	2		5	11

for the Australian species; and C. Gibeaux for the European and Afrotropical species.

Descriptive works in Yponomeutidae and the other four families once included in Yponomeutidae sensu Kyrki (1990) have overall increased after the 1820s, although there was a dip during the 1940s and 50's (fig. 3). The highest peak in total can be found during the 1910s which corresponds to the highest peak for Argyresthiidae. The second highest peak after 2000 corresponds to the highest peak for Yponomeutidae, although this period includes three more years than other decades. Overall patterns in total described species numbers follow the patterns of the two most speciose families, Argyresthiidae and Yponomeutiidae. Scythropiidae (not shown in fig. 3) remains monobasic since 1767.

Argyresthiidae: Most species were described during the 1910s due to the works of two prolific researchers, E. Meyrick and A. Busck. Two almost equally high peaks are the 1870s and 1960s due to Zeller and Chambers for the first decade, and the Japanese species described by Moriuti (1969a).

Attevidae: The two highest peaks in the number of described species are found during the 1870s and 1900s, followed by a close second highest peak in the 2000s, when the Neotropical species were reviewed (Becker, 2009). Species descriptions of Attevidae have been few or absent at other times than those three peaks.

Praydidae: The first and highest peak is found between 1900 and 1919 when many species were described by Meyrick, mainly in his *Exotic Microlepidoptera*

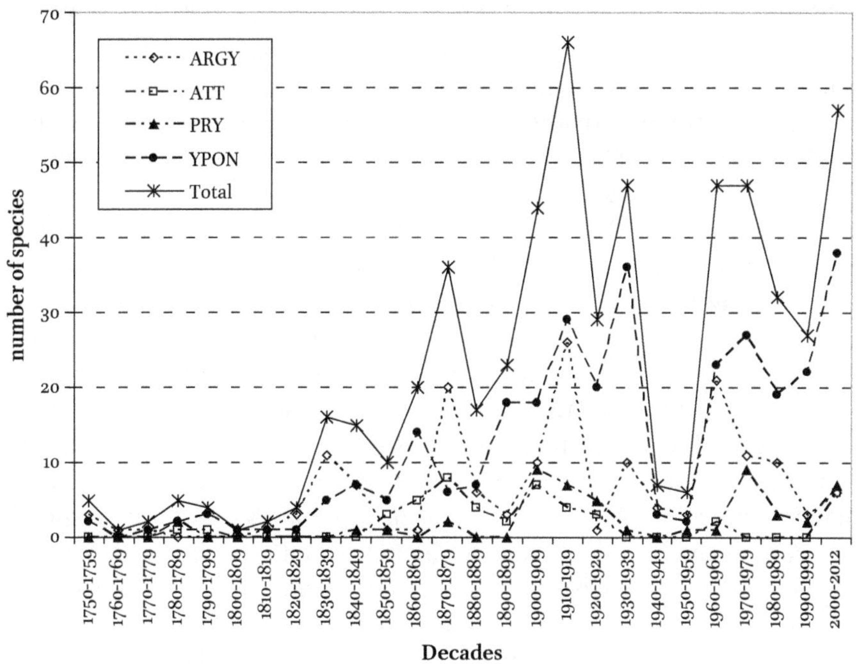

FIGURE 3 *The number of new, valid species-group names per decade of Argyresthiidae (*ARGY*, open rhomboids), Attevidae (*ATT*, open squares), Praydidae (*PRY*, closed triangles), and Yponomeutidae (*YPON*, closed circles).*

series. Another peak happened when Moriuti described several Japanese species in his *Fauna Japonica* monograph. Recently (after 2000), several new species have been described, especially from China (Yu & Li, 2004; Sohn & Wu, 2011).

Yponomeutidae: The highest number of new species was described after 2000, especially in the late 2000s when many new species of *Thecobathra* were described by H. Li and his research team. The second highest peak corresponds to Matsumura's and Meyrick's works during the 1930s. The third highest peak, in the 1910s is also mainly due to Meyrick.

Methods and Conventions

Data for this catalogue have come primarily from the *Zoological Record* until 2012 and a previous world catalogue by Meyrick (1914c). Other important data sources includes Hemming, 1937 (2 volumes) on the works of Hübner, local catalogues and checklists such as Kyrki (1990), Gershenson & Ulenberg (1998), Moriuti (1977a), and online data bases such as Beccaloni et al. (2005) and Pitkin & Jenkins (2004). All described genera and species of Yponomeutidae were compiled from these and other data sources, originally using the family definition by Kyrki (1990) and then rearranged according to the latest working classification system. Genera and species once misassigned to Yponomeutidae were also considered. For the poorly-known taxa, we estimated their relationships with the families treated in this catalogue. Although the families are not monophyletic, they are commonly lacking ocelli on the head. Therefore, genera possessing ocelli were excluded from these families. However, we do not rule out the possibility that they belong to Yponomeutoidea. Other characteristics useful for estimating taxonomic placement include the presence/absence of scales on the proboscis or bristles on the labial palpi. All excluded taxa are listed in a separate section, along with historically confused non-yponomeutoids. Some genera, such as *Ethmia* and *Cydonia*, were associated with Yponomeutidae briefly in older literature, but these family assignments were soon dismissed, have seldom reappeared, and are not included in our catalogue. We also do not treat taxa that have been referred to Yponomeutidae in an expanded definition of the family, for example plutellid and ypsolophid genera in Moriuti (1977a) or gelechioid genera in Walsingham (1907a). In the section "Genera and their component species removed from the families treated here," descriptions of species currently placed in the excluded genera are not cited in the references because they are not treated in this catalogue.

All nomenclatural issues involved in this catalogue follow the provisions of the 4th edition of the International Code of Zoological Nomenclature (ICZN, 1999), except for the endings of the specific names which are given as presented from the original description to avoid complications according to subsequent changes in generic assignment. Families, genera and species are listed alphabetically, using their original spelling. Valid taxon names of included species are given in bold. All junior synonyms and misspellings are listed alphabetically, indented under their senior synonyms. Misspellings are separated from their author by a semicolon in order to distinguish them from true synonyms, for example, *fabricella*; Wallengren, 1861. Original generic names are given in parentheses following the author and year of publication. Singly dated references which appeared in print on the year indicated in the publication are given without the year being enclosed in brackets. For references and citations with two years, one enclosed in brackets, the year without brackets (usually an older year) indicates the year when given within the original work; the year enclosed in brackets (usually a more current year) is when known from other sources that the work first appeared in print. For example,

Hübner, 1816 [1825] was dated 1816 but did not appear in print until 1825. References and citations only with a year enclosed in brackets is given when an undated work or portion of a larger work is known from sources other than the publication to have first appeared in print. For example, Hübner, [1813] is given for independently printed plates without text published in 1813, as part of the work by Hübner, entitled "Sammlung europäischer Schmetterlinge," spanning the period 1796–1838. Dates and details of publications for Hübner's works were obtained from Hemming, (1937), comprised of two volumes.

Excluded taxa and non-lepidopteran homonyms are given only by taxon names without further documentation. Information with taxon names is given as follows.

For example,

abies Freeman, 1972 (*Argyresthia*), Can. Ent. 104(5): 696. (CNC: Holotype ♂);
①   ②

described from 6♂, 9♀ (CNC).
③

(1) Literature source of original description. For the long titles, abbreviations were used (see the abbreviations below for the full titles). The full citations of the references can be found in the 'References' section of this catalogue.

(2) Deposition and status of primary types in parentheses. Specimen depositories are given as abbreviations (also see below). Primary types comprise holotypes, lectotypes, neotypes and syntypes. The sex of the primary type specimens is given when known. See "Notes" for the uncertainties or other corrections such as primary type designations and lectotypes designated in the present catalogue.

(3) Deposition and status of type series. Based on the original descriptions, all specimens comprising the type series are mentioned with their sexes, when known. In cases where the original description did not specify them, the number of type specimens are recorded as "not given." When the number and the sex of the type specimens become secondarily available, such information is given with the statement "in type series." All depositories for the type series are listed in parentheses, when known.

The type locality is given progressively from general to specific following the country name. The locality names in the original descriptions are mentioned, but additional details or new names are also given, when known. The country names follow Wikipedia (www.wikipedia.org). For some land masses such as islands, specific localities are given instead of country names in spite of their political dependence.

Distribution is given primarily by countries, but some specific locality names (e.g. some islands) are included in alphabetical order as well. For large countries including multiple ecozones, the state/province-level or equivalent divisions are given in parentheses. Distributions due to anthropogenic introduction are indicated. All these localities were sorted according to zoogeographic ecozones. The definition of each zoogeographic regions follows:

Palearctic Region: Europe including Azores, Canary Islands, Madeira, Africa north of the Sahara (Morocco, Algeria, Tunisia, Libya and Egypt), Russia, Near and Middle East including the Arabian Peninsula, Turkey, Iran and Afghanistan, Central Asia (Azerbaijan, Turkmenistan, Tajikistan, Uzbekistan, Kyrgyzstan), Nepal, Pakistan, temperate East Asia (Mongolia, northern China, Korea, Japan).

Oriental Region: India including Kashmir, Sri Lanka, southern China (Fujian, Guangdong, Guangxi, Guizhou, Hainan, Hunan, Jiangxi, Shanghai, Sichuan, Yunnan, Zhejiang), Taiwan, Japan (Ryukyu Islands, Nansei Archipelago), Southeast Asia (Myanmar, Thailand, Laos, Cambodia, Vietnam, Malaysia, the Philippines, Indonesia).

Ethiopian Region: Africa south of the Sahara, Cape Verde Islands, St. Helena, Madagascar, Mascarene Islands, Comores, Seychelles, and the islands in the southern Indian Ocean.

Australasian Region: Australia, Tasmania, New Zeland, New Guinea including Irian Jaya, the Bismarck and Solomon Archipelagos, the islands in the Pacific Ocean including Easter Island, the Hawaiian and the Juan Fernandez Islands.

Nearctic Region: North America north of Mexico including Greenland and the Bermuda Islands.

Neotropical Region: Mexico, Central America, Caribbean Islands, South America including Galápagos Islands and the Falkland Islands.

Host plants are given, if known. The data sources are primarily the original descriptions and other taxonomic papers, but records from electronic sources are also considered. Rough statements of host plants such as oaks and pines are given in formal scientific names. Scientific names and classifications of plants follow primarily two online databases: the International Plant Name Index (http://www.ipni.org) and the Plant List (http://www.theplantlist.org). Host plant species descriptions are not included in references. Records of possible host plants due to adult-habitat association are excluded. Cecidophagies are given as originally stated in the references. Plant names are sorted into current family groups and arranged in alphabetical order. Plants belonging to the same genus are listed with the first or first two letters of the genus after the statement of full genus name for the first species. Citations of the host records are given in parentheses. We did not distinguish between original host records and secondary citations. We simply list as many host references as possible to recover the whole host plant range. Suspicious or possibly mistaken host records are discussed.

Verbatim label data are given for the lectotypes and the neotypes designated in this catalogue. The line breaks in the labels are marked with ' | '. When there is a line dividing two rows in the label, it is marked with ' || '. The labels colored with other than white and/or shaped as other than rectangular are indicated by the color and the shape. When the label content is written by hand, it is stated so.

List of Changes

Lectotypes are Designated for the Following Species

albistria Haworth, 1828 (*Erminea*) [Argyresthiidae]
antistatica Meyrick, 1931a (*Hyponomeuta*) [Yponomeutidae]
aphrospora Meyrick, 1892 [1893] (*Zelleria*) [Yponomeutidae]
araeodes Meyrick, 1892 [1893] (*Zelleria*) [Yponomeutidae]
argentea Busck, 1912 (*Calantica*) [Yponomeutidae]
atmoriella Bankes, 1896 (*Argyresthia*) [Argyresthiidae]
autocasis Meyrick, 1907d (*Prays*) [Praydidae]
calcarata Meyrick, 1924 (*Hyponomeuta*) [Yponomeutidae]
callidoxa Meyrick, 1892 [1893] (*Zelleria*) [Yponomeutidae]
calycias Meyrick, 1907d (*Prays*) [Praydidae]
citrina Meyrick, 1892 [1893] (*Zelleria*) [Yponomeutidae]
conspicua Walsingham, 1900 (*Atteva*) [Attevidae]
cuprina Felder & Rogenhofer, 1875 (*Amblothridia*) [Attevidae]
cynetica Meyrick, 1892 [1893] (*Zelleria*) [Yponomeutidae]
endocarpa Meyrick, 1919 (*Prays*) [Praydidae]
enneacentra Meyrick, 1925c (*Hyponomeuta*) [Yponomeutidae]
ergatica Walsingham, 1914 (*Atteva*) [Attevidae]
flavivitta Walker, 1866b (*Carthara*) [Attevidae]
hemileuca Meyrick, 1933 (*Hyponomeuta*) [Yponomeutidae]
idiograpta Meyrick, 1935a (*Argyresthia*) [Argyresthiidae]
inscripta Meyrick, 1907d (*Prays*) [Praydidae]
lustralis Meyrick, 1911a (*Argyresthia*) [Argyresthiidae]
mathewi Butler, 1887 (*Corinea*) [Attevidae]
megalastra Meyrick, 1907d (*Atteva*) [Attevidae]

memorella Meyrick, 1892 [1893] (*Zelleria*) [Yponomeutidae]
monerythra Meyrick, 1926b (*Atteva*) [Attevidae]
monoplanetis Meyrick, 1910b (*Atteva*) [Attevidae]
mystarcha Meyrick, 1892 [1893] (*Zelleria*) [Yponomeutidae]
nivosa Meyrick, 1938a (*Zelleria*) [Yponomeutidae]
nubeculella Tengström, 1848 (*Tinea*) [Yponomeutidae]
obtorta Meyrick, 1923a (*Xyrosaris*) [Yponomeutidae]
paurodes Meyrick, 1907d (*Hyponomeuta*) [Yponomeutidae]
perficitellus Walker, 1863 (*Hyponomeuta*) [Yponomeutidae]
porphyris Meyrick, 1907d (*Atteva*) [Attevidae]
praecocella Zeller, 1839 (*Argyresthia*) [Argyresthiidae]
proterospila Meyrick, 1892 [1893] (*Zelleria*) [Yponomeutidae]
puncticornis Walsingham, 1891b (*Hyponomeuta*) [Yponomeutidae]
pustulellus Walker, 1863 (*Hyponomeuta*) [Yponomeutidae]
reticulata Staudinger, 1877 (*Argyresthia*) [Argyresthiidae]
retinella Zeller, 1839 (*Argyresthia*) [Argyresthiidae]
sigillata Meyrick, 1892 [1893] (*Zelleria*) [Yponomeutidae]
subaurata Durrant, 1900 (*Atteva*) [Attevidae]
teratias Meyrick, 1907d (*Atteva*) [Attevidae]
trifasciata Staudinger, 1871 (*Argyresthia*) [Argyresthiidae]
variegata Tengström, 1869 (*Swammerdamia*) [Yponomeutidae]

Neotype is Designated for the Following Species

glaphyropis Meyrick, 1908c (*Hyponomeuta*) [Yponomeutidae]

New Synonyms

heliodoxa Meyrick, 1910b (*Atteva*) = ***basalis*** Vollenhoven, 1863 (*Oeta*)
Syncallia Guérin-Méneville, [1844] = ***Atteva*** Walker, 1854
stellata Guérin-Méneville, 1844 (*Syncallia*) = ***carteri*** Walsingham, 1891b (*Oeta*)
tonseana Tams, 1935 (*Atteva*) = ***cuprina*** Felder & Rogenhofer, 1875 (*Amblothridia*)
triplex Diakonoff, 1967 (*Atteva*) = ***apicalis*** Vollenhoven, 1863 (*Oeta*)

New Combinations

pyrothorax Meyrick, 1928b (*Atteva*) to *Lactura.*
stratellus Zeller, 1877 (*Prays*) to *Atemelia.*

Catalogue of the Yponomeutoidea

ARGYRESTHIIDAE Bruand, 1849 [1850].
Mém. Soc. Emul. Doubs 3(3): 49. Note: Cited as: Tribus XIV. *Argyrestidæ*.

ARGYRESTHIADAE Meyrick, 1889.
Trans. N. Z. Inst 21: 162. emendation of Argyresthiidae.

ARGYRESTHIA Hübner, [1825]. Verz. bekannt. Schmett.: 422.
Type species: *Phalaena goedartella* Linnaeus, 1758, by subsequent designation by Busck, 1907, Proc. USNM 32: 5.

ARGYROSETIA Stephens, 1829. Syst. Cat. Brit. Ins. Part 2: 205. A junior objective synonym of *Argyresthia*.
Type species: *Phalaena goedartella* Linnaeus, 1758, by subsequent designation by Westwood, 1840, Introd. Mod. Classif. Insects 2 (Synopsis Genera Br. Insects): 112.

BLASTOTERE Ratzeburg, 1840. Forst.-Ins. 2: 246. [valid as a subgenus name (Moriuti, 1965b)].
Type species: *Phalaena bergiella* Ratzeburg, 1840, by monotypy.

BLASTOTETE; Csóka, 2001. In: Liebhold et al., Proc. Integ. Man. Dyn. For. Defol. Ins.: 35. [misspelling of *Blastotere*].

EDERESA Curtis, 1833. Ent. Mag. 1(2): 191. A junior objective synonym of *Oligos* Treitschke, 1830.
Type species: *Phalaena pruniella* Clerck, 1759, by original designation.

EDERESSA; Stephens, 1835. Ill. Brit. Ent. Haust. 4: 397. [misspelling of *Ederesa*].

ISMENE Stephens, 1834. Ill. Brit. Ent. Haust. 4: 247. A junior homonym of *Ismene* Savigny, 1816. [Lepidoptera, Crambidae].
Type species: *Phalaena pruniella* Clerck, 1759, by subsequent designation by Fletcher, 1928, Cat. Indian Insects 17: 2.

OLIGOS Treitschke, 1830. In: Ochsenheimer & Treitschke, Schmett. Eur. 8: 299. An objective replacement name for *Ismene* Stephens, 1834.

 | DOI 10.1163/9789004264267_003

Type species: *Phalaena pruniella* Clerck, 1759, by subsequent designation by Fletcher, 1929.

PARAARGYRESTHIA Moriuti, 1969a. Bull. Univ. Osaka Pref. (B) 21: 30.
Type species: *Paraargyresthia japonica* Moriuti, 1969a, by original designation.

abdominalis Zeller, 1839 (*Argyresthia*), Isis Oken 32(3): 205. (BMNH: Syntype); not given (1 ♂ in BMNH).
Type locality: Poland – Glogów.
Distribution: Palearctic – Austria, Belgium, Bulgaria, Czech Republic, Denmark, England, Estonia, Finland, France, Germany, Hungary, Italy, Latvia, Netherlands, Norway, Poland, Romania, Russia (Murmansk), Slovakia, Sweden, Switzerland, Turkey, [former] Yugoslavia.
Host: Cupressaceae – *Juniperus communis* L. (Baraniak, 1993; Agassiz, 1996).

helvetica Heinemann, 1876 (*Argyresthia abdominalis*), in Heinemann & Wocke, Schmett. Deutsch. 2(2)2: 652. (Type not found); not given.
Type locality: Germany.

inauratella Tengström, 1848 (*Argyresthia*), Not. Sällsk. Faun. Fenn. Förh. 1: 137. (ZMH: Syntype); not given (1 ♂ in ZMH).
Type Locality; Finland – Kuusamo.

inaurata; Sharp, 1893, Zool. Rec. 30: 301. [misspelling of *inauratella*].

abies Freeman, 1972 (*Argyresthia*), Can. Ent. 104(5): 696. (CNC: Holotype ♂); described from 6♂, 9♀ (CNC).
Type locality: Canada – Ontario, Hoyle.
Distribution: Nearctic – Canada (Alberta, Ontario, Québec, Saskatchewan).
Host: Pinaceae – *Abies balsamea* (L.) Mill. (Freeman, 1972).

affinis Braun, 1940 (*Argyresthia*), Trans. Am. Ent. Soc. 66: 276. (ANSP: Holotype ♂); described from 4♂, 1♀ (ANSP; USNM).
Type locality: USA – Ohio, Clermont Co., Cabin Run.
Distribution: Nearctic – Canada (Ontario), USA (Kentucky, Ohio).
Host: Cupressaceae – *Juniperus virginiana* L. (Braun, 1940)

albicomella Moriuti, 1969a (*Argyresthia*), Bull. Univ. Osaka Pref. (B) 21: 27. (UOP: Holotype ♀); described from 5♀ (UOP; USNM).

Type locality: Japan – Hokkaido, Onuma.
Distribution: Palearctic – Japan (Hokkaido, Honshu).

albistria Haworth, 1828 (*Erminea*), Lep. Brit. 4: 517. (BMNH: Lectotype ♀ [designated here]); not given (1♂, 1♀ in BMNH).
Type locality: [England: London and Kent].
Distribution: Palearctic – Austria, Belgium, Bulgaria, Corsica, Czech Republic, Denmark, England, Finland, France, Germany, Hungary, Ireland, Italy, Latvia, Lithuania, Netherlands, Norway, Poland, Romania, Russia (Kaliningrad, Middle Volzhsky, Zabaykalsk), Sardinia, Slovakia, Spain, Sweden, Switzerland, [former] Yugoslavia.
Hosts: Rosaceae – *Crataegus* spp., *Prunus cerasus* L., *P. domestica* L., *P. insititia* L., *P. persica* (L.) Batsch, *P. spinosa* L., *P. syriaca* Borkh., *Pyrus communis* L. (Friese, 1969; Agassiz, 1996; Baraniak, 1993; Alford, 2007; Robinson et al. 2010). Records from *Betula, Corylus* and *Fagus* are suspicious (Friese, 1969).
Note: In the species description, two color variations, α and β are described. One female of the type series was selected as lectotype by M.S. Parson but this lectotypy was never published. We validate the type designation here. The lectotype female has no label indicating locality but subsequently has added labels: "B.M. | Genitalia 7154," "unpublished Lectotype designation | teste MS Parson 1948."

> *fagetella* Zeller, 1839 (*Argyresthia*), Isis Oken 32(3): 204. (unknown); not given.
> Type locality: Poland – Glogów.
> Notes: Moritz was given as the author of *fagetella* by Zeller (1839). However, this authorship is invalid because it has never been published.

> *tarnis* Réal, 1990 (*Argyresthia albistria* ssp.), Bull. Soc. Ent. Mulh. (1–3): 1. (P. Réal collection [private]: Holotype ♀); described from 4♀ (Réal collection).
> Type locality: France – Midi-Pyrénées, Verdalle.

alpha Friese & Moriuti, 1968 (*Argyresthia*), Trans. Lep. Soc. Japan 19(1): 13. (UOP: Holotype ♂); described from 4♂, 6♀ (UOP; DEIB).
Type locality: Japan – Honshu, Sinano [=Nagano], Kamikoti.
Distribution: Palearctic – Japan (Hokkaido, Honshu).
Host: Rosaceae – *Sorbus commixta* Hedl. (Friese & Moriuti, 1968).

alternatella Kearfott, 1908 (*Argyresthia*), Jour. N. Y. Ent. Soc. 16: 182. (USNM: Lectotype: ♂ [designated by Sohn & Lewis, 2012: 4]); described from 7♂, 1♀ (BMNH; USNM).
Type locality: USA – New Jersey, Essex Co., Caldwell and Watchug Mountains.
Distribution: Nearctic – Canada (Ontario, Québec), USA (Kentucky, Maine, Massachusetts, New Jersey, New York, Oklahoma).
Host: Cupressaceae – *Juniperus virginiana* L. (Forbes, 1923).

altissimella Chambers, 1877 (*Argyresthia*), Bull. U. S. Geol. Surv. 3: 130 (MCZ: Syntype); not given (1♂ in MCZ).
Type locality: USA – Colorado, Lake Co., Mount Elbert.
Distribution: Nearctic – USA (California, Colorado).

amiantella Zeller, 1847b (*Blastotere*), Linn. Ent. 2: 294. (TLMF: Neotype ♀ [designated by Bengtsson & Johansson, 2011b: 271]); described from more than one specimen but destroyed (Bengtsson & Johansson, 2011b).
Type locality: Austria – Tirol, Vennatal.
Distribution: Palearctic – Austria, Czech Republic, Finland, France, Germany, Italy, Poland, Russia?, Slovakia, Switzerland, [former] Yugoslavia.
Host: Pinaceae – *Picea excelsa* (Lam.) Lk. (Baraniak, 1993)
Note: Some authors have cited Kollar as the author of the description. In fact, Zeller cited Kollar as species collector, not author.

andrianella Gibeaux, 1983b (*Argyresthia*), Revue Fr. Ent. (N. S.) 5(3): 124. (MNHN: Holotype ♂); described from 2♂, 3♀ (MNHN).
Type locality: Madagascar – Centre, Soaindrana Plateau.
Distribution: Ethiopian – Madagascar.

angusta Moriuti, 1969a (*Argyresthia*), Bull. Univ. Osaka Pref. (B) 21: 12. (USNM: Holotype ♂); described from 2♂, 2♀ (UOP; USNM).
Type locality: Japan – Kyushu, Oita Pref., Mt. Kurodake.
Distribution: Palearctic – Japan (Honshu, Kyushu).

annettella Busck, 1907 (*Argyresthia*), Proc. USNM 32: 12. (USNM: Syntype); not given (1♀ in USNM).
Type Locality; USA – Ohio, Cincinnati.
Distribution: Nearctic – Canada (Newfoundland, Ontario, Québec), USA (Connecticut, Maine, Ohio).
Host: Cupressaceae – *Juniperus communis* L. (Braun, 1912).

anthocephala Meyrick, 1936b (*Argyresthia*), Exot. Micr. 4(20): 622. (BMNH: Lectotype ♂ [designated by Clarke, 1965: 280]); described from 1♂, 1♀ (BMNH).
Type locality: Japan – Honshu, Kyoto.
Distribution: Palearctic – Japan (Honshu, Kyushu, Shikoku).
Host: Cupressaceae – *Cryptomeria japonica* (Thunb. ex L. f.) D. Don (Moriuti, 1969b).

aphoristis Meyrick, 1938a (*Argyresthia*), In: Caradja & Meyrick, Dt. Ent. Z. Iris 52(1): 22. (BMNH: Lectotype ♂ [designated by Clarke, 1965: 280]); described from 4 specimens.
Type locality: China – Yunnan, Likiang [=Lijiang], Yülingshanmass [possibly Yulongxueshan].
Distribution: Oriental – China (Yunnan).

apicimaculella Chambers, 1874 (*Argyresthia*), Can. Ent. 6(1): 11. (Type not found); not given.
Type locality: USA – Kentucky, "in oak woods."
Distribution: Nearctic – Canada (Québec), USA (Florida, Kentucky, Maine, Massachusetts, Michigan, New Jersey, Ohio, Pensylvannia).
Note: The type specimen of *apicimaculella* was erroneously labeled as "*A. visaliella*" by Chambers. The name *visaliella* was published in Chambers (1875b) without description and hence it is invalid (Covell, 1999). This problem was first noticed by Busck (1907) who concluded that the specimen labelled as "*A. visaliella*" is the actual type specimen of *apicimaculella*.

arceuthina Zeller, 1839 (*Argyresthia*), Isis Oken 32(3): 205. (BMNH: Syntype); type series stated as "many" specimens (1 ♀ in BMNH).
Type locality: Poland – Glogów.
Distribution: Palearctic – Austria, Belgium, Czech Republic, Denmark, England, Finland, France, Germany, Hungary, Ireland, Italy, Latvia, Netherlands, Norway, Poland, Portugal, Romania, Russia (European Northwest), Slovakia, Spain, Sweden, Switzerland, [former] Yugoslavia.
Host: Cupressaceae – *Juniperus* sp.; *J. communis* L. (Baraniak, 1993; Agassiz, 1996).

argentella Linnaeus, 1761 (*Phalaena Tinea*), Fauna Suecica: 367, No. 1431. (LSL: Lectotype ♂ [designated by Robinson & Nielsen, 1983: 202]); not given.
A junior homonym of *Phalaena argentella* Clerck, 1759 [Elachistidae].
Type locality: Sweden.

arceuthobiella Busck, 1916 (*Argyresthia*), Proc. Ent. Soc. Wash. 18(3): 153. (USNM: Syntype); not given (1 ♀ in USNM).
Type locality: USA – Oregon, Mistletoe.
Distribution: Nearctic – USA (California, Oregon).
Host: Cupressaceae – *Calocedrus decurrens* (Torr.) Florin (Busck, 1916).

assimilis Moriuti, 1977a (*Argyresthia*), Fauna Jap. Ypon. s. l.: 266. (UOP: Holotype ♂); described from 1 specimen.
Type locality: Japan – Honshu, Nara Pref., Mt. Yosinoyama.
Distribution: Palearctic – China (Gansu, Shaanxi, Xinjiang, Xizang), Japan (Honshu).
Host: Rosaceae – *Malus domestica* Borkh. (Ma & Sun, 1982).

atlanticella Rebel, 1940 (*Argyresthia*), Soc. Scient. Fenn., Comm. Biol. 8 (1): 37. (NHMW: Syntypes); described from more than one specimen (1♂ in NHMW).
Type locality: Azores.
Distribution: Palearctic – Azorean Islands.
Hosts: Ericaceae – *Erica azorica* Hochst. ex Seub., *Vaccinium cylindraceum* Smith; Myricaceae – *Myrica faya* Dryand. (Gardner et al. 1988; Silva et al. 1995).

aureoargentella Brower, 1953 (*Argyresthia*), Ann. Ent. Soc. Amer. 46(1): 97. (USNM: Holotype ♂); described from 60♂, 38♀ specimens (CNC; CUIC; MUIC; USNM).
Type Locality; USA – Maine, Hancock Co., Mariaville, Ellsworth Falls.
Distribution: Nearctic – Canada (New Brunswick, Ontario, Québec), USA (Maine).
Host: Cupressaceae – *Thuja occidentalis* L. (Silver, 1951; Browne, 1968).

aurulentella Stainton, 1849b (*Argyresthia*), Zoologist 7: [in Appendix] xl. (Type not found); not given.
Type locality: England – Sanderstead, Surrey.
Distribution: Palearctic – Austria, Belgium, Czech Republic, Denmark, England, Estonia, Finland, France, Germany, Italy, Latvia, Lithuania, Netherlands, Norway, Poland, Russia (European Northwest, Kaliningrad), Sardinia, Slovakia, Sweden, Switzerland, [former] Yugoslavia.
Host: Cupressaceae – *Juniperus communis* L. (Baraniak, 1993; Agassiz, 1996)

austerella Zeller, 1873 (*Argyresthia*), Verh. Zool.-Bot. Ges. Wien 23(3): 305. (MCZ: Syntypes); described from 6 specimens (2♂, 1♀ in MCZ).
Type locality: USA – Texas, Dallas.

Distribution: Nearctic – Canada (Québec); USA (Florida, Illinois, Iowa, Kentucky, Maine, Michigan, Missouri, New Hampshire, Maryland, Ohio, Oklahoma, Texas).

belangerella Chambers, 1875b (*Argyresthia*), Can. Ent. 7(8): 145. (MCZ & USNM: Syntypes); not given (1♂ in MCZ; 1♀ in USNM [not found]).
Type locality: Canada – [Québec].
Distribution: Nearctic – Canada (Ontario, Québec), USA (Maine).

bergiella Ratzeburg, 1840 (*Tinea Blastotere*), Forst.-Ins. 2: 246. (Boje collection [not traced]: Syntypes); not given.
Type locality: Germany – Thuringia, Thuringian Forest.
Distribution: Palearctic – Austria, Czech Republic, Denmark, Estonia, Finland, France, Germany, Italy, Latvia, Lithuania, Luxembourg, Netherlands, Norway, Poland, Romania, Russia (European Northwest, Murmansk), Slovakia, Sweden, Switzerland.
Host: Pinaceae – *Picea abies* (L.) H. Karst. (Friese, 1969; Baraniak, 1993).
Note: The authorship of *P. bergiella* was attributed to Saxesen by Ratzeburg. This attribution is invalid according to the ICZN (Article 50.1).

> *certella* Zeller, 1847b, (*Argyresthia*), Linn. Ent. 2: 289. (BMNH: Holotype ♀); described from 1 specimen.
> Type locality: Poland – Reinerz.
> Note: The species description erroneously stated that the type is a male; JCS examined the specimen and found it to be a female.

beta Friese & Moriuti, 1968 (*Argyresthia*), Trans. Lep. Soc. Japan 19(1): 15. (UOP: Holotype ♂); described from 9♂, 12♀ (UOP; DEIB).
Type locality: Japan – Honshu, Osaka Pref., Izumi, Mt. Makinosan.
Distribution: Palearctic – Japan (Hokkaido, Honshu, Kyushu).
Hosts: Ericaceae – *Rhododendron kaempferi* Planch., *R. macrosepalum* Maxim., *R. quinquefolium* Bisset & S. Moore (Friese & Moriuti, 1968)

biruptella Zeller, 1877 (*Argyresthia*), Hor. Soc. Ent. Ross. 13: 249. (BMNH: Holotype ♀); described from 1 specimen.
Type locality: Colombia – Bogota.
Distribution: Neotropical – Colombia.

bobyella Gibeaux, 1983b (*Argyresthia*), Revue Fr. Ent. (N. S.) 5(3): 120. (MNHN: Holotype ♀); described from 1 specimen.
Type locality: Madagascar – Centre, Boby Cirque.
Distribution: Ethiopian – Madagascar.

bolliella Busck, 1907 (*Argyresthia*), Proc. USNM 32: 21. (USNM: Lectotype ♀ [designated by Sohn & Lewis, 2012: 5]); not given (2♀ in USNM).
Type locality: USA – Texas.
Distribution: Nearctic – USA (Texas).

bonnetella (Linnaeus, 1758) (*Phalaena Tinea*), Syst. Nat. 10(1): 541 . (LSL: Lectotype ♂ [designated by Robinson & Nielsen, 1983: 204]); not given (3 specimens in LSL).
Type locality: not given.
Distribution: Palearctic – Austria, Belgium, Bulgaria, Czech Republic, Denmark, England, Estonia, Finland, France, Germany, Hungary, Ireland, Italy, Latvia, Lithuania, Netherlands, Norway, Poland, Romania, Russia (European Central, European Northwest, Kaliningrad), Slovakia, Spain, Sweden, Switzerland, [former] Yugoslavia.
Hosts: Rosaceae – *Crataegus laevigata* (Poiret) DC., *Prunus spinosa* L. (Friese, 1969; Baraniak, 1993).

denudatella Zeller, 1847b (*Argyresthia nitidella* var.), Linn. Ent. 2: 250. Nomen nudum.
Note: *denudatella* "F. R. [Fisher von Röslerstamm] in litt." was cited by Zeller. It is a manuscript name referred to by Zeller in his 1847 article, which Fischer von Röslerstamm never published (Ole Karsholt, pers. comm.).

maritella Zeller, 1847b (*Argyresthia*), Linn. Ent. 2: 250. Nomen nudum.
Note: *maritella* "F. R. [Fisher von Röslerstamm] in litt." was cited by Zeller. It is a manuscript name referred to by Zeller in his 1847 article, which Fischer von Röslerstamm never published (Ole Karsholt, pers. comm.).

nitidella Denis & Schiffermüller, 1775 (*Tinea*), Schmett. Wien: 137. (Type lost); not given.
Type locality: not given.
Note: Zeller (1847b: 250) attributed the authorship to Fabricius, 1794, which is incorrect. Fabricius was referencing *Tinea nitidella* Denis & Schiffermüller, 1775.

ossea Haworth, 1828 (*Erminea bonnetella* ab.), Lep. Brit. 4: 517. (BMNH & OUM: Syntypes); not given (1 ♂ in BMNH; 1♀ in OUM).
Type locality: England – London.

purpurascentella Stainton, 1849a (*Argyresthia*), Zoologist 7: [in Appendix] v. (BMNH: Holotype ♂); described from 2 specimens (BMNH).
Type locality: England – Stirlingshire, near Carron.

tetrapodella Linnaeus, 1767 (*Phalaena Tinea*), Syst. Nat. (ed. 12) 1(2): 890 [no. 388]. (NRM: Lectotype ♂ [designated by Robinson & Nielsen, 1983: 234]); not given (2 specimens in NRM).
Type locality: Italy – Salerno Prov., Ceraso.

brockeella Hübner, [1813] (*Tinea*), Samml. Eur. Schmett. Vol. 8, pl. 53, Tineæ 2, Pyraliformes B, fig. 262 [in Hübner, 1796–1838]. (Type lost); not given.
Type locality: not given.
Distribution: Palearctic – Austria, Belgium, China (Northeast), Czech Republic, Denmark, England, Estonia, Finland, France, Germany, Hungary, Ireland, Israel, Italy, Japan (Hokkaido, Honshu), Latvia, Lithuania, Netherlands, Norway, Poland, Romania, Russia, Slovakia, Spain, Sweden, Switzerland.
Hosts: Betulaceae – *Alnus glutinosa* (L.) Gaertn., *Alnus* spp., *Betula* spp. (Agassiz, 1996).

aurivittella Haworth, 1828 (*Tinea*), Lep. Brit. 4: 570. (type not found); described from 2 specimens.
Type locality: [England].

brockella; Stephens, 1834 (*Argyrosetia*), Ill. Brit. Ent. 4: 252. [misspelling of *brockeella*].

brokeella; Gómez-Bustillo, 1979 (*Argyresthia*), SHILAP Revta. lepid. 7(28): 240. [misspelling of *brockeella*].

debrockeel Duponchel, 1838 [1840] (*Oecophora*), In: Godart & Duponchel, Hist. Nat. Lép. France, 11: 464. (MNHN: Syntypes); not given.
Type locality: France – Northern Department.

fulguralis Matsumura, 1931 (*Argyresthia*), 6000 Ill. Ins. Jap.: 1096. (EIHU: Syntype); not given (1 ♂ in EIHU).
Type locality: Japan – Hokkaido, Sapporo.

iwella Haworth, 1828 (*Tinea*), Lep. Brit. 4: 569. (Type not found); not given.
Type locality: England – Coomb Wood.

buvati Gibeaux, 1993 (*Blastotere*), Entomol. Gall. 3(4): 189. (Gibeaux collection [private]: Holotype ♂); described from 8♂, 4♀ (Gibeaux collection).
Type locality: France – Hautes-Alpes, Saint-Crépin.
Distribution: Palearctic – France.

calliphanes Meyrick, 1913a (*Argyresthia*), Exot. Micr. 1(5): 135. (BMNH: Lectotype ♂ [designated by Clarke, 1965: 280]); described from 50 specimens, 14 of them extant in BMNH (Clarke, 1965: 280).
Type locality: Canada – Ontario, Toronto.
Distribution: Nearctic – Canada (British Columbia, Ontario), USA (Alaska, California, Maine, New York).
Note: This species has been confused with *A. goedartella* L. It is still unclear if both *A. calliphanes* and *A. goedartella* are present in North America. Therefore, some Nearctic records of *A. goedartella* may be due to confusion with *A. calliphanes*. A careful revision of these records is necessary.

canadensis Freeman, 1972 (*Argyresthia*), Can. Ent. 104(5): 692. (CNC: Holotype ♂); described from 13♂, 3♀ (CNC).
Type locality: Canada – Québec, Aylmer.
Distribution: Nearctic – Canada (Manitoba, New Brunswick, Nova Scotia, Ontario, Québec), USA (Northeast: Connecticut).
Host: Cupressaceae – *Thuja occidentalis* L. (Freeman, 1972).

carcinomatella Zeller, 1877 (*Argyresthia*), Hor. Soc. Ent. Ross. 13: 247. (BMNH: Syntypes); described from 5 specimens (1♂, 1♀ in BMNH).
Type locality: Colombia – Bogota.
Distribution: Neotropical – Colombia.

castaneella Busck, 1915 (*Argyresthia*), Proc. Ent. Soc. Wash. 17(2): 90. (USNM: Lectotype ♂ [designated by Sohn & Lewis, 2012: 5]); not given (1♂, 1♀ in USNM).
Type locality: USA – New Hampshire, Hampton.
Distribution: Nearctic – USA (Maine, New Hampshire, Virginia).
Host: Fagaceae – *Castanea dentata* (Marshall) Borkh. [bark] (Busck, 1915)

chalcocausta Meyrick, 1935a (*Argyresthia*), Mat. Microlep.-Fauna Chin. Prov. Kiangsu, Chekiang, Hunan: 91. (BMNH: Lectotype ♀ [designated by Clarke, 1965: 280]); described from 3 specimens.
Type locality: China – Zhejiang, Mt. Tienmushan.
Distribution: Oriental – China (Zhejiang).

chalcochrysa Meyrick, 1918a (*Argyresthia*), Exot. Micr. 2(6): 187. (BMNH: Holotype ♂); described from 1 specimen.
Type locality: Canada – Ontario, Parry Sound.
Distribution: Nearctic – Canada (Ontario), USA (Maine).

chalybaeella Costa, 1836 (*Plutella*), Fau. Reg. Napoli. Lepid.1836: [263] (Type not found); not given.
Type locality: Italy – Citra Prov. [=Salerno Prov.], Luglio.
Distribution: Palearctic – Italy.

chamaecypariae Moriuti, 1965b (*Argyresthia*), Bull. Univ. Osaka Pref. (B) 16: 69. (Issiki collection: Holotype ♂ [not traced]); described from 7♂, 4♀ (USNM & UOP).
Type locality: Japan – Honshu, Yamaguchi Pref., Takanoko.
Distribution: Palearctic – Japan (Honshu, Shikoku).
Hosts: Cupressaceae – *Chamaecyparis obtusa* (Siebold & Zucc.) Endl., *C. pisifera* (Siebold & Zucc.) Endl. (Moriuti, 1977a).

chionochrysa Meyrick, 1931a (*Argyresthia*), Exot. Micr. 4(3): 92. (BMNH: Holotype ♀); described from 1 specimen.
Type locality: China – Tibet, Yatung.
Distribution: Palearctic – China (Tibet).

chrysidella Peyerimhoff, 1877 (*Argyresthia*), Pet. Nouv. Ent. 2: 102. (Holotype lost, J. Minet, pers. comm.); described from 1 specimen.
Type locality: France – Côte d'Azur, Golfe Juan.
Distribution: Palearctic – France.
Host: Cupressaceae – *Juniperus oxycedrus* L. (Silva et al. 1995).

> *walsinghamella* Millière 1880 [1882] (*Argyresthia*), Lépidoptérologie 8: 7. (MNHN: Possibly lost); may have been destroyed in World War II, (G. Luquet via pers. comm. to J. Minet); not given.
> Type locality: France – Cannes.

columbia Freeman, 1972 (*Argyresthia*), Can. Ent. 104(5): 696. (CNC: Holotype ♂); described from 3♂, 4♀ (CNC).
Type locality: Canada – British Columbia, Christina Lake near Fife.
Distribution: Nearctic – Canada (Alberta, British Columbia).
Host: Pinaceae – *Larix occidentalis* Nutt. (Freeman, 1972).

communana Moriuti, 1969a (*Argyresthia*), Bull. Univ. Osaka Pref. (B) 21: 22. (UOP: Holotype ♂); described from 8♂, 7♀ (UOP; USNM).
Type locality: Japan – Kyushu, Kagosima Pref., Mt. Kirisimayama.
Distribution: Palearctic – Japan (Honshu, Kyushu).

conjugella Zeller, 1839 (*Argyresthia*), Isis Oken 32(3): 204. (BMNH: Syntypes); described from 5♂, 4♀ (1♀ in BMNH).
Type locality: Germany – Ulmen, Boehmen (Bohemia) & Glogau.
Distribution: Nearctic – Canada (Alberta, British Columbia, Newfoundland, Nova Scotia, Ontario, Québec, Saskatchewan), USA (California, Florida, Maine, New York); Oriental – Bhutan; Palearctic – Austria, Belgium, China (Shandong, Shanxi), Czech Republic, Denmark, England, Estonia, Finland, France, Germany, Hungary, Ireland, Israel, Italy, Japan (Hokkaido, Honshu), Latvia, Lithuania, Luxembourg, Netherlands, Norway, Poland, Russia, Slovakia, South Korea, Spain, Sweden, Switzerland, Turkey.
Hosts: Rosaceae – *Crataegus* spp., *Malus sylvestris* (L.) Mill., *Prunus domestica* L., *P. padus* L., *Prunus* sp., *Sorbus aucuparia* L., *S. commixta* Hedl. (Krämer, 1960; Friese, 1969; Moriuti, 1969a; Mirzoyan & Grigoryan, 1987; Sharma et al. 1988; Agassiz, 1996).
Note: Zeller (1839) did not specify the type series of *conjugella*. One specimen, probably the only one remaining from the type series, is indicated as "Type." This specimen was erroneously treated as the holotype of *conjugella* by Moriuti (1969a, 1977a). He examined the specimen and reported the sex as male; JCS reexamined the type specimen holding the same labels as Moriuti mentioned and found that it is actually female, based on the number of frenular bristles. A female paratype Moriuti (1969a, 1977) mentioned is actually a topotype as indicated on a label.

aerariella Stainton, 1871 (*Argyresthia*), Ent. Ann. 1871: 100. (not found); not given, but recorded as "several specimens."
Type locality: England – near Staleybridge, Brushes.

maculosa Tengström, 1848 (*Argyresthia conjugella* var.), Not. Sällsk. Faun. Fenn. Förh. 1: 136. (ZMH: Syntype); not given (1♂ in ZMH).
Type locality: Finland – Northern Ostrobothnia, Uleåborg [=Oulu].

conspersa Butler, 1883 (*Argyresthia*), Trans. Ent. Soc. Lond.: 85. (BMNH: Syntype); not given (1♂ in BMNH).
Type locality: Chile – Mountains of the hacienda of Cauquenes.
Distribution: Neotropical – Chile.

cupressella Walsingham, 1891a (*Argyresthia*), Insect Life, 3: 118. (BMNH: Syntypes); described from 4 specimens (1♂, 1♀ in BMNH; 2♂ in USNM).
Type locality: USA – California, Los Angeles.
Distribution: Nearctic – Canada (British Columbia), USA (California, Oregon, Washington); Palearctic – England (introduced).
Hosts: Cupressaceae – *Chamaecyparis lawsoniana* (A. Murray bis) Parl., *Cupressus guadaloupensis* S. Watson, *C.* x *leylandii* A.B. Jacks et Dallim, *C. macrocarpa* Hartweg, *C. sempervirens* L., *Juniperus chinensis* L., *J. scopulorum* Sarg. (Coquillett, 1891; Rosenthal et al. 1969; Agassiz & Tuck, 1999; Goodey, 2000; Powell & Opler, 2009); records from *Thuja* and *Sequoia* by Furniss & Carolin (1977) need verification.

curvella Linnaeus, 1761 (*Tinea*), Fauna Suecica: 359. (LSL: Syntypes); not given (2 specimens in LSL).
Type locality: [Sweden].
Distribution: Palearctic – Austria, Belgium, Bulgaria, Corsica, Czech Republic, Denmark, England, Finland, France, Germany, Hungary, Ireland, Israel, Italy, Latvia, Lithuania, Netherlands, Norway, Poland, Romania, Russia (European Central, European Northwest, Kaliningrad, Middle Volzhsky, South Ural, Volga-Donskoy, West Caucasus, Zabaykalsk), Slovakia, Sweden, Switzerland.
Hosts: Cornaceae – *Cornus* sp.; Rosaceae – *Malus domestica* Borkh., *M. sylvestris* Mill., *Crataegus*, *Prunus*; Cornaceae – *Cornus* (Gershenson, 1981; Baraniak, 1993; Küppers, 2008; Bengtsson & Johansson, 2011a).

> *arcella* Fabricius, 1776 (*Tinea*), Gen. Ins. 1: 295. (ZMUC: Lectotype [designated by Karsholt & Nielsen, 1976: 244; sex undeterminable, O. Karsholt, pers. comm.]); not given.
> Type locality: Germany.

> *curva* Haworth, 1828 (*Erminea*), Lep. Brit. 4: 516. [unjustified emendation of *curvella*].

deletella Zeller, 1873 (*Argyresthia*), Verh. Zool.-Bot. Ges. Wien 23(3): 305. (BMNH & MCZ: Syntypes); mentioned as "several specimens" (1♂, 1♀ in BMNH; 2♂, 1♀ in MCZ).
Type locality: USA – Texas.
Distribution: Nearctic – USA (Texas).

diffractella Zeller, 1877 (*Argyresthia*), Hor. Soc. Ent. Ross. 13: 245. (BMNH: Holotype ♂); described from 1 specimen.

Type locality: Colombia – Bogota.
Distribution: Neotropical – Colombia, Cuba.

dilectella Zeller, 1847b (*Argyresthia*), Linn. Ent. 2: 272. (Holotype ♂ [not found]); described from 1 specimen.
Type locality: Poland – Glogòw.
Distribution: Palearctic – Austria, Belgium, Czech Republic, Denmark, England, Estonia, Finland, France, Germany, Hungary, Ireland, Italy, Latvia, Lithuania, Netherlands, Norway, Poland, Russia (European Northwest, Kaliningrad), Slovakia, Sweden, Switzerland.
Hosts: Cupressaceae – *Juniperus communis* L., *Juniperus* spp.; *Chaemaecyparis* spp. (Baraniak, 1993; Agassiz, 1996).

dilutella; Dallas, 1866, Zool. Rec. 3: 495. [misspelling of *dilectella*].

dislocata Meyrick, 1914f (*Argyresthia*), Jour. Bomb. Nat. Hist. Soc. 23(1): 123. (BMNH: Lectotype ♂ [designated by Clarke, 1965: 283]); described from 12 specimens (BMNH).
Type locality: India – Assam, Khasi Hills.
Distribution: Oriental – India (Assam).

eugeniella Busck, 1916 (*Argyresthia*), Proc. Ent. Soc. Wash. 18(3): 153. (USNM: Syntype); not given (1♂ in USNM).
Type locality: USA – Florida, Key West.
Distribution: Nearctic – USA (Florida).
Hosts: Myrtaceae – *Eugenia foetida* Pers., *Psidium guajava* L., *Psidium* sp. (Busck, 1916; Sarwar, 2006).

festiva Moriuti, 1969a (*Argyresthia*), Bull. Univ. Osaka Pref. (B) 21: 14. (UOP: Holotype ♂); described from 3♂, 1♀ (UOP; NIAS)
Type locality: Japan – Honshu, Nara/Mie Pref., Mt. Kunimidake.
Distribution: Palearctic – Japan (Honshu).

flavicomans Moriuti, 1969a (*Argyresthia*), Bull. Univ. Osaka Pref. (B) 21: 16. (UOP: Holotype ♂); described from 3♂, 5♀ (UOP; USNM).
Type locality: Japan – Hokkaido, Hidaka, Mt. Apoidake.
Distribution: Palearctic – Japan (Hokkaido, Honshu).

flavipes Gibeaux, 1983b (*Argyresthia*), Revue Fr. Ent. (N. S.) 5(3): 121. (MNHN: Holotype ♂); described from 4♂, 2♀ (MNHN).

Type locality: Madagascar – Centre, Imaitso Forest, Anjavidilava.
Distribution: Ethiopian – Madagascar.

flexilis Freeman, 1960 (*Argyresthia*), Can. Ent. 92, Supp. 16: 28. (CNC: Holotype ♂); described from 1♂, 1♀ (CNC).
Type locality: USA – Montana, St. Marys.
Distribution: Nearctic – USA (Montana).
Host: Pinaceae – *Pinus flexilis* E. James. (Freeman, 1960).

franciscella Busck, 1915 (*Argyresthia*), Proc. Ent. Soc. Wash. 17(2): 90. (USNM: Holotype ♂); described from 1 specimen.
Type locality: USA – California, San Francisco.
Distribution: Nearctic – USA (California).
Host: Cupressaceae – *Cupressus macrocarpa* Hartweg (Keifer, 1930; Rosenthal et al. 1969; Powell & Opler, 2009)

freyella Walsingham, 1891a (*Argyresthia*), Insect Life 3: 119. (BMNH: Holotype ♂); described from 1 specimen.
Type locality: USA – Texas, Dallas.
Distribution: Nearctic – Canada (British Columbia, Manitoba, New Brunswick, Nova Scotia, Ontario, Québec), USA (Arkansas, Kentucky, Maine, Minnesota, Mississippi, Missouri, New York, Ohio, Tennessee, Texas).
Host: Cupressaceae – *Juniperus virginiana* L. (Silver, 1957).

friulii Huemer, 2009 (*Argyresthia*), Gorta. Botan. Zool. 31: 138. (TLMF: Holotype ♂); described from 10♂, 2♀ (MFSN, TLMF).
Type locality: Italy – Udine Province, Monte Canin (northern slope).
Distribution: Palearctic – Italy.

fujiyamae Moriuti, 1969a (*Argyresthia*), Bull. Univ. Osaka Pref. (B) 21: 11. (UOP: Holotype ♀); described from 1 specimen.
Type locality: Japan – Honshu, Yamanashi Pref., Mt. Huzisan.
Distribution: Palearctic – Japan (Honshu).
Host: Pinaceae – ?*Larix leptolepis* (Siebold & Zucc.) Gordon [reared from a pupa collected on the suspected host] (Moriuti, 1969a).

fundella Fischer von Röslerstamm, 1834 (*Oecophora*), Abb. Ber. Erg. Schmett. 1: 24. (NHMW: Syntypes); given as "in numbers."
Type Localities: Germany – Reichstadt; Poland – Glogòw.

Distribution: Palearctic – Austria, Belgium, Bulgaria, Czech Republic, Denmark, France, Germany, Italy, Netherlands, Norway, Poland, Romania, Russia (European Northwest, Kaliningrad), Slovakia, Sweden, Switzerland, [former] Yugoslavia.
Hosts: Pinaceae – *Abies alba* Miller, *A. nordmanniana* (Steven) Spach. (Friese, 1969; Baraniak, 1993).

furcatella Busck, 1916 (*Argyresthia*), Proc. Ent. Soc. Wash. 18(3): 154. (USNM: Lectotype ♂ [designated by Sohn & Lewis, 2012: 8]); not given (4♂, 1♀ in USNM).
Type locality: USA – Colorado, Colorado Springs Co., Cheyenne Canon.
Distribution: Nearctic – USA (Colorado).
Host: Bred from a cynipid gall on Oak (*Quercus* sp.) (Busck, 1916).

gephyritis Meyrick, 1938a (*Argyresthia*), In: Caradja & Meyrick, Dt. Ent. Z. Iris 52(1): 22. (BMNH: Holotype ♀); described from 1 specimen.
Type locality: China – Yunnan, Likiang [=Lijiang], Yülingshanmass [possibly Yulongxueshan].
Distribution: Oriental – China (Yunnan).

glabratella Zeller, 1847b (*Blastotere*), Linn. Ent. 2: 293. (BMNH: Lectotype ♀ [designated by Bengtsson & Johansson, 2011b: 270]); not given (3 specimens in BMNH).
Type locality: Poland – Walbrzych, Reinerz.
Distribution: Palearctic – Austria, Belgium, Bulgaria, Czech Republic, Denmark, Estonia, Finland, France, Germany, England, Italy, Latvia, Lithuania, Netherlands, Norway, Poland, Romania, Russia (European Central, Murmansk, South Ural), Slovakia, Sweden, Switzerland, [former] Yugoslavia.
Hosts: Pinaceae – *Picea abies* (L.) H. Karst., *P. excelsa* (Lam.) Lk. (Baraniak, 1993; Agassiz, 1996).

> *oleaginella* Standfuss, 1851 (*Argyresthia*), Bresl. Zeitschr. Ent. 5: 60. (RMNH: Holotype ♂); described from 1 specimen.
> Type locality: Poland – Reinerz, Wagnerberg.

glaucinella Zeller, 1839 (*Argyresthia*), Isis Oken 32(3): 205. (Type not found); described from 7 specimens.
Type locality: Germany – Schleben (=Torgau-Oschatz, Sachsen); ?Poland – Espengestrauch [=Glogów?].
Distribution: Palearctic – Austria, Belgium, Bulgaria, Czech Republic, Denmark, England, Estonia, Finland, France, Germany, Greece, Hungary, Ireland, Latvia,

Lithuania, Netherlands, Poland, Romania, Russia (Kaliningrad), Spain, Sweden, Switzerland, [former] Yugoslavia.
Hosts: Betulaceae – *Betula* sp. [bark]; Fagaceae – *Quercus* sp. [bark]; Sapindaceae – *Aesculus hippocastanum* L. [bark] (Friese, 1969; Agassiz, 1996).

dzieduszyckii Nowicki, 1860 (*Argyresthia*), Enum. Lep. Hal.: 204. (LMVA: Holotype [sex unknown]); described from 1 specimen.
Type locality: Austria.

goedartella Linnaeus, 1758 (*Phalaena Tinea*), Syst. Nat. 10(1): 541. (LSL: Lectotype ♀ [designated by Robinson & Nielsen, 1983: 214]); not given.
Type locality: not given.
Distribution: Nearctic – Canada (British Columbia, Québec), USA (Alaska, California, Colorado, Maine, Massachusetts, Michigan, New Mexico, Ohio, Pennsylvania, West Virginia); Palearctic – Austria, Belgium, Bulgaria, Czech Republic, Denmark, England, Estonia, Finland, France, Germany, Hungary, Ireland, Israel, Italy, Latvia, Lithuania, Luxembourg, Netherlands, Norway, Poland, Portugal, Romania, Russia (European Central, European Northwest, European South Taiga, Kaliningrad, Karelia, Middle Volzhsky, Murmansk, Primorsky, South Khabarovsk, South Ural, Volga-Donskoy, West Caucasus, Zabaykalsk), Slovakia, Spain, Sweden, Switzerland, [former] Yugoslavia.
Hosts: Betulaceae – *Betula verrucosa* Ehrh., *Betula* sp., *Alnus glutinosa* (L.) Gaertn., *Alnus* sp. [catkin]; Fagaceae – *Quercus* sp.; Ulmaceae – *Ulmus* sp. (Friese, 1969; Baraniak, 1993).
Note: Powell & Opler (2009) raised the possibility that *Argyresthia goedartella* in North America may represent a complex of several species, including *A. calliphanes* Meyrick.

aurescentella Uffeln, 1930 (*Argyresthia goedartella* f.), Sond. Abhand. West. Prov. Nat. 1: 68. (WPNM: Type not found); not given.
Type locality: Germany.

divisella Dufrane, 1960 (*Argyresthia goedartella* f.), Bull. Inst. Roy. Sci. nat. Belg. 36 (29): 14 (IRSN: Holotype ♂); described from 1 specimen.
Type locality: Belgium – Hainaut Prov., Frameries.

goedastella; Chambers, 1875c (*Argyresthia*), Cinc. Quart. Journ. Sci. 2: 294. [misspelling of *goedartella*].

goedertella; Dyar, 1903 (*Argyresthia*), Bull. U. S. Nat. Mus. 52: No. 6458. [misspelling of *goedartella*].

literella Haworth, 1828 (*Tinea*), Lep. Brit. 4: 570. (BMNH: Holotype ♀); described from 1 specimen.
Type locality: [England].

semiargentella Donovan, 1793 (*Phalaena*), Nat. Hist. Brit. Ins. 2: 76 & pl. 65, figs. 2 & 3. (Type not found); not given.
Type locality: England – Highgate.

splendida Reutti, 1898 (*Argyresthia goedartella* ab.), Verh. Natur. Vereins Karlsruhe: 295. (LNK: Syntypes); not given.
Type locality: Germany – Überlingen; Hinterzarten; Freiburg; Karlsruhe; Wertheim.

uniformella Dufrane, 1960 (*Argyresthia goedartella* f.), Bull. Inst. Roy. Sci. nat. Belg. 36 (29): 14 (IRSN: Holotype ♂); described from 2♂, 2♀ (IRSN).
Type locality: Belgium – Hainaut Prov., Ghlin.

hilfiella Rebel, 1910a (*Argyresthia*), Verh. Zool.-Bot. Ges. Wien 60: 429. (NHMW: Holotype ♀); described from 1 specimen.
Type locality: Greece – Cephalonia (=Kefalonia).
Distribution: Palearctic – Greece.

huguenini Frey, 1882 (*Argyresthia*), Mitt. Schweiz. Ent. Ges. 6: 369. (BMNH: Holotype ♂); described from 1 specimen.
Type locality: [Switzerland].
Distribution: Palearctic – Switzerland.

icterias Meyrick, 1907b (*Argyresthia*), Jour. Bomb. Nat. Hist. Soc. 17(4): 984. (BMNH: Lectotype ♂ [designated by Clarke, 1965: 284]); described from 16 specimens (BMNH).
Type locality: Sri Lanka – Central Prov., Maskeliya.
Distribution: Oriental – Sri Lanka.

idiograpta Meyrick, 1935a (*Argyresthia*), Mat. Microlep.-Fauna Chin. Prov. Kiangsu, Chekiang, Hunan: 91. (BMNH: Lectotype ♀ [designated here]); described from 2 specimens (BMNH).
Type locality: China – Zhejiang, Mt. Tienmushan.

Distribution: Oriental – China (Zhejiang).
Note: Only one type specimen of *Argyresthia idiograpta* is still available in the BMNH collection. We disignate this specimen as the lectotype to preserve the species identity. The type labels read: "Lecto- | type" [circular label with red borders], "Type" [circular label with red borders],"Tian-Mu-Shan | China | H. 5300. 5. 32" [hand-writing], "idiograpta Meyr." [hand-writing], "Argyresthia | idiograpta | 1/1 Meyr. | E. Meyrick det. | in Meyrick Coll.," "LECTOTYPE | *Argyresthia idiographa* | Meyrick | teste J. Sohn, 2013."

illuminatella Zeller, 1839 (*Argyresthia*), Isis Oken 32(3): 205. (BMNH: Lectotype ♂ [designated by Bengtsson & Johansson, 2011b: 271]); described from at least 2♂, 1♀ (only 1♂ still present in BMNH).
Type locality: Austria – [Mt] Spitzberge.
Distribution: Palearctic – Austria, Belgium, Bulgaria, Czech Republic, Estonia, France, Germany, Italy, Lithuania, Netherlands, Poland, Romania, Russia (European Northwest, Kaliningrad), Slovakia, Spain, Switzerland, Turkey.
Hosts: Pinaceae – *Abies alba* Mill., *Larix decidua* Mill. (Friese, 1969; Baraniak, 1993; Baraniak et al. 2003).

> *tatrica* Baraniak, Kulfan & Patočka, 2003 (*Argyresthia*), Mitt. Mus. Nat. Berl., Dtsch. entomol. Z. 50(2): 231. (SNMB: Holotype ♂); described from 5♂, 4♀ (AMUP, RMBB, SNMB, SASZ, ZMB).
> Type locality: Slovakia – High Tatra Mts. (Vysoké Tatry), Velická dolina.

impura Staudinger, 1880 (*Zell*[*eria*]?), Hor. Soc. Ent. Ross. 25: 280. (ZMB: Lectotype ♂ [designated by Friese, 1963: 408]); described from 7♂, 3♀ (ZMB).
Type locality: Turkey – Amasia, Jenikeui, Hochebene.
Distribution: Palearctic – Macedonia, Turkey, Uzbekistan.

inexpectella Gibeaux, 1983b (*Argyresthia*), Revue Fr. Ent. (N. S.) 5(3): 124. (MNHN: Holotype ♂); described from 2♂, 3♀ (MNHN).
Type locality: Madagascar – Centre, Soandrana Plateau.
Distribution: Ethiopian – Madagascar.

inscriptella Busck, 1907 (*Argyresthia*), Proc. USNM 32: 15. (USNM: Syntype); not given (1♂ in USNM).
Type locality: USA – Arizona, Williams.
Distribution: Nearctic – USA (Arizona).

iopleura Meyrick, 1918a (*Argyresthia*), Exot. Micr. 2(6): 187. (BMNH: Holotype ♂); described from 1 specimen.
Type locality: India – Almora, Kumaon.
Distribution: Oriental – India.
Host: Pinaceae – *Pinus roxburghii* Sarg. (Meyrick, 1918a).

italaviana Gibeaux, 1983b (*Argyresthia*), Revue Fr. Ent. (N. S.) 5(3): 122. (MNHN: Holotype ♂); described from 3♂, 6♀ (MNHN).
Type locality: Madagascar – Amoron'i Mania Region, Italaviana.
Distribution: Ethiopian – Madagascar.

ivella Haworth, 1828 (*Tinea*), Lep. Brit. 4: 570. (Type not found); not given.
Type locality: England – Kent, Darn Wood [=Darenth Wood?].
Distribution: Palearctic – Austria, Belgium, China (Northeast), Czech Republic, Denmark, England, Estonia, France, Germany, Hungary, Japan (Hokkaido, Honshu), Latvia, Lithuania, Norway, Poland, Romania, Russia (Kamchatka, Primorsky, South Khabarovsk, Zabaykalsk), South Korea, Sweden, Switzerland.
Hosts: Betulaceae – *Corylus avellana* L., *C. heterophylla* Fischer & Trautvetter; Rosaceae – *Malus domestica* Borkh., *M. sieboldii* Regel, *M. sylvestris* (L.) Mill., *Pyrus communis* L.; Salicaceae – *Salix* sp. (Friese, 1969; Moriuti, 1977a; Baraniak, 1993; Hua, 2005).

> *andereggiella* Duponchel, 1838 [1840] (*Oecophora*), In: Godart & Duponchel, Hist. Nat. Lép. France, 11: 469 pl. 305: 9. (MNHN: Syntype); not given (1♂ in MNHN).
> Type locality: Switzerland – Valais, vicinity of Gamsen.
>
> *anduegiella*; Chambers, 1877 (*Argyresthia*), Bull. U. S. Geol. Surv. 3: 131. [misspelling of *andereggiella*].
>
> *kuwayamella* Matsumura, 1931 (*Argyresthia andereggiella* f.), 6000 Ill. Ins. Jap.: 1095. (EIHU: Holotype ♂); described from 1 specimen.
> Type locality: Japan – Hokkaido.
>
> *robustella* Caradja, 1920 (*Argyresthia andereggiella* var.), Dt. Ent. Z. Iris 34: 92. (MGAB: Lectotype ♂ [designated by Popescu-Gorj, 1992: 135]); described from 3 specimens (MGAB).
> Type locality: Russia – Primorsk Krai, Kasakewitsch [=Kazakevich].

japonica Moriuti, 1969a (*Paraargyresthia*), Bull. Univ. Osaka Pref. (B) 21: 30. (UOP: Holotype ♂); described from 1♂, 2♀ (UOP).

Type locality: Japan – Shikoku, Ehime Pref., Mt. Isizutiyama.
Distribution: Palearctic – Japan (Honshu, Shikoku).

juniperivorella Kuznetzov, 1958 (*Argyresthia*), Dokl. Akad. Nauk Arm. SSR 27: 54. (ZMAS: Syntypes); described from 6 ♂, 12♀ [stated as "20 specimens"] (ZMAS; ZMB).
Type locality: Armenia – Alaverdskii, Akhtala.
Distribution: Palearctic – Armenia.
Host: Cupressaceae – *Juniperus* sp. (Kuznetzov, 1958).

kasyi Friese, 1963 (*Argyresthia*), Öst Akad. Wiss. Abt. I 172 (9/10): 408. (NHMW: Holotype ♂); described from 3♂, 5♀ (NHMW).
Type locality: Macedonia – Treskaschluct, Skopje.
Distribution: Palearctic – Macedonia.

kulfani Bengtsson & Johansson, 2011b (*Argyresthia*), Ent. Tidskr. 132(4): 269. (MZLU: Holotype ♀); described from 2♂, 1♀ (MZLU; NRM; ZMUC).
Type locality: Germany – Saxony, Upper Lusatia [=Oberlausitz].
Distribution: Palearctic – Germany, Slovakia.
Host: Pinaceae – *Abies alba* Mill. (Bengtsson and Johansson, 2011b).

kurenzovi Gershenson, 1988 (*Argyresthia*), Vestnik Zoologii 5: 84. (ZMAS: Holotype ♂); described from 1♂, 6♀ (ZMAS).
Type locality: Russia – Primorsk Krai, Ussuriysk.
Distribution: Palearctic – Russia (Primorsky).

laevigatella Herrich-Schäffer, [1855] (*Tinea*), Syst. Bearb. Schmett. Eur. 5: 271. [see Herrich-Schäffer, 1853–1855]. (Type lost); not given.
Type locality: Austria/Switzerland – Rhätischen Alpen [=now ?Rhätikon Mts. (Agassiz, 1996)]
Distribution: Palearctic – Austria, Czech Republic, Denmark, England, Finland, France, Germany, Hungary, Ireland, Italy, Japan (Honshu), Latvia, Netherlands, Norway, Poland, Russia (European Northeast, Irkutsk, Primorsky, Zabaykalsk), Slovakia, South Korea, Sweden, Switzerland.
Hosts: Pinaceae – *Larix decidua* Miller, *L. leptolepis* (Siebold & Zucc.) Gordon, *L. kaempferi* (Lamb.) Carrière (Moriuti, 1977a; Agassiz, 1996).
Note: According to O. Karsholt (pers. comm. to JAL) Herrich-Schäffer described a species in 1855 as *Tinea laevigatella* v Heyd. (H. v Heyden had not published the name, making Herrich-Schäffer the author) and credited H. v Heyden [not Heydenreich] as the source of the specimen(s) in the description at the end,

confusing the authorship for subsequent writers, who credit the authorship to Heydenreich erroneously. However, in 1851, Heydenreich did describe a species as *Brachmia laevigatellum* (family Tineidae, currently known as *Rhodobates laevigatellus* Heydenreich, 1851) found listed on page Nachtrag at the end of his Lepidopterorum Europaeorum Catalogus. Heydenreich cited Herrich-Schäffer as the source of the specimen in his description and referenced Herrich-Schäffer's figure 270 as an illustration of the species for the nomenclaturally unavailable [no genus] *laevigatellus* (Herrich-Schäffer, [1855], vol. 5, pl. 39, figure 270). Heydenreich's reference to Herrich-Schäffer's figure makes his name nomenclaturally available so he is the author of a species in the family Tineidae, not Argyresthiidae, and is not the author of *Argyresthia laevigatella.*

atmorella Tutt, 1896 (*Argyresthia*), Ent. Rec. J. Var. 7: 201 (Type not found); not given.
Type locality: England – King's Lynn.
Distribution: England
Note: It is obvious that Tutt (1896) and Bankes (1896) described the same species since they mentioned specimens collected by the same collector and from the same locality and host plant. See Beirne (1942) for further details on this issue.

atmoriella Bankes, 1896 (*Argyresthia*), Ent. M. Mag. 32: 25. (BMNH: Lectotype ♀ [designated here]); not given (3♂, 6♀ in BMNH).
Type locality: England – Norfolk, Kings Lynn.
Note: Moriuti (1969a: 10) stated that he had examined the holotype (♂), allotype (♀) and two male paratypes of *atmoriella*. His statement was, however, misguided because Bankes (1896) did not specify the type series in his description and specimen labels. Consequently, a female specimen was selected as lectotype of *atmoriella* by K. Sattler, as evidenced by a pin label, but his selection was not published. Because the male holotype of Moriuti cannot be located, and the female is undoubtedly the allotype he reported, to ensure the preservation and integrity of this taxon, we designate the female that was labelled by K. Sattler as lectotype here. The pin data labels read: "LECTO- | TYPE" [circular label with indigo borders], "Type" [circular label with red borders], "Kings Lynn; Nfk | VI. 1893 | Atmore," "June 1893 | King's Lynn, | ATMORE" [elongate octagonal label], "*Argyresthia* | *atmoriella*, Bankes. | E. A. Atmore Coll. King's Lynn, | June 1893. ♀. II TYPE E.M.M. Ser. 2, vii, 25" [hand-written except "TYPE"; with black-lined margins], "E.R.Bankes | Collection. | B.M. 1928-208," "B.M. ♀ |

Genitalia Slide | No. 15456," "Lectotype ♀ | *Argyresthia* | *atmoriella* Bankes | teste K. Sattler 1968."

majorella Müller-Rutz, 1934 (*Argyresthia laevigat*[*ella*] f.), Mitt. Schweiz. Ent. Ges. 16: 127. (NHMS: Syntypes); described from 6 specimens (2♂, 1♀ in NHMS). Type locality: Switzerland – Graubünden, Ochesenalp.

lamiella Bradley, 1965 (*Argyresthia*), Ruwenzori Exped. 1952 2(12): 107. (BMNH: Holotype ♂); described from 1♂, 5♀ (BMNH).
Type locality: Uganda – Ruwenzori, Lami Valley.
Distribution: Ethiopian – Uganda.

laricella Kearfott, 1908 (*Argyresthia*), Jour. N. Y. Ent. Soc. 16: 182. (USNM: Lectotype ♂ [designated by Sohn & Lewis, 2012: 9]); described from 6 specimens (3♂, 2♀ in USNM).
Type locality: Canada – Ontario, Mer Bleue [Bog].
Distribution: Nearctic – Canada (Alberta, British Columbia, Manitoba, New Brunswick, Nova Scotia, Ontario, Prince Edward Island, Québec, Saskatchewan), USA (Maine, Michigan, Vermont).
Hosts: Pinaceae – *Larix decidua* P. Miller, *L. laricina* (Du Roi) K. Koch, *L. occidentalis* Nuttall (Kearfott, 1908; Forbes, 1923; Eidt & Sippell, 1961).

leuculias Meyrick, 1914f (*Argyresthia*), Jour. Bomb. Nat. Hist. Soc. 23(1): 123. (BMNH: Lectotype ♂ [designated by Clarke, 1965: 284]); described from 6 specimens (BMNH).
Type locality: Sri Lanka – Central Prov., Maskeliya.
Distribution: Oriental – Sri Lanka.

libocedrella Busck, 1916 (*Argyresthia*), Proc. Ent. Soc. Wash. 18(3): 153. (USNM: Lectotype ♂ [designated by Sohn & Lewis, 2012: 10]); not given (3♂, 1♀, 1 of sex undetermined in USNM).
Type locality: USA – Oregon, Jackson Co., Ashland.
Distribution: Nearctic – USA (California. Oregon).
Hosts: Cupressaceae – *Calocedrus decurrens* (Torr.) Florin, *Chamaecyparis lawsoniana* (A. Murray bis) Parl. (Busck, 1916; Furniss & Carolin, 1977).

liparodes Meyrick, 1914e (*Argyresthia*), Exot. Micr. 1(7): 205. (BMNH: Holotype ♂); described from 1 specimen.
Type locality: Malawi – Nyassaland, Mlanje plateau.
Distribution: : Ethiopian – Malawi, South Africa.

lustralis Meyrick, 1911a (*Argyresthia*), Trans. Linn. Soc. Lond. (2) 14: 292. (BMNH: Lectotype ♂ [designated here]); described from 11 specimens (1♂ in BMNH).
Type locality: Seychelles – Silhouette.
Distribution: Ethiopian – Seychelles.
Note: Only one male specimen of the type series of *lustralis* can be found in the BMNH collection. We designate it as the lectotype to preserve the unique identity of the species. The pin labels read: "Lecto- | type" [circular label with indigo borders], "Type" [circular label with red borders], "Seychelles: | Silhouette, | Mare-aux-Cochons. | 1000H. IX. 1908, | H.Scott.," "Sihouette 1908. | Seychelles Exp. | 121," "Brit. Mus. | 1913-170," "*Argyresthia* | *lustralis* | Meyr | TYPE ♂" [rectangular label with black margins], "LECTOTYPE | *Argyresthia lustralis* | Meyrick | teste J. Sohn, 2014."

magna Moriuti, 1969a (*Argyresthia*), Bull. Univ. Osaka Pref. (B) 21: 23. (UOP: Holotype ♂); described from 3♂, 4♀ (UOP; USNM).
Type locality: Japan – Honshu, Nagano Pref., Sigakogan.
Distribution: Palearctic – Japan (Honshu, Shikoku), South Korea.

magnificella Karisch, 2002 (*Argyresthia*), Lambillionea 102(2): 169. (Karisch collection [private]: Holotype ♂); described from 1 specimen.
Type locality: Equatorial Guinea – Bioko Island, Gran Caldera de Luba u Ose.
Distribution: Ethiopian – Equatorial Guinea (Bioko Island).

mariana Freeman, 1972 (*Argyresthia*), Can. Ent. 104(5): 695. (CNC: Holotype ♂); described from 9♂, 10♀ (CNC).
Type locality: Canada – Ontario, North Bay.
Distribution: Nearctic – Canada (Alberta, Ontario, Québec, Yukon Territory).
Host: Pinaceae – *Picea mariana* (Mill.) Britton, Sterns & Poggenburg (Freeman, 1972).

media Braun, 1914 (*Argyresthia*), Ent. News 25: 113. (Braun collection: Holotype [sex unknown, not found]); described from 1 specimen.
Type locality: USA – Ohio, Cincinnati.
Distribution: Nearctic – USA (Maine, Ohio).

melitaula Meyrick, 1918a (*Argyresthia*), Exot. Micr. 2(6): 187. (BMNH: Lectotype ♂ [designated by Clarke, 1965: 291]); described from 3 specimens (1♂ in BMNH).
Type locality: Colombia – Valle del Cauca Dept., La Crumbre.
Distribution: Neotropical – Colombia.

mesocausta Meyrick, 1913a (*Argyresthia*), Exot. Micr. 1(5): 135. (BMNH: Holotype: ♂); described from 1 specimen.
Type locality: Canada – Ontario, Toronto.
Distribution: Nearctic – Canada (Ontario).

metallicolor Moriuti, 1969a (*Argyresthia*), Bull. Univ. Osaka Pref. (B) 21: 16. (UOP: Holotype ♂); described from 4♂, 1♀ (UOP).
Type locality: Japan – Honshu, Nagano Pref., Mt. Ontake.
Distribution: Palearctic – Japan (Honshu).

minusculella Rebel, 1940 (*Argyresthia*), Soc. Scient. Fenn., Comm. Biol. 8(1): 38. (NHMW: Syntype); described from more than one specimen according to the original description, however, only 1♀ found in NHMW.
Type locality: Azores.
Distribution: Palearctic – Azores.

mirabiella Toll, 1847 (1848) (*Argyresthia*), Ztschr. Wien. Entomol. Ges. 32: 114. (ISEZ: Holotype ♀); described from one specimen.
Type locality: Iran – Kuh -i- Mirabi Mts.
Distribution: Palearctic – Iran.

monochromella Busck, 1921 (*Argyresthia*), Can. Ent. 53: 279. (USNM: Holotype ♀); described from 2 specimens (1♂, 1♀ in USNM)
Type locality: Canada – British Columbia, Victoria.
Distribution: Nearctic – Canada (Alberta, British Columbia), USA (Utah).
Host: Pinaceae – ?*Pseudopsuga menziesii* (Mirbel) (Braun, 1925).

montana Fissetchko, 1970 (*Argyresthia*), Zool. Zhur. 49: 794. (ZMAS: Holotype ♂); described from 7♂, 13♀ (ZMAS).
Type locality: Kazakhstan – Reserve Aksu Dhabagly [=Aksu Zhabagly Nature Reserve].
Distribution: Palearctic – Kazakhstan.

montella Chambers, 1877 (*Argyresthia*), Bull. U. S. Geol. Surv. 3: 130. (MCZ: Syntype); not given (1♂ in MCZ).
Type locality: USA – Colorado, Edgerton.
Distribution: Nearctic – USA (Colorado).

nemorivaga Moriuti, 1969a (*Argyresthia*), Bull. Univ. Osaka Pref. (B) 21: 11. (UOP: Holotype ♂); described from 6♂, 5♀ (UOP).
Type locality: Japan – Shikoku, Ehime Pref., Mt. Isizutiyama.

Distribution: Palearctic – Japan (Hokkaido, Honshu, Shikoku).
Host: Pinaceae – *Abies sachalinensis* (F. Schmidt) Mast. (Moriuti, 1977a).

niphospora Meyrick, 1938a (*Argyresthia*), In: Caradja & Meyrick, Dt. Ent. Z. Iris 52(1): 23. (BMNH: Holotype ♀); described from 1 specimen.
Type locality: China – Yunnan, Likiang [=Lijiang], Yülingshanmass [possibly Yulongxueshan].
Distribution: Oriental – China (Yunnan).

nivifraga Diakonoff, 1955 (*Argyresthia*), Verh. K. ned. Akad. Wet. (2) 50(3): 36. (RMNH: Holotype ♂); described from six specimens (RMNH).
Type locality: Netherlands New Guinea – Scree Valley Camp.
Distribution: Australasian – Papua.

notoleuca Turner, 1913 (*Zelleria*), Proc. Linn. Soc. N.S. Wales 38(1): 194. (ANIC: Holotype ♀); described from 1 specimen.
Type locality: Australia – Queensland, Kuranda, near Cairns.
Distribution: Australasian – Australia (Queensland).

nymphocoma Meyrick, 1919 (*Argyresthia*), Exot. Micr. 2(8): 234. (BMNH: Holotype ♀); described from 1 specimen.
Type locality: Canada – Ontario, Muskoka Lake.
Distribution: Nearctic – Canada (Ontario).

ochridorsis Zeller, 1877 (*Argyresthia*), Hor. Soc. Ent. Ross. 13: 250. (BMNH: Holotype: ♂); described from 1 specimen.
Type locality: Colombia – Bogota.
Distribution: Neotropical – Colombia.

oreasella Clemens, 1861a (*Argyresthia*), Proc. Acad. Nat. Sci. Phil. 12: 7. (ANSP and Stainton collections [not found]: Syntypes); not given (2 specimens according to Busck, 1903).
Type locality: [North America – specific locality unknown]
Distribution: Nearctic – Canada (Alberta, British Columbia, Manitoba, Newfoundland, New Brunswick, Nova Scotia, Ontario, Québec, Saskatchewan), USA (California, Colorado, Idaho, Kentucky, Maine, Michigan, Missouri, New Mexico, New York, Ohio, Pensylvania, West Virginia).
Hosts: Rosaceae – *Prunus pensylvanica* L. f., *P. serotina* Ehrhart, *P. virginiana* L.; Oak excluded (Prentice, 1965).

oreadella Meyrick, 1914c (*Argyresthia*), Lep. Cat. 19: 6. [unjustified emendation of *oreasella*].

ornatipennella Moriuti, 1974b (*Argyresthia*), Trans. Lep. Soc. Japan 25(4): 113. (UOP: Holotype ♂); described from 1 specimen.
Type locality: Taiwan – Nantow Co., Rushan-Unchen.
Distribution: Oriental – Taiwan.

pallidella Braun, 1918 (*Argyresthia*), Can. Ent. 50: 231. (ANSP: Holotype [sex unknown, not found]); described from 2 specimens (ANSP).
Type locality: USA – California, San Bernardino Mts.
Distribution: Nearctic – USA (California).

pedmontella Chambers, 1877 (*Argyresthia*), Bull. U. S. Geol. Surv. 3: 131. (MCZ: Syntypes); not given (2♀ in MCZ).
Type locality: USA – Colorado, Edgerton.
Distribution: Nearctic – USA (Colorado).

pedemontella; Meyrick, 1914c (*Argyresthia*), Lep. Cat. 19: 10. [misspelling of *pedmontella*].

pentanoma Meyrick, 1913b (*Argyresthia*), Ann. Transv. Mus. 3: 318. (TMP: Holotype ♂); described from 1 specimen.
Type locality: South Africa – Northern Cape, Three Sisters.
Distribution: Ethiopian – Comoros, South Africa (Northern Cape).

perbella Moriuti, 1969a (*Argyresthia*), Bull. Univ. Osaka Pref. (B) 21: 28. (UOP: Holotype ♀); described from 1 specimen.
Type locality: Japan – Honshu, Hyogo Pref., Mt. Oginosen.
Distribution: Palearctic – Japan (Honshu).

percussella Zeller, 1877 (*Argyresthia*), Hor. Soc. Ent. Ross. 13: 246. (BMNH: Syntypes); not given (1♂, 1♀ in BMNH).
Type locality: Colombia – Bogota.
Distribution: Neotropical – Colombia.

perezi Vives Moreno, 2001 (*Argyresthia*), SHILAP Revta. lepid. 29(114): 167. (MNCN: Holotype ♂); described from 3♂, 9♀ (BMNH; MNCN; MNHN; ZMB).
Type locality: Spain – Mora de Rubielos, Teruel.

Distribution: Palearctic – Spain.
Host: Cupressaceae – *Juniperus oxycedrus* L. (Vives Moreno, 2001).

picea Freeman, 1972 (*Argyresthia*), Can. Ent. 104(5): 695. (CNC: Holotype ♂); described from 14♂, 12♀ (CNC).
Type locality: Canada – Ontario, North Bay.
Distribution: Nearctic – Canada (Alberta, Ontario, Québec, Yukon), USA (Alaska).
Host: Pinaceae – *Picea glauca* (Moench) Voss. (Freeman, 1972).

pilatella Braun, 1910 (*Argyresthia*), Ent. News 21: 171. (ANSP: Holotype [sex unknown, not found]); described from 9 specimens (ANSP; MCAS; USNM).
Type locality: USA – California, Oakland, Mills College.
Distribution: Nearctic – USA (California).
Host: Pinaceae – *Pinus radiata* D. Don. (Keifer, 1930).

plectrodes Meyrick, 1913a (*Argyresthia*), Exot. Micr. 1(5): 135. (BMNH: Holotype ♀); described from 1 specimen.
Type locality: Malaysia – Borneo, Kuching.
Distribution: Oriental – Malaysia (Borneo).

plicipunctella Walsingham, 1891a (*Argyresthia*), Insect Life 3: 119. (BMNH: Syntypes); not given (1♂, 1♀ in BMNH).
Type Localities: USA – California; Oregon.
Distribution: Nearctic – USA (California, Oregon).

praecocella Zeller, 1839 (*Argyresthia*), Isis Oken 32(3): 205. (BMNH: Lectotype ♂ [designated here]); described from 12 specimens (1♂ in BMNH).
Type locality: Poland – Glogòw.
Distribution: Palearctic – Austria, Belgium, Czech Republic, Denmark, England, Estonia, Finland, France, Germany, Hungary, Italy, Japan (Honshu), Latvia, Lithuania, Netherlands, Norway, Poland, Russia (European Central, European Northeast, European Northwest, Kaliningrad, Primorsky), Slovakia, Spain, Sweden, Switzerland.
Hosts: Cupressaceae – *Juniperus communis* L., *J. rigida* Siebold & Zucc. (Moriuti, 1977a; Baraniak, 1993; Agassiz, 1996).
Note: Moriuti (1977a) reported a type series of *praecocella* Zeller, which he claimed consisted of a 'holotype' and 20 paratypes, located in the BMNH collection. However, Zeller (1839) did not distinguish among his types, and included only 12 specimens in his description. JCS examined all those reported by Moriuti and found that the cited 'holotype' is the only specimen marked

unambiguously as a type. Therefore, we designate this male specimen as the lectotype to stabilize and preserve the species identity. The type data labels read: "Lecto- | type" [circular label with indigo borders], "Type" [circular label with red borders], "Gross Glogau | Silesia | b. v. ante 1839," "Walsingham Collection | 1910-427," "praecocella 7/2. 33," "Argyresthia | praecocella, Z. | Is. 1839. 205 | TYPE," "LECTOTYPE | *Argyresthia praecocella* | Zeller | teste J. Sohn. 2014."

praecosella; Strand, 1920 (*Argyresthia*), Archiv Naturgesch. (A) 85(4): 60. [misspelling of *praecocella*].

prenjella Rebel, 1901a (*Argyresthia*), Verh. Zool.-Bot. Ges. Wien 51(10): 802. (NHMW: Syntypes); described from 1♂, 1♀ (NHMW).
Type locality: Herzegovina – [Mt.] Prenj.
Distribution: Palearctic – Herzegovina.

pretiosa Staudinger, 1880 (*Argyresthia*), Hor. Soc. Ent. Ross. 15: 281. (ZMB: Syntypes); described from 1♂, 1♀ (ZMB).
Type locality: Turkey – Amasia, Jenikeui, Hochebene.
Distribution: Palearctic – Turkey.

pruniella Clerck, 1759 (*Phalaena*), Icon. Ins. Rar.: pl. 11, fig. 4. (Type not found); not given.
Type locality: [Europe] (not given).
Distribution: Nearctic – Canada (British Columbia, Nova Scotia); Palearctic – Albania, Austria, Belgium, Bulgaria, Czech Republic, Denmark, England, Estonia, Finland, France, Germany, Greece, Hungary, Ireland, Israel, Italy, Latvia, Lithuania, Luxembourg, Netherlands, Norway, Poland, Romania, Russia (European Central, European Northwest, European South Taiga, Kaliningrad, Middle Volzhsky, Ural, Volga-Donskoy, West Caucasus), Sardinia, Sicily, Slovakia, Spain, Sweden, Switzerland, [former] Yugoslavia.
Hosts: Rosaceae – *Prunus avium* (L.) L., *P. cerasus* L., *P. persica* (L.) Batsch, *Pyrus* sp., *Sorbus* sp. (Friese, 1969; Baraniak, 1993; Agassiz, 1996; Küppers, 2008). Records from *Corylus* sp. (Betulaceae) are suspicious.
Note: Clerck illustrated and named *Phalaena pruniella* in 1759. Linnaeus published a description of the species in 1761 (Fauna Suecica: 359, species number 1386), citing it as *Phalaena Tinea pruniella.* Then, Hübner (1796) published an illustration, cited as *Tinea pruniella* Clerck, 1759, in Samml. Eur. Schmett. Vol. 8, pl. 26, Tineæ 5, Ignobiles B, fig. 175 [in Hübner, 1796–1838]. The authorship of the species was subsequently erroneously attributed to Hübner by Treitschke

(1830: 299), and to Linnaeus by Curtis (1833: 191) and Stephens (1834: 247). Further, Fletcher (1928: 2) erred when he cited Stephens as the author.

comella Scopoli, 1763 (*Phalaena*) Ent. Carn.: 253. (Type lost); not given.
Type locality: not given.
ephippella Fabricius, 1776 (*Alucita*), Gen. Ins. I: 297. (ZMUC: Syntype); not given (1♀ in ZMUC).
Type locality: not given.

granonensis Gibeaux, 1983a (*Argyresthia pruniella* ssp.), Linneana Belgica 9(3): 146. (Gibeaux collection: Holotype ♂); described from 6♂, 3♀ (Gibeaux collection).
Type locality: France – Hautes-Alpes, Granon Pass.

pruni Haworth, 1828 (*Erminea*), Lep. Brit. 4: 516. [unjustified emendation of *pruniella*].

semialbella Haworth, 1802 (*Tinea*), Prod. Lepidopt. Brit.: 38. (Type not found); not given.
Type locality: [probably England].

psamminopa Meyrick, 1932c (*Argyresthia*), Exot. Micr. 4(11): 339. (BMNH: Lectotype ♂ [designated by Clarke, 1965: 288]); described from 13 specimens (BMNH).
Type locality: India – Kashmir, Gulmarg.
Distribution: Oriental – India (Kashmir).

pseudotsuga Freeman, 1972 (*Argyresthia*), Can. Ent. 104(5): 697. (CNC: Holotype ♂); described from 2♂, 1♀ (CNC).
Type Locality; Canada – British Columbia, Sooke.
Distribution: Nearctic – Canada (British Columbia, Ontario), USA (California).
Host: Pinaceae – *Pseudotsuga menziesii* (Mirb.) Franco (Freeman, 1972).

pulchella Lienig & Zeller, 1846 (*Argyresthia*), Isis Oken 39(3/4): 293. (BMNH: Syntypes); described from 1♂, 1♀ (BMNH).
Type locality: Austria – Salzburg, Gastein.
Distribution: Palearctic – Austria, Czech Republic, Estonia, Finland, France, Germany, Italy, Latvia, Netherlands, Norway, Poland, Romania, Russia (European Central, European Northwest, Karelia, Middle Volzhsky), Slovakia, Sweden, Switzerland.
Hosts: Rosaceae – *Malus sylvestris* Mill., *Sorbus aucuparia* L. (Friese, 1969). The record of *Corylus avellana* L. (Betulaceae) by Osthelder (1951) is suspicious.

pumilella Gibeaux, 1983b (*Argyresthia*), Revue Fr. Ent. (N. S.) 5(3): 128. (MNHN: Holotype ♂); described from 1 specimen.
Type locality: Madagascar – Andranotobaka, Ambatolampy.
Distribution: Ethiopian – Madagascar.

purella Chrétien, 1907 [1908] (*Argyresthia*), Bull. Soc. Ent. Fr. 20: 340. (MNHN: Holotype ♂); described from 1 specimen.
Type locality: Algeria – Biskra.
Distribution: Palearctic – Algeria.

pusiella Gibeaux, 1983b (*Argyresthia*), Revue Fr. Ent. (N. S.) 5(3): 126. (MNHN: Holotype ♂); described from 3♂ (MNHN).
Type locality: Madagascar – Amoron'i Mania Region, Italaviana.
Distribution: Ethiopian – Madagascar.

pygmaeella Denis & Schiffermüller, 1775 (*Tinea*), Schmett. Wien: 141. (Type lost); not given.
Type locality: Austria – Vienna district.
Distribution: Nearctic – Canada (Alberta, British Columbia, Manitoba, Newfoundland, New Brunswick, Ontario, Québec, Saskatchewan), USA (Alaska, Maine, Michigan, Washington); Palearctic – Austria, Belgium, Czech Republic, Denmark, England, Estonia, Finland, France, Germany, Hungary, Ireland, Italy, Latvia, Lithuania, Luxembourg, Netherlands, Norway, Poland, Romania, Russia (European Central, European Northwest, Kaliningrad, Murmansk), Slovakia, Spain, Sweden, Switzerland, Turkey, [former] Yugoslavia.
Hosts: Salicaceae – *Salix caprea* L., *Salix* spp. (Friese, 1969; Baraniak, 1993; Agassiz, 1996).

> *alpina* Müller-Rutz, 1920 (*Argyresthia pygmaeella* f.), Mitt. Ent. Zurich 5: 345 (NHMS: Holotype ♀); described from 1 specimen.
> Type locality: Switzerland – Graubünden, Calfreisen.

> *capilella* Strand, 1901 (*Argyresthia pygmaeella* var.), Nyt Mag. Nat. 39(1): 70 (ZMO: Syntypes); described from 3 specimens (ZMO).
> Type locality: Norway – Nordland Co., Tysfjorden.

> *hyperboreella* Strand, 1920 (*Argyresthia pygmaeella* var.), Archiv Naturgesch. (A) 85(4): 60. (ZMO: Syntypes); described from 3 specimens.
> Type locality: Norway – Nordland Co., Ranen.

> *rudolphella* Esper, 1791 (*Tinea*), Naturforscher 25: 43. (ZSM [Type not found]); not given.

Type locality: Germany – possibly Erlangen.
Note: Koçak (1981) erroneously treated *rudolphella* as a valid name over *pygmaeella*. He confused *pygmaeella* Denis & Schiffermüller, 1775 and *pygmaeella* Hübner, [1813]. The latter was a citation of *pygmaeella* Denis & Schiffermüller, 1775.

semifasciella Haworth, 1828 (*Tinea*), Lep. Brit. 4: 570. (Type not found); not given.
Type locality: England.

quadristrigella Zeller, 1873 (*Argyresthia*), Verh. Zool.-Bot. Ges. Wien 23: 304. (MCZ: Holotype ♂); described from 1 specimen.
Type locality: USA – Texas.
Distribution: Nearctic – USA (Maine, Oklahoma, Texas).
Host: Cupressaceae – ?*Juniperus virginiana* L. [adult association] (Busck, 1907).

quercicolella Chambers, 1877 (*Argyresthia*), Bull. U. S. Geol. Surv. 3: 130. (MCZ: Syntypes); not given (2♂ in MCZ).
Type locality: USA – Colorado, Edgerton.
Distribution: Nearctic – USA (Colorado).
Hosts: Fagaceae – *Quercus gambelii* Nuttall, *Quercus* spp. (Chambers, 1877; Busck, 1907).

quercicalella; Chambers, 1879, *Argyresthia*; Can. Ent. 11: 144. [misspelling of *quercicolella*].

rara Moriuti, 1969a (*Argyresthia*), Bull. Univ. Osaka Pref. (B) 21: 22. (UOP: Holotype ♂); described from 3♂, 4♀ (UOP; USNM).
Type locality: Japan – Honshu, Nagano Pref., Sigakogan.
Distribution: Palearctic – Japan (Honshu, Shikoku).

resplenderella Gibeaux, 1983b (*Argyresthia*), Revue Fr. Ent. (N. S.) 5(3): 121. (MNHN: Holotype ♂); described from 5♂, 3♀ (MNHN).
Type locality: Madagascar – Centre, Boby Cirque.
Distribution: Ethiopian – Madagascar.

reticulata Staudinger, 1877 (*Argyresthia*), Stett. Ent. Zeit. 38: 205. (ZMB: Lectotype ♂ [designated here]); not given [both sexes mentioned] (4♂, 4♀ in ZMB).
Type locality: Switzerland – Graubünden, Muotas Valley, near Samaden.

Distribution: Palearctic – Belgium (introduced), Czech Republic, France, Netherlands, Switzerland.
Host: Cupressaceae – *Juniperus thurifera* L. (Bigot et al. 1988).
Note: One male specimen of *Argyresthia reticulata* in the type series holds a label which reads "lectotype." This designation (probably by Gerrit Friese) has never been published (Wolfram May, 2013, pers. comm. to JCS). This selection is validated here. The type labels read: "Muottas | 8-74" [hand-writttten], "Engadin | Turati" [hand-written], "281," "Origin." [pink label], "Reticulata Stgn." [hand-writttten], "Staudinger.," "*Blastotere* ♂ | *reticulata* Stgr. | nov. comb. | Dr. G. Friese det. 1968 | Lectotype [red label attached]" [label with red margins].

marmorata Frey, 1880 (*Argyresthia*), Lep. Schweiz: 384. (Type not found); described from 2 specimens.
Type locality: Switzerland – Valais.

retinella Zeller, 1839 (*Argyresthia*), Isis Oken 32(3): 205. (BMNH: Lectotype ♀ [designated here]); given as "many" specimens (1♀ in BMNH).
Type locality: [Poland – Glogów].
Distribution: Palearctic – Austria, Belgium, Czech Republic, Denmark, England, Estonia, Finland, France, Germany, Hungary, Ireland, Italy, Japan (Honshu), Latvia, Lithuania, Netherlands, Norway, Poland, Romania, Russia (European Central, European Northeast, European Northwest, European South Taiga, Kaliningrad, Karelia, Middle Volzhsky, Primorsky, South Khabarovsk, South Ural, Volga-Donskoy, Zabaykalsk), Slovakia, Spain, Sweden, Switzerland.
Hosts: Betulaceae – *Betula pubescens* Ehrh., *B. verrucosa* Ehrh., *Betula* sp. (Werner, 1958; Baraniak, 1993; Agassiz, 1996; Tenow et al. 1999). Records from *Quercus* sp. (Fagaceae) and *Salix caprea* L. (Salicaceae) by Werner (1958) are questionable.
Note: Zeller (1839) did not specifically record the type series of *retinella* except for stating that there were many specimens. There is only one type specimen still available in the BMNH collection and Moriuti (1977a) erroneously referred to it as the holotype. We designate the unique specimen as lectotype, correcting the status. The pin label data reads as follows: "Lecto- | type" [circular label with indigo borders], "retinella Z. | Is. 1839.8.209," "Zeller Coll. | Walsingham Collection | 1910-427," "TYPE in BMNH," "LECTOTYPE | *Argyresthia retinella* | Zeller | teste J. Sohn, 2014."

albicornis Caradja, 1920 (*Argyresthia fundella* ab.), Dt. Ent. Z. Iris 34: 92. (MGAB: Holotype ♂); described from 1 specimen.
Type locality: Russia – Primorsk Krai, Kasakewitsch [=Kazakevich]

ocellea Stephens, 1834 (*Ismene*), Ill. Brit. Ent. Haust. 4: 251. (Type not found); described from 2 specimens.
Type locality: England – Darenth Wood.

rileiella Busck, 1907 (*Argyresthia*), Proc. USNM 32: 20. (USNM: Syntype); not given (1♂ in USNM).
Type locality: USA – District of Columbia, Washington.
Distribution: Nearctic – USA (District of Columbia, Maine).

rileyella; Brower, 1984, List Lepid. Maine, Pt. 2: 27. [misspelling of *rileiella*].

ruidosa Braun, 1940 (*Argyresthia*), Trans. Am. Ent. Soc. 66: 275. (ANSP: Holotype ♂.); described from 15♂, 6♀ (ANSP, USNM)
Type locality: USA – New Mexico, Lincoln National Forest, Ruidoso Canyon.
Distribution: Nearctic – USA (New Mexico).
Host: Pinaceae – *Abies concolor* (Gordon) Lindley ex Hildebrand (Braun, 1940).

sabinae Moriuti, 1965b (*Argyresthia*), Bull. Univ. Osaka Pref. (B) 16: 71. (UOP: Holotype ♂); described from 16♂, 5♀ (UOP; USNM).
Type locality: Japan – Kyushu, Kumamoto.
Distribution: Palearctic – Japan (Kyushu).
Host: Cupressaceae – *Juniperus procumbens* (Siebold ex Endl.) Miq. (Moriuti, 1965b).

semiflavella Christoph, 1882 (*Argyresthia*), Bull. Soc. Imp. Nat. Mosc. 57: 12. (ZMB: Holotype ♀); described from 1 specimen.
Type locality: Russia – Primorsky Krai, Vladivostok.
Distribution: Palearctic – Russia (Primorsky Krai).
semifusca Haworth, 1828 (*Erminea*), Lep. Brit. 4: 517. (Type not found); not given.
Type locality: England – vicinity of London.
Distribution: Palearctic – Austria, Belgium, Czech Republic, Denmark, England, Estonia, Finland, France, Germany, Hungary, Ireland, Italy, Latvia, Lithuania, Netherlands, Norway, Poland, Romania, Russia (European Central, European Northwest, European South Taiga, Kaliningrad, Middle Volzhsky, Volga-Donskoy), Slovakia, Spain, Sweden, Switzerland, [former] Yugoslavia.
Hosts: Rosaceae – *Crataegus* spp., *Prunus padus* L., *P. spinosa* L., *Sorbus aucuparia* L. (Friese, 1969; Agassiz, 1996; Baraniak, 1993).

semipurpurella Stephens, 1834 (*Ismene*), Ill. Brit. Ent. Haust. 4: 249. (Type not found); described from 2 specimens.
Type locality: England – New Forest.

spiniella Zeller, 1839 (*Argyresthia*), Isis Oken 32(3): 204. (Type not found); described from 2♂, 1♀.
Type locality: Czech Republic – Bohemia.

semitestacella Curtis, 1833 (*Ederesa*), Ent. Mag. 1(2): 191. (DEMV: Syntype); not given (1♀ in DEMV).
Type locality: England – New Forest.
Distribution: Palearctic – Austria, Belgium, Czech Republic, Denmark, England, Estonia, France, Germany, Hungary, Ireland, Italy, Luxembourg, Netherlands, Norway, Poland, Romania, Russia (Kaliningrad, West Caucasus), Slovakia, Sweden, Switzerland, [former] Yugoslavia.
Host: Fagaceae – *Fagus sylvatica* L. (Baraniak, 1993; Agassiz, 1996).
Note: The type specimen of *Ederesa semitestacella* holds no labels attached. The sex of the specimen has been determined by the exposed genitalia.

semitrunca Meyrick, 1914f (*Argyresthia*), Jour. Bomb. Nat. Hist. Soc. 23(1): 123. (BMNH: Lectotype ♀ [designated by Clarke, 1965: 288]); described from 5 specimens.
Type locality: India – Assam, Khasi Hills.
Distribution: Oriental – India (Assam).

semitruncata; Clarke, 1965 (*Argyresthia*), Cat. Type Specimens Microlepid. Br. Mus. Nat. Hist. Descr. E. Meyrick: 5: 288. [misspelling of *semitrunca*].

sorbiella Treitschke, 1833 (*Oecophora*), in Ochsenheimer & Treitschke, Schmett. Europa 9(2): 160. (Type not found); not given.
Type locality: Czech Republic/Germany – borders between Saxony and Bohemia.
Distribution: Palearctic – Austria, Belgium, Czech Republic, Denmark, England, Estonia, Finland, France, Germany, Hungary, Ireland, Italy, Lithuania, Netherlands, Norway, Poland, Romania, Russia (European Central, European Northeast, European Northwest, European South Taiga, Kaliningrad, Karelia, Middle Volzhsky, Murmansk, North Ural, Volga-Donskoy), Slovakia, Sweden, Switzerland, [former] Yugoslavia.

Hosts: Rosaceae – *Cotoneaster* sp., *Prunus* sp., *Sorbus aria* (L.) Crantz, *S. aucuparia* L., *S. intermedia* (Ehrh.), *S. torminalis* (L.) Crantz (van Deurs, 1965; Bengtsson & Johansson, 2011a).

esperella de Villers, 1789 (*Phalaena Tinea*), Caroli Linn. Ent. 526, species 1073. (Type not found); not given.
Type locality: [probably Sweden].

spinosella Stainton, 1849a (*Argyresthia*), Zoologist 7: [in Appendix] ix. (Type not found); not given.
Type locality: England.
Distribution: Palearctic – Austria, Belgium, Bulgaria, Czech Republic, Denmark, England, Estonia, Finland, France, Germany, Greece, Hungary, Ireland, Italy, Latvia, Luxembourg, Netherlands, Norway, Poland, Romania, Russia (European Central, European Northwest, Kaliningrad, Middle Volzhsky, West Caucasus), Sardinia, Sicily, Slovakia, Spain, Sweden, Switzerland, [former] Yugoslavia.
Hosts: Rosaceae – *Prunus domestica* L., *P. spinosa* L. (Baraniak, 1993; Küppers, 2008).

mendica Haworth, 1828 (*Erminea*), Lep. Brit. 4: 517. [misidentification and unnecessary emendation of *Tinea mendicella* Hübner, 1796].

stilpnota Meyrick, 1913b (*Argyresthia*), Ann. Transv. Mus. 3: 318. (TMP: Syntypes); described from 4 specimens.
Type locality: South Africa – Barberton, Waterval Onder.
Distribution: Ethiopian – South Africa.

submontana Frey, 1870 [1871] (*Argyresthia*), Mitt. Schweiz. Ent. Ges. 3: 255. (BMNH: Syntype); not given (1♂ in BMNH).
Type locality: Switzerland – Uetliberg.
Distribution: Palearctic – Austria, Finland, France, Germany, Italy, Latvia, Russia (?), Sweden, Switzerland.
Hosts: Rosaceae – *Amelanchier ovalis* Medik., *Sorbus aria* (L.) Crantz. (Frey, 1870 [1871]; Mitterberger, 1912).

subreticulata Walsingham, 1882 (*Argyresthia*), Trans. Am. Ent. Soc. Phila. 10(3): 173. (ANSP: Holotype ♀); described from 1 specimen.
Type locality: USA – [?Pennsylvania].
Distribution: Nearctic – Canada (Ontario, Québec), USA (Florida, Maine, Massachusetts, New Hampshire, New Jersey, New York, North Carolina, Oklahoma, Pennsylvania, Tennessee).

Note: The type specimen in the ANSP does not hold a label indicating the collecting locality (Jason Weintraub, 2009, pers. comm.), but Busck (1907) stated he examined a specimen from Pennsylvania (perhaps a specimen in the collection of the American Entomological Society in Philadelphia) then stated the type is in the ANSP, so the type locality as given herein is submitted as questionable.

subrimosa Meyrick, 1932a (*Argyresthia*), Exot. Micr. 4(8): 227. (BMNH: Holotype: ♂); described from 1 specimen.
Type locality: China – Sichuan, Kwanhsien.
Distribution: Oriental – China (Sichuan); Palearctic – Japan (Hokkaido, Honshu).

> *mutuurai* Moriuti, 1964 (*Argyresthia*), Trans. Lep. Soc. Japan 15(1): 20. (UOP: Holotype ♂); described from 2♂, 1♀, 1 specimen of unknown sex. (UOP).
> Type locality: Japan – Hokkaido, Isikari, Sapporo.

svenssoni Bengtsson & Johansson, 2011b (*Argyresthia*), Ent. Tidskr. 132(4): 261. (MZLU: Holotype ♀); described from 25♂, 17♀ [stated as 41 specimens] (B.Å. Bengtsson collection [private]; MZLU; R. Johansson collection [private]; TLMF; ZMUC).
Type locality: Sweden – Jämtland, Vällista.
Distribution: Palearctic – Finland, Italy, Norway, Slovakia, Sweden.

taiwanensis Moriuti, 1968 (*Argyresthia*), Trans. Lep. Soc. Japan 19(3–4): 85. (USNM: Holotype ♂); described from 2♂ (USNM).
Type locality: Taiwan – Mt. Tattaka [=Sungkang].
Distribution: Oriental – Taiwan.

talassica Fissetchko, 1970 (*Argyresthia*), Zool. Zhur. 49: 792. (ZMAS: Holotype ♂); described from 10♂, 9♀ (ZMAS).
Type locality: Kazakhstan – Aksu Zhabagly Nature Reserve.
Distribution: Palearctic – Kazakhstan.

tarmanni Gibeaux, 1993 (*Argyresthia*), Entomol. Gall. 4(4): 161. (TLMF: Holotype ♂); described from 5♂ (TLMF).
Type locality: Austria – East Tyrol.
Distribution: Palearctic – Austria.
Hosts: Cupressaceae – *Juniperus communis* L., *J. nana* Willdenow (Gibeaux, 1993).

thoracella Busck, 1907 (*Argyresthia*), Proc. USNM 32: 9. (USNM: Syntype); not given (1♂ in USNM).
Type locality: USA – Arizona, Williams.
Distribution: Nearctic – USA (Arizona, California, Nevada).

thuiella Packard, 1871 (*Bucculatrix*), 1st Ann. Rpt. Mass.: 24. (MCZ: Syntypes); not given (1♂, 2♀ in MCZ).
Type locality: USA – Maine, Brunswick.
Distribution: Nearctic – Canada (New Brunswick, Nova Scotia, Ontario, Québec), USA (Maine, Massachusetts, Michigan, Missouri, New Jersey, Pennsylvania); Palearctic (introduced) – Austria, Belgium, Czech Republic, Germany, Hungary, Netherlands, Poland, Switzerland.
Host: Cupressaceae – *Thuja occidentalis* L. (Packard, 1871; Busck, 1907; Freeman, 1967).

thujella; Povolný & Zacha, 1990 (*Blastotere*), Ochr. Rost. 26: 67. [misspelling of *thuiella*].

thuriferana Gibeaux, 1992 [1993] (*Blastotere*), Entomol. Gall. 3(4):189 (MNHN: Holotype ♂); described from 6♂, 1♀ (MNHN; Gibeaux collection).
Type locality: France – Hautes Alpes, Saint-Crépin.
Distribution: Palearctic – France.
Host: Cupressaceae – *Juniperus thurifera* L. (Gibeaux, 1992 [1993]).

trifasciae Braun, 1910 (*Argyresthia*), Ent. News 21: 171. (ANSP: Syntypes); not given.
Type locality: USA – California, San Francisco, Ocean Beach.
Distribution: Nearctic – USA (California).
Host: Cupressaceae – *Cupressus macrocarpa* Hartw. (Powell & Opler, 2009).

triplicata Meyrick, 1914c (*Argyresthia*), Lep. Cat. 19: 4. [unnecessary replacement name for *trifasciae*].

trifasciata Staudinger, 1871 (*Argyresthia*), in Staudinger & Wocke, Cat. Lep. Eur. Faun.: 425. (ZMB: Lectotype ♂ [designated here]); not given (3♂, 3♀ in ZMB).
Type locality: Switzerland – Valais.
Distribution: Palearctic – Austria, Belgium, Czech Republic, Denmark, England, France, Hungary, Italy (?), Netherlands, Poland, Spain, Sweden, Switzerland.

Hosts: Cupressaceae – *Chamaecyparis* spp., *Cupressus* x *leyandii* (A.B. Jacks. & Dallim), *Cupressocyparis* sp., *Juniperus chinensis* L., *J. communis* L., *J. phoenicea* L., *J. sabina* L., *J. thurifera* L., *J. scopulorum* Sarg., *Thuja* spp. (Lhomme, 1946; Emmet, 1982; Agassiz, 1996; Gal & Szeoke, 2001; Baraniak & Walczak, 2003; Bengtsson & Johansson, 2011a).
Note: One male specimen of *Argyresthia trifasciata* in the type series bears a label reading "lectotype." This designation (probably by Gerrit Friese) has never been published (Wolfram May, 2013, pers. comm. with JCS). This selection is validated here. The type labels read: "Valesia Ag. | [leg. Anderegg]" [hand-writtten], "340," "Origin." [pink label], "Staudinger.," "*Blastotere* ♂ | *trifasciata* Stgr. | nov. comb. | Dr. G. Friese det. 1968 | Lectotype [red label attached]" [label with red margins].

tristella Gibeaux, 1983b (*Argyresthia*), Revue Fr. Ent. (N. S.) 5(3): 126. (MNHN: Holotype ♂); described from 1 specimen.
Type locality: Madagascar – Andranotobaka, Ambatolampy.
Distribution: Ethiopian – Madagascar.

trochaula Meyrick, 1938a (*Argyresthia*), In: Caradja & Meyrick, Dt. Ent. Z. Iris 52(1): 22. (BMNH: Lectotype ♀ [designated by Clarke, 1965: 288]); described from 2 specimens (BMNH).
Type locality: China – Yunnan, Likiang [=Lijiang], Yülingshanmass [possibly Yulongxueshan].
Distribution: Oriental – China (Yunnan); Palearctic – Nepal.

tsuga Freeman, 1972 (*Argyresthia*), Can. Ent. 104(5): 697. (CNC: Holotype ♂); described from 1♂, 1♀ (CNC).
Type Locality; Canada – British Columbia, Vancouver Island, Holberg.
Distribution: Nearctic – Canada (British Columbia, Ontario), USA (Alaska).
Host: Pinaceae – *Tsuga heterophylla* (Raf.) Sarg. (Freeman, 1972).

tutuzicolella Moriuti, 1969a (*Argyresthia*), Bull. Univ. Osaka Pref. (B) 21: 26. (UOP: Holotype ♂); described from 9♂, 9♀ (UOP; USNM).
Type locality: Japan – Honshu, Wakayama Pref., Osima.
Distribution: Palearctic – Japan (Honshu, Kyushu).
Host: Ericaceae – *Rhododendron dilatatum* Miq. (Moriuti, 1969a).

undulatella Chambers, 1874 (*Argyresthia*), Can. Ent. 6: 10. (MCZ & USNM: Syntypes); not given (3♂, 1♀, 2 specimens of unknown sex in MCZ; 1♂ in USNM).
Type locality: USA – Kentucky.

Distribution: Nearctic – USA (Kansas, Kentucky, Maine, Maryland, Michigan, New York, North Carolina, Ohio, Texas).
Host: Ulmaceae – *Ulmus rubra* Muhl. (Braun, 1921).

uralensis Baraniak & Junnilainen, 2011 (*Argyresthia* (*Blastotere*)), Pol. J. Ent. 80: 500. (ZMH: Holotype ♂); described from 2♂ (ZMH).
Type locality: Russia – South Ural, Chelyabinsk Dist., Iremel Mountain Reserve.
Distribution: Palearctic – Russia (South Ural).

ATTEVIDAE Mosher, 1916.
Bull. Illinois Lab. Nat. Hist. 12: 71.

ATTEVA Walker, 1854, List Lepid. Ins. BM 2: 526.
Type species: *Atteva niveigutta* Walker, 1854, by monotypy.

AMBLOTHRIDIA Wallengren, 1861, K. Svenska Fregatten Eugenies Resa 2(1)(10): 385.
Type species: *Phalaena fabriciella* Swederus, 1787, by subsequent designation by Fletcher, 1929.
Note: Fletcher (1928) designated *Atteva niveigutta* Walker, 1854 as the type species but this is unavailable. *Atteva niveigutta* Walker, 1854, a nominal species was not originally included in *Amblothridia* and not linked in synonymy with one of the originally included nominal species (Nye & Fletcher (eds.), 1991).

CARTHARA Walker, 1866b, List Lepid. Ins. BM 35: 1871. A junior homonym of *Carthara* Walker, 1865, the objective replacement name is *Synadia* Walker, 1866b.
Type species: *Carthara flavivitta* Walker, 1866b, by monotypy.

CORINEA Walker, 1863, List Lepid. Ins. BM 28: 542.
Type species: *Corinea niviguttella* Walker, 1863, by subsequent designation by Fletcher, 1928
oeta Grote, 1865, Proc. Ent. Soc. Philad. 5: 230. An objective replacement name for *Poeciloptera* Clemens, 1861b, a junior homonym.
Type species: *Poeciloptera compta* Clemens, 1861b, by monotypy.

POECILOPTERA Clemens, 1861b, Proc. Acad. Nat. Sci. Phil. 12: 546. A junior homonym of *Poeciloptera* Loew, 1846 (Diptera). The replacement name is *Oeta* Grote, 1865.
Type species: *Poeciloptera compta* Clemens, 1861b, by monotypy.

SCINTILLA Guenée, 1879, Ann. Soc. Ent. Fr. (Série 5) 9(18–19): 287. A junior homonym of *Scintilla* Deshayes, 1856 (Mollusca).
Type species: *Tinea pastulella* Fabricius, 1787, by monotypy.

SYBLIS Guenée, 1879, Ann. Soc. Ent. Fr. (Série 5) 9(18–19): 288.
Type species: *Syblis glaucopidella* Guenée, 1879, by monotypy.

SYNADIA Walker, 1866b, List Lepid. Ins. BM 35: 1984. The objective replacement name for *Carthara* Walker, 1866b.
Type species: *Carthara flavivitta* Walker, 1866b, by monotypy.

SYNCALLIA Guérin-Méneville, [1844], Icon. Règne Anim. Cuvier 3: 497.
Type species: *Syncallia stellata* Guérin-Méneville, [1844], by monotypy. **Syn. nov.**

albiguttata Zeller, 1873 (*Oeta*), Verh. Zool.-Bot. Ges. Wien 23: 230. (BMNH: Syntypes); not given (1♂, 1♀ in BMNH).
Type locality: Australia – Queensland
Distribution: Australasian – Australia (Queensland, New South Wales).
Host: Simaroubaceae – *Quassia* sp. (Britton & Couper, 2009).

albitarsis Felder & Rogenhofer, 1875 (*Amblothridia*), Reise öst. Fregatte Novara (Zool.) (2) 2: legend for pl. 139 [in footnote]. (NHMW: Holotype ♂); described from 1 specimen.
Type locality: [Indonesia].
Distribution: Oriental – Indonesia.

aleatrix Meyrick, 1922a (*Atteva*), Exot. Micr. 2(18): 551. (BMNH: Holotype ♀); described from 1 specimen.
Type locality: Fiji Islands – Lumi.
Distribution: Australasian – Fiji Islands.

anisochrysa Meyrick, 1928b (*Atteva*), Exot. Micr. 3(13): 415. (BMNH: Holotype ♀); described from 1 specimen.
Type locality: Papua New Guinea – Talesea Island.
Distribution: Australasian – Papua New Guinea (Talesea Island).

apicalis Vollenhoven, 1863 (*Oeta*), Tijdschr. v. Ent. 6: 139. (RMNH: Syntypes); not given.
Type locality: Indonesia – Java
Distribution: Oriental – Indonesia (Java), Philippines (Calayan Island).

triplex Diakonoff, 1967 (*Atteva*), USNM Bull. 257: 240. (USNM: Holotype ♀); described from 1 specimen. **Syn. nov.**
Type locality: Philippines – Calayan Island.

aurata Butler, 1882 (*Corinea*), Ann. Mag. Nat. Hist. 5(10): 238. (BMNH: Syntype); not given (1♂ in BMNH).
Type locality: Papua New Guinea – Duke of York Islands.
Distribution: Australasian – Papua New Guinea, Solomon Islands.

aurea Fitch, 1856 (*Deiopeia*), Trans. N. Y. State Agr. Soc. 16: 1856: 486. (CNC: Neotype ♀ [designated by Wilson et al. 2010: 54]); not given.
Type locality: USA – Florida, Marion Co., Ocala National Forest.
Distribution: Nearctic – Canada (Nova Scotia, Ontario, Québec), USA (Alabama, Arkansas, California, Connecticut, Delaware, Florida, Georgia, Illinois, Indiana, Iowa, Kansas, Kentucky, Louisiana, Maine, Maryland, Massachusetts, Michigan, Mississippi, Missouri, Nebraska, New Hampshire, New Jersey, New Mexico, New York, North Carolina, Ohio, Oklahoma, Pennsylvania, Rhode Island, South Carolina, Tennessee, Texas, Vermont, Virginia, West Virginia, Wisconsin). Neotropical – Belize, Costa Rica, Guatemala, Honduras, Mexico (Coahuila, Galeana, Puebla, Vera Cruz), Nicaragua.
Hosts: Simaroubaceae – *Ailanthus altissima* (Mill.) Swingle, *Castela emoryi* (A. Gray) Moran & Felger, *C. peninsularis* Rose, *C. polyandra* Moran & Felger, *Simarouba amara* Aubl, *S. glauca* de Candolle (Delgado & Couturier, 2008; Powell & Opler, 2009; Becker, 2009).

aurera; Becker, 2009 (*Oeta*), Rev. Bras. Ent. 53(3): 350. [a wrong citation (misspelling) of *aurea* in Stretch, 1873: 240].

compta Clemens, 1861b (*Poeciloptera*), Proc. Acad. Nat. Sci. Phil. 12: 547. (possibly lost); not given.
Type locality: USA – Texas.

edithella Busck, 1908 (*Atteva*), Proc. Ent. Soc. Wash. 9: 85. (USNM: Syntype); not given (1♀ in USNM).
Type locality: USA – Texas, Maverick Co.

ergatica Walsingham, 1914 (*Atteva*), Biol. Centr. Amer. 4: 328. (BMNH: Lectotype ♀ [designated here]); described from 10 specimens (BMNH; USNM).
Type locality: Belize – Toledo Dist., Rio Sarstoon.

Note: This species was described based on 10 specimens from several Neotropical countries. Walsingham (1914) roughly sorted the type series ("Type" vs "PTT [Paratypes]") but still included more than two specimens referred to "Type." Therefore, this description was actually based on syntypes. Becker (2009) mistakenly reported one as the holotype. We correct the type status, stabilize the type series, and preserve the identity of the name by designating the same specimen as the lectotype. The pin data labels read: "Lecto- | type" [circular label with indigo borders], "R. Sarstoon | B. Honduras | Blancaneau.," "R. Sarstoon | BR. HONDURAS | Blancaneau | Godm-Salv. Coll. | 66064 Wlsm," "Walsingham | Collection | 1910-427," "Atteva ♀ | ergatica | Wlsm. | TYPE ♀ deau[?]" [rectangular label with black margins], "B.M. ♀ | Genitalia slide | No. 31548," "Database # | CNC LEP | 00060676" [with barcode], "Barcode of Life Project | DNA extracted" [pale blue, rectangular label], "LECTOTYPE | *Atteva ergatica* | Walsingham | teste J. Sohn, 2014."

exquisita Busck, 1912 (*Atteva*), Proc. Ent. Soc. Wash. 14: 86. (USNM: Lectotype ♂ [designated by Sohn & Lewis, 2012: 7]); described from more than 2 specimens (1♂, 1♀ in USNM).
Type locality: Mexico – Coahuila, Mobano.

floridana Neumoegen, 1891 (*Oeta compta* var.), Can. Ent. 23: 122. (USNM: Lectotype ♂ [designated by Sohn & Lewis, 2012:8]); described from about 50 specimens (1♂ in USNM).
Type locality: USA – Florida, Upper Indian River.

microsticta Walsingham, 1914 (*Atteva*), Biol. Centr. Amer. 4: 330. (USNM: Holotype ♀); described from 1 specimen.
Type locality: Mexico – Puebla, Tehuacan.

balanota Meyrick, 1910b (*Atteva*), Trans. Roy. Ent. Soc. Lond. 1910: 471. (BMNH: Holotype ♀); described from 1 specimen.
Type locality: Philippines – Manila.
Distribution: Oriental – Philippines (Manila).

basalis Vollenhoven, 1863 (*Oeta*), Tijdschr. v. Ent. 6: 140. (RMNH: Syntypes); not given.
Type locality: Indonesia – North Maluku, Morotai Island.
Distribution: Australasian – New Guinea (Jobi Island), Solomon Islands (Shortland). Oriental – Indonesia (Buru, Luzon, Maluku, Morotai).

conspicua Walsingham, 1900 (*Atteva*), in Swinhoe et al., Cat. East. Aust. Lepid. Het. Oxford. Univ. Mus. 2: 559. (BMNH: Lectotype ♂ [desognated here]); described from 7 specimens (BMNH; OUM).
Type locality: Indonesia – Maluku, Buru Island.
Note: One male specimen in the type series of *conspicua* Walsingham is designated as lectotype here. The labels read: "LECTO- | TYPE" [circular label with indigo margins], "BURU | (below 1000 ft.) | Doherty, 1892 | No. 41151," "Walsingham | Collection. | 1910–427.," "*CORINEA* | *CONSPICUA*, Wlsm. | Cat.East. & Austral. Lep. Het. | Oxf. Mus. II. 559–560. No 3661 (1900) || TYPE ♂ ders." [label with double black-lined margins, outer lines bold], "LECTOTYPE | *Atteva conspicua* | Walsingham | teste J. Sohn, 2014."

heliodoxa Meyrick, 1910b (*Atteva*), Trans. Ent. Soc. Lond. 1910: 471. (BMNH: Holotype ♂); described from 1 specimen. **Syn. nov.**
Type locality: New Guinea – Jobi Island.

carteri Walsingham, 1891b (*Oeta*), Trans. Ent. Soc. Lond. 1891: 90. (BMNH: Holotype ♀); described from 1 specimen.
Type locality: Gambia – Bathurst [=Banjul].
Distribution: Ethiopian – Gambia, Nigeria, South Africa.
Host: Simaroubaceae – *Quassia silvestris* Check & Jongkind (Eidt, 1967).

stellata Guérin-Méneville, 1844 (*Syncallia*), Icon. Règne Anim. Cuvier 3: 497. (BMNH: Syntype); not given (1♀ in BMNH). **Syn. nov.**
Type locality: Senegal – Casamance River.

charopis Turner, 1903 (*Atteva*), Proc. Linn. Soc. N.S. Wales 28: 80. (BMNH: Syntypes); not given [stated as "a series including the type"] (17 specimens in BMNH).
Type locality: Australia – Queensland, Cedar Bay.
Distribution: Australasian – Australia (Queensland).

chionosticta Durrant, 1916 (*Atteva*), Rep. Brit. Ornith. Uni. Exped. 2(15): 159. (BMNH: Holotype ♀); described from 1 specimen.
Type locality: Papua New Guinea – Utakwa River.
Distribution: Australasian – Papua New Guinea.

cosmogona Meyrick, 1931a (*Atteva*), Exot. Micr. 4(3): 87. (NHMW: Holotype ♀); described from 1 specimen.
Type locality: Brazil – Rio de Janeiro, Nova Friburgo.
Distribution: Neotropical – Brazil (Rio de Janeiro).

covelli Heppner, 2009b (*Atteva*), Lepid. Novae 2(2): 125. (MCGF: Holotype ♀); described from 1 specimen.
Type locality: Ecuador – Quito, Yasuni Research Station, Dept. Napo.
Distribution: Neotropical – Ecuador.

cuprina Felder & Rogenhofer, 1875 (*Amblothridia*), Reise öst. Fregatte Novara (Zool.) (2) 2: pl. 139, fig. 21. (BMNH: Lectotype ♀ [designated here]); not given (1♀ in BMNH; 1♂ in NHMW).
Type locality: Indonesia – Maluku Island.
Distribution: Oriental – Indonesia (Maluku, Sulawesi).
Note: The type series is comprised by one male in the NHMW and one female in the BMNH. To stabilize and preserve the unique species, the female specimen is designated here as lectotype. The pin data labels read: "Lecto- | type" [a circular label with indigo borders], "Lorq[uin]," "219," "Novara CXXXIX | f 21. Amblothridia | cuprina n. | Moll. 1," "Felder Coll. | Rothschild | 1913-86," "FELDER'S | TYPE," "LECTOTYPE | *Atteva cuprina* | F. & R. | teste J. Sohn, 2014."

> *tonseana* Tams, 1935 (*Atteva*), Mém. Mus. H. n. Belg., hors sér. 4 (12): 64, pl. 2, fig. 6. (IRSN: Holotype ♂); described from 2 specimens (IRSN). **Syn. nov.**
> Type locality: Indonesia – Sulawesi, Menado, Tonsea Lama.

emissella Walker, 1863 (*Corinea*), List Lepid. Ins. BM 28: 543. (BMNH: Holotype ♂); described from 1 specimen.
Type locality: Malaysia – Borneo, Sarawak.
Distribution: Oriental – Bhutan, Malaysia (Borneo)

fabriciella Swederus, 1787 (*Phalaena Tinea*), K. Svenska. Vetensk. Akad. Handl. 8: 277. (USNM: Neotype ♀[designated by Sohn & Wu, 2013: 5]); not given.
Type locality: India.
Distribution: Oriental – China (Guangxi, Sichuan, Yunnan), India (Gujarat, Orissa), Indonesia (Borneo, Java, Sulawesi, Sumatra), Philippines (Luzon), Sri Lanka, Thailand.
Hosts: Simaroubaceae – *Ailanthus excelsa* Roxb., *A. triphysa* (Dennst.) Alston., *Brucea sumatrana* Roxb., *Quassia indica* (Gaertn.) Nooteboom (Horsfield & Moore, 1859; Fletcher, 1914; Mishra and Pandey, 1966; Mohanadas and Verma, 1984; Verma, 1992). Note: Records from *Acacia* spp. (Fabacae), *Boswellia serrata* Roxb. & Colebr.. (Burseraceae) and *Santalum album* L. (Santalaceae) are questionable (Beeson, 1941; Browne, 1968; Robinson et al. 1994).

brucea Moore, 1859 (*Atteva*), in Horsefield & Moore, Cat. Lep. Ins. E. Ind. Co. 2: 300. (BMNH: Lectotype ♂ [designated by Sohn & Wu, 2013: 5]); described from 6 specimens (1♂, 3♀ in BMNH).
Type locality: Indonesia – Java

bruceella Guenée, 1879 (*Atteva*), Ann. Soc. Ent. Fr. (Série 5) 9(18–19): 290. [unjustified emendation of *brucea*]

fabricella; Wallengren, 1861 (*Amblothridia*), K. Svenska Fregatten Eugenies Resa 2(1)(10): 385. [misspelling of *fabriciella*].

niveiguttella; Berg, 1880 (*Atteva*), An. Soc. Cient. Argent. 10: 104. [misspelling of *niviguttella*].

niviguttella Walker, 1863 (*Corinea*), List Lepid. Ins. BM 28: 542. (BMNH: Lectotype ♀ [designated by Sohn & Wu, 2013: 5]); described from 5 specimens (2♀ in BMNH).
Type locality: India – South.

flavivitta Walker, 1866b (*Carthara*), List Lepid. Ins. BM 35: 1872. (BMNH: Lectotype ♀ [designated here]); described from 1♂, 1♀, 1 sex undetermined (BMNH).
Type locality: Colombia – Magdalena Dept., Santa Marta.
Distribution: Neotropical – Colombia, Venezuela.
Note: Walker (1866) did not give the number of type specimens in his description. Becker (2009) incorrectly reported the existence of a holotype. In fact, there are three syntypes still available, one of which bears a "lectotype" label. This lectotypy, however, has never been published. We stabilize the type series by designating the unpublished lectotype. The pin labels read: ""Lecto- | type" [circular label with indigo margins], "Sta | Marta" [hand-written] [circular label] "a. Santa Marta," "*Carthara* | *flavivitta* | Wlkr. || TYPE ♂" [rectangular label with black margins], "LECTOTYPE | *Atteva flavivitta* | Walker | teste J. Sohn, 2014."

fulviguttata Zeller, 1873 (*Oeta*), Verh. Zool.-Bot. Ges. Wien 23: 231. (ZMB: Holotype ♀); described from 1 specimen.
Type locality: "Neuholland" [Antilles?].
Distribution: Neotropical – Dominican Republic, Jamaica.
Note: Becker (2009: 352) cites the type locality as "[Antilles?]," and further states Zeller mentioned Australia in the text, Becker referred to it as a mislabelling of the species. However, the last line of the species description reads:

"Vaterland: wahrscheinlich Neuholland.," interpreted here as: 'fatherland [Type locality]: probably Neuholland'.

fulviguttella; Walsingham, 1891c [1892] (*Oeta*), Proc. Zool. Soc. Lond. 1891: 532. [misspelling of *fulviguttata*].

glaucopidella Guenée, 1879 (*Syblis*), Ann. Soc. Ent. Fr. (Série 5) 9(18–19): 289. (BMNH: Lectotype ♂ [designated by Becker, 2009: 352]); not given (2♂ in BMNH).
Type locality: Jamaica.

glaucopodella; Kirby, 1879, Zool. Rec. (Ins.) 16: 188. [misspelling of *glaucopidella*].

gemmata Grote, 1873 (*Oeta*), Bull. Buff. Soc. N.S. 1.: 93. (Type not found); described from 1 specimen.
Type locality: Cuba.
Distribution: Neotropical – Cuba.

fastuosa Zeller, 1877 (*Oeta*), Hor. Soc. Ent. Ross. 13: 225. (ZMB: Holotype ♀); described from 1 specimen.
Type locality: Cuba.

hermosa Heppner, 2009a (*Atteva*), Lepid. Novae 2(2): 105. (UNAM: Holotype ♂); described from 16 specimens (MCGF, NHMW, USNM).
Type locality: Peru – near San Ramon, Pampa Hermosa Lodge.
Distribution: Neotropical – Peru.

holenopla Diakonoff, 1967 (*Atteva*), USNM Bull. 257: 241. (BMNH: Holotype ♀); described from 1 specimen.
Type locality: Philippines – Luzon Island, Benguet, Panai.
Distribution: Oriental – Philippines (Luzon).

hysginiella Wallengren, 1861 (*Amblothridia*), K. Svenska Fregatten Eugenies Resa 2(1)(10): 386. (NRM: Syntype); not given (1♀ in NRM).
Type locality: Panama [probably an error: according to B. Landry, this species is endemic to the Galápagos Islands].
Distribution: Neotropical – Ecuador (Galápagos Islands).
Host: Simaroubaceae – *Castela galapageia* Hooker *f.* (Landry & Landry, 1998)

monerythra Meyrick, 1926b (*Atteva*), Trans. Ent. Soc. Lond. 74: 278. (BMNH: Lectotype ♀ [designated here]); described from 3 specimens (BMNH).
Type locality: Ecuador – Galápagos Islands, Charles Island.
Note: Landry & Landry (1998) stated the holotype is in the BMNH, but subsequently, JCS found two other specimens with "paratype" labels as well. Since Meyrick (1926) did not designate a holotype in the description, the three specimens are actually syntypes. We correct the type status and designate as lectotype the specimen which Landry & Landry (1998) reported. Label data is repeated here verbatim following Landry & Landry (1998): "Holo- | type" [circular, red-bordered, white [label], printed]; "Charles Island, Galapagos, | at light, | 31.7.24. | St. George Expedn. | C. L. Collenette." [rectangular, grey [label], handwritten]; "Cydosia | sylpharis | Butler Type" [rectangular, white [label], handwritten]; "sylpharis | Butler." [rectangular, white [label], handwritten, with printed horizontal line near bottom]; "Type" [rectangular, white [label], printed, black-bordered, with thin horizontal line at basal third]; "genitalia slide ♀ | JFL 1468" [rectangular, pale green [label], printed (except for ♀)] (BMNH). The two other specimens in the BMNH collection which bear paratype labels are designated as paralectotypes.

sylpharis Butler, 1877 (*Cydosia*), Proc. Zool. Soc. Lond. 1877: 87. (BMNH: Holotype ♀); described from 1 specimen.
Type locality: Ecuador – Galápagos Islands, Albemarle Island.

impariguttata Zeller, 1877 (*Atteva*), Hor. Soc. Ent. Ross. 13: 224. (ZMB: Holotype ♀); described from 1 specimen.
Type locality: Sri Lanka – Colombo.
Distribution: Oriental – Sri Lanka.

impariguttella Walsingham, 1887 (*Atteva*), in Moore, Lep. Ceyl. 3: 505 [unjustified emendation of *impariguttata*].

impunctella Ritsema, 1875 (*Atteva*), Pet. Nouv. Ent. 7: 479. (RMNH: Holotype ♀); described from 1 specimen.
Type locality: Indonesia – Sumatra.
Distribution: Oriental – Indonesia (Sumatra).

intermedia Becker, 2009 (*Atteva*), Rev. Bras. Ent. 53(3): 352. (USNM: Holotype ♂); described from 8♂, 1♀ (USNM).
Type locality: Antigua Island – St. John's, Coolidge Airport.
Distribution: Neotropical – Antigua Island.

iris Felder & Rogenhofer, 1875 (*Amblothridia*), Reise öst. Fregatte Novara (Zool.) (2) 2: pl. 139, fig. 25. (BMNH: Holotype ♀); described from 1 specimen.
Type locality: Indonesia – Maluku Island.
Distribution: Oriental – Indonesia (Maluku).

mathewi Butler, 1887 (*Corinea*), Ann. Mag. Nat. Hist. 20(5): 414. (BMNH: Lectotype ♂ [designated here]); not given [stated to be based on both sexes] (2 specimens in BMNH).
Type locality: Solomon Islands – Malayta Island.
Distribution: Australasian – Solomon Islands.
Note: To eliminate possible future confusion and stabilize the species identity, the lectotype of *Corinea mathewi* is designated here. The data labels read: : "LECTO- | TYPE" [circular label with indigo margins], "Malayta | 83. 3 |·| (Woodford)" [hand-written], "*CORINEA* | *MATHEWI*. Btlr. | Ann.Mag.N.H. 5S. XIX.P414.1887 || TYPE ♂" [label with double black-lined margins, outer lines bold], "JDB [vertical] B.M. | Genitalia slide | No. 1674 [in red]," "LECTOTYPE | *Atteva mathewi* | Butler | teste J. Sohn, 2014."

megalastra Meyrick, 1907d (*Atteva*), Proc. Linn. Soc. N.S. Wales 32(1): 81. (BMNH: Lectotype ♀ [designated here]); described from 2 specimens (BMNH).
Type locality: Australia – Queensland, Port Douglas.
Distribution: Australasian – Australia (Queensland), New Guinea.
Note: A female specimen in the type series of *megalastra* Meyrick bears a "Lectotype" label which indicates that it was selected as such by I. F. B. Common in 1966. This lectotypy is however known only from the specimen, not from any publication. We formalize the lectotype designation by publishing the data for the specimen. The pin label data reads: "LECTO- | TYPE" [circular label with indigo margins], "Port Douglas | Queensland | TPL. ba [?] /85" [cream-colored, rectangular label] [hand-written], "Meyrick Coll. | B.M. 1938–290.," "megalastra Meyr." [hand-written], "Atteva | megalastra | 1/11 Meyr. | E. Meyrick det. | in Meyrick Coll." "LECTOTYPE ♀ | Atteva | megalastra Meyr.1907 | selected by | I.F.B. Common, 1966."

monoplanetis Meyrick, 1910b (*Atteva*), Trans. Ent. Soc. Lond. 1910: 470. (BMNH: Lectotype ♀ [designated here]); described from 2 specimens (BMNH).
Type locality: Papua New Guinea – Bismark Archipelago, New Hanover.
Distribution: Australasian – Papua New Guinea (Bismark Archipelago).
Note: One of the type specimens of *monoplanetis* bears a lectotype label, however, we have not found any evidence that this lectotype has been published. To stabilize the species and formalize the lectotypy, we designate the specimen

here. The pin data labels read: "LECTO- | TYPE" circular label with indigo margin], "New Hanores | Bismark Is. | R. 0.3.97" [hand-written on a cream-colored, rectangular label], "Meyrick Coll. | B.M. 1938–290.," "*monoplanetis* Meyr." [hand-written], "*Atteva* | *monoplanetis* | 1/2 Meyr. | E. Meyrick det. | in Meyrick Coll.," "Box: N11 | HYPON." [blue rectangular label], "LECTOTYPE | *Atteva monoplanetis* | Meyrick | teste J. Sohn, 2014."

niphocosma Turner, 1903 (*Atteva*), Proc. Linn. Soc. N.S. Wales 28: 79. (ANIC: Syntypes); described from 2 specimens (ANIC).
Type Localities: Australia – Queensland, Brisbane & Townsville.
Distribution: Australasian – Australia (Queensland).
Host: Araliaceae – *Polyscias murrayi* (F. Muell) Harms (Common, 1990).

> *myriastra* Meyrick, 1907d (*Atteva*), Proc. Linn. Soc. N.S. Wales 32(1): 82. (BMNH: Holotype ♀); described from 1 specimen.
> Type locality: Australia – Queensland, Maryborough.

niveigutta Walker, 1854 (*Atteva*), List Lepid. Ins. BM 2: 526. (BMNH: Holotype ♀); described from 1 specimen.
Type locality: Bangladesh – Sylhet.
Distribution: Oriental – Bangladesh, China (?), India (Assam), Nepal, Taiwan (introduced?).
Host: Simaroubaceae – *Ailanthus excelsus* Roxb. (Moore, 1867: 669).

> *niveiguttata*; Berg, 1880 (*Atteva*), An. Soc. Cient. Argent. 10: 103 [misspelling of *niveigutta*].

numeratrix Meyrick, 1930a (*Atteva*), Ann. Naturhist. Mus. Wien 44: 262. (NHMW: Holotype ♂); described from 1 specimen.
Type locality: Brazil – Amazonas, Taperinha.
Distribution: Neotropical – Brazil (Amazonas).

porphyris Meyrick, 1907d (*Atteva*), Proc. Linn. Soc. N.S. Wales 32(1): 79. (BMNH: Lectotype ♀ [designated here]); described from 2 specimens (BMNH).
Type locality: Solomon Islands – Bougainville.
Distribution: Australasian – Solomon Islands.
Note: The lectotype of *Atteva porphyris* is designated here to eliminate confusion and stabilize the species. The lectotype labels read: "LECTO- | TYPE" [circular label with indigo margins], "Bougainvile | Solomon Is. | ASM. 0.05" [cream-colored rectangular label] [hand-written], "Meyrick Coll. | B.M.

1938–290.," "*porphyris* Meyr." [hand-written], "*Atteva* | *porphyris* | 1/2 Meyr. | E. Meyrick det. | in Meyrick Coll.," "LECTOTYPE | *Atteva porphyris* | Meyrick | teste J. Sohn, 2014."

pulchella Moore, 1879 [1888] (*Atteva*), In: Hewitson & Moore, Descr. Lep. Atk. 3: 282. (ZMB: Syntype); not given (1♀ in ZMB).
Type locality: [India].
Distribution: Oriental – India.

pustulella Fabricius, 1794 (*Tinea*), Ent. Syst. Suppl. 3(2): 292. An objective replacement name of *punctella* Stoll, and justified emendation of *pastulella*.
Distribution: Neotropical – Argentina, Brazil, Costa Rica, Dominica, Haiti, Jamaica, Martinique, Peru, Suriname, Uruguay.
Hosts: Simaroubaceae – *Ailanthus altissima* (Mill.) Swingle, *Simarouba amara* Aubl. (Berg, 1880; Becker, 2009); Records from *Cinnamomum porphyrium* (Griseb.) Kosterm. (Lauraceae) by Schreiter (1943) and *Laurus nobilis* L. (Lauraceae) by Biezanko et al. (1957) need further verification. The record from *Melia azedarach* L. (Meliaceae) by Hayward (1969) is possibly an error (Becker, 2009).

> *pastulella*; Fabricius, 1787 (*Tinea*), Mant. Ins. 2: 241, no. 16. [misspelling of *pustulella*].

> *punctella* Stoll, 1782 (*Phalaena*), in Cramer, Pap. Exot. 4: 164, pl. 372, fig. L. (USNM: Neotype ♀ [designated by Wilson et al. 2010: 47]); not given. A junior homonym of *Phalaena punctella* Linnaeus, 1761.
> Type locality: Costa Rica – Guanacaste, Area de Conservacion.
> Note: On page 164, last line, is the following statement with an asterisk: [approximately translated] "These are in my collection. C. Stoll." establishing C. Stoll as the author, not P. Cramer.

> *pustulata* Fabricius, 1798 (*Lithosia*), Suppl. Entom. Sist.: 462. [unjustified emendation of *pustulella*].

> *subtilis* Hübner, 1819 [1822] (*Crameria*), Verz. bekannt. Schmett.: 168. [unnecessary replacement name for *punctella*].

rawlinsi Becker, 2009 (*Atteva*), Rev. Bras. Ent. 53(3): 351.(CUIC: Holotype ♂); described from 54♂, 47♀ (CUIC).
Type locality: Dominican Republic – Barahona Prov., Barahona.
Distribution: Neotropical – Dominican Republic.

rex Butler, 1887 (*Corinea*), Ann. Mag. Nat. Hist. 20(5): 414. (BMNH: Syntype); not given (1♂ in BMNH).
Type locality: Philippines – Sulu, Alu [volcano].
Distribution: Oriental – Philippines (Sulu).

sciodoxa Meyrick, 1908a (*Atteva*), Rec. Ind. Mus. 2, Part 4(39): 398. (BMNH & IMCK: Syntypes); described from 4 specimens (BMNH, IMCK).
Type Localities: Malaysia – Borneo, Labuan & Penang Island; Myanmar – Dawna Hills.
Distribution: Oriental – Malaysia (Borneo, Penang Island), Myanmar.

scolecias Meyrick, 1928c (*Atteva*), Exot. Micr. 3(14): 419. (BMNH: Lectotype ♀ [designated by Clarke, 1965: 292]); described from 2 specimens (BMNH).
Type locality: Philippines – Luzon, Mt. Makiling.
Distribution: Oriental – Philippines (Luzon).

siderea Walsingham, 1891c [1892] (*Oeta*), Proc. Zool. Soc. Lond. 1891: 533. (BMNH: Holotype ♀); described from 1 specimen.
Type locality: Dominican Republic – San Domingo.
Distribution: Neotropical – Dominican Republic, US Virgin Islands (St. Thomas).

sidereoides Becker, 2009 (*Atteva*), Rev. Bras. Ent. 53(3): 352. (USNM: Holotype ♂); described from 1 specimen.
Type locality: Jamaica – St. Ann, near Runaway Bay Cave.
Distribution: Neotropical – Jamaica.

sphaerodoxa Meyrick, 1918a (*Atteva*), Exot. Micr. 2(6): 189. (BMNH: Holotype ♂); described from 1 specimen.
Type locality: New Guinea – Upper Setekwa River, Snow Mountains.
Distribution: Australasian – New Guinea.

sphaerotrocha Meyrick, 1936c (*Atteva*), Exot. Micr. 5(2): 41 (NHMW: Holotype ♂), described from 1 specimen.
Type locality: New Caledonia.
Distribution: Australasian – New Caledonia.

subaurata Durrant, 1900 (*Atteva*), Cat. Het. 2: 560. (BMNH: Lectotype ♂ [designated here]); described from 5 specimens (BMNH).
Type locality: India – Sikkim.
Distribution: Oriental – India (Assam, Sikkim).

Note: One specimen in the type series of *subaurata* Durrant bears a "Lectotype" label. To our knowledge, there is no evidence that this has ever been published. We formalize the lectotypy by designating the lectotype here. The pin data labels read: "LECTO- | TYPE" [circular label with indigo margins], "SIKKIM | 1889 | Möller" [cream-colored rectangular label] [hand-written] "Walsingham | collection | 1910-427," "ATTEVA | SUBAURATA | Drnt | Swinh. Cat. Het. II. p.560, (1900) || TYPE ♂" [rectangular label with black margins], "LECTOTYPE | *Atteva subaurata* | Durrant | teste J. Sohn, 2014."

modesta Snellen, 1901 (*Atteva*), Tijdschr. v. Ent. 44: 79. (RMNH: Syntypes); described from 9 specimens (RMNH).
Type locality: India – Assam.

teratias Meyrick, 1907d (*Atteva*), Proc. Linn. Soc. N.S. Wales 32(1): 79. (BMNH: Lectotype ♀ [designated here]); described from 2 specimens (BMNH).
Type locality: New Guinea – Sariba Island.
Distribution: Australasian – New Guinea (Sariba Island).
Note: One specimen in the type series of *teratias* Meyrick bears a Lectotype label. To our knowledge, there is no evidence that this has ever been published. We formalize the selection by designating the lectotype here. The pin data labels read: "LECTO- | TYPE" [circular label with indigo borders], "Sariba I. | New Guinea | ASN. 05" [cream-colored rectangular label] [hand-written] "Meyrick Coll. | B.M. 1938-290," "*teratias* Meyr." [hand-written], "*Atteva* | *teratias* | 3/6 Meyr. | E. Meyrick det. | In Meyrick Coll.," "LECTOTYPE | *Atteva teratias* | Meyrick | teste J. Sohn, 2014."

wallengreni Sohn & Wu, 2013 (*Atteva*), J. Ins. Sci. 13(66): 7. (USNM: Holotype ♀); described from 3♀ (USNM).
Type Locality: China – Guangzhou, Canton.
Distribution: Oriental – China (Guangdong, Hainan, Hongkong, Shaanxi), Indonesia (Bali), Malaysia (Perak), Thailand, Vietnam.

yanguifella Sohn & Wu, 2013 (*Atteva*), J. Ins. Sci. 13(66): 11. (IZAS: Holotype ♂); described from 4♂, 1♀ (IZAS).
Type Locality: China – Xizang, Motuo [=Medog].
Distribution: Oriental – China (Xizang).

zebra Duckworth, 1967 (*Atteva*), Proc. Ent. Soc. Wash. 69(1): 71. (USNM: Holotype ♂); described from 7♂, 33♀ (USNM).
Type locality: Panama – Canal Zone, Barro Colorado Island.

Distribution: Neotropical – Costa Rica, Panama.
Host: Simaroubaceae – *Simarouba amara* Aubl. (Wilson et al. 2010).

zebrina Becker, 2009 (*Atteva*), Rev. Bras. Ent. 53(3): 354. (VOBC: Holotype ♂); described from 2♂, 1♀ (VOBC).
Type locality: Brazil – Espírito Santo, Linhares.
Distribution: Neotropical – Brazil (Espírito Santo).

PRAYDIDAE Moriuti, 1977a.
Fauna Jap. Ypon. s. l.: 32.

ATEMELIA Herrich-Schäffer, [1853], Syst. Bearb. Schmett. Eur. 5: 9 (key), 33 [see Herrich-Schäffer, 1853–1855].
Type species: *Oecophora torquatella* Lienig & Zeller, 1846, by subsequent designation by Fletcher, 1929, Mem. Dep. Agric. India (Ent) 11: 26.

> *ORINYMPHA* Meyrick, 1927a, Exot. Micr. 3(12): 360.
> Type species: *Orinympha aetherias* Meyrick, 1927a, by monotypy.

aetherias Meyrick, 1927a (*Orinympha*), Exot. Micr. 3(12): 361. (BMNH: Holotype ♂); described from 1 specimen.
Type locality: USA – Texas: Alpine.
Distribution: Nearctic – USA (Texas).

mahonivora Sohn & Peralta, 2014 (*Atemelia*), Ann. Ent. Soc. Amer. 107(2): 340. (USNM: Holotype ♂); described from 8♂, 2♀ (USNM).
Type locality: Chile, San Antonio, Comuna de Santo Domingo.
Distribution: Neotropical – Chile (Bio-Bio, Curico, San Antonio).
Host: Berberidaceae – *Mahonia* sp. (Sohn & Alba, 2014).

stratellus Zeller, 1877 (*Prays*), Hor. Soc. Ent. Ross. 13: 243. (BMNH: Holotype ♂); described from 1 specimen. **comb. nov.**
Type locality: Colombia – Bogota.
Distribution: Neotropical – Colombia.

torquatella Lienig & Zeller, 1846 (*Oecophora*), Isis Oken 39(3/4): 280. (BMNH: Holotype ♂); described from 1 specimen.
Type locality: Germany – Pomerania [=Stettin Co.], Grünhof [from specimen label].
Distribution: Palearctic – Austria, Belgium, Czech Republic, England, Estonia, Finland, France, Germany, Hungary, Italy, Latvia, Lithuania, Netherlands,

Norway, Poland, Russia (European Northwest, Irkutsk, Middle Volzhsky, Murmansk, South Ural, South Western Siberia), Slovakia, Sweden, Switzerland, [former] Yugoslavia.
Hosts: Betulaceae – *Betula pubescens* Ehrn., *Betula* spp.; Myricaceae – *Myrica gale* L.; Ulmaceae – *Ulmus* spp. (Hering, 1951, 1957; Agassiz, 1996; Beavan & Heckford, 2012).

compressella Herrich-Schäffer, [1855] (*Atemelia*), Syst. Bearb. Schmett. Eur. 5: 112 [see Herrich-Schäffer, 1853–1855]. (type not found); described from 1 specimen.
Type locality: Germany – Mannheim.

fuscifrontella Stainton, 1849c (*Oecophora*), Syst. Cat. Brit. Tin. & Pteroph. 8: 13. (type not found); described from 1 specimen.
Type locality: England – Scotland.

DICTYOPRAYS Sohn, 2012, Tinea 22(2): 121.
Type species: *Dictyoprays kyutekparki* Sohn, 2012, by monotypy.

kyutekparki Sohn, 2012 (Dictyoprays), Tinea 22(2): 122. (MCGF: Holotype ♂); described from 1 specimen.
Type locality: Vietnam – Ninh Binh Prov., Cucphuong National Park.
Distribution: Oriental – North Vietnam.

EUCATAGMA Busck, 1900a, Jour. N. Y. Ent. Soc. 8: 247.
Type species: *Eucatagma amyrisella* Busck, 1900a, by original designation.

amyrisella Busck, 1900a (*Eucatagma*), Jour. N. Y. Ent. Soc. 8: 247. (USNM: Lectotype ♀ [designated by Sohn & Lewis, 2012: 4]); described from 2 specimens (USNM; BMNH [not found]).
Type locality: USA – Florida, Palm Beach.
Distribution: Nearctic – USA (Florida).
Host: Burseraceae – *Amyris floridana* Nuttall (Busck, 1900a)

amyridella Meyrick, 1914c (*Prays*), Lep. Cat. 19: 16. [unjustified emendation of *amyrisella*].

PRAYS Hübner, 1825 [1826], Verz. bekannt. Schmett.: 413.
Type species: *Tinea caenobitella* Hübner, [1813] 1816, by subsequent designation by Walsingham, 1907c.

Note: Walsingham (1907c) (in Sharp, Fauna Hawaiiensis (1): 651) cited the type species as *Phalaena curtisella* Donovan. When Walsingham designated as type species *Phalaena curtisella* Donovan, 1793, a nominal species not originally included in *Prays*, he also placed *caenobitella* (but cited as *coenobitella*, an incorrect subsequent spelling by Hübner, 1825 [1826]) as a nominal species that was originally included in *Prays*, as a junior subjective synonym of *curtisella*. Under the Code (Edition 3), Article 69(a)(v), this designation constitutes the fixation of the originally included nominal species as the type species.

PEPILLA Guenée, 1845a, Ann. Soc. Ent. Fr. 2(3): 344. A junior objective synonym of *Prays* Hübner, 1825 [1826].
Type species: *Tinea caenobitella* Hübner, [1813] 1816, by monotypy.

PRATYS; Desmarest, 1857, in Chenu, Encycl. Hist. Nat. (Papillons nocturnes): 303. [misspelling of *Prays*].

acmonias Meyrick, 1914f (*Prays*), Jour. Bomb. Nat. Hist. Soc. 23(1): 125. (BMNH: Lectotype: ♀ [designated by Clarke, 1965: 372]); described from 2 specimens (BMNH).
Type locality: India – Punjab, Rawalpindi.
Distribution: Oriental – India (Punjab), Pakistan.
Hosts: Adoxaceae – *Viburnum cotinifolium* D. Don, *V. grandiflorum* Wall. ex DC., *V. mullaha* Buch.-Ham. ex Don (Ahmad, 1976).

alpha Moriuti, 1977a (*Prays*), Fauna Jap. Ypon. s. l.: 119. (UOP: Holotype ♂); described from 4♂, 4♀ (UOP).
Type locality: Japan – Hokkaido, Kamikawa, Ashikawa.
Distribution: Palearctic – China (Heilongjiang), Japan (Hokkaido).
Hosts: Juglandaceae – *Juglans ailanthifolia* Carrière; Oleaceae – *Fraxinus mandshurica* var. *japonica* Maxim. (Moriuti, 1977a).

amblystola Turner, 1923 (*Prays*), Trans. R. Soc. S. Austr. 47: 168. (ANIC: Syntypes); described from 2 specimens (ANIC).
Type locality: Australia – New South Wales, Mt. Kosciusko.
Distribution: Australasian – Australia (New South Wales).

atomocella Dyar, 1902b (*Hyponomeuta*), Jour. N. Y. Ent. Soc. 10: 208. (USNM: Syntypes); described from 2 specimens (1♂, 1♀ in USNM).
Type locality: USA – Texas, LeBaron.

Distribution: Nearctic – USA (Illinois, Michigan, Ohio, Texas).
Host: Rutaceae – *Ptelea trifoliata* L. (Forbes, 1923).

atomosella; Meyrick, 1914c (*Yponomeuta*), Lep. Cat. 19: 17. [misspelling of *atomocella*].

diaphorus Walsingham, 1907a (*Hyponomeuta*), Proc. USNM 33: 220. (BMNH: Holotype ♀); decribed from 3 specimens (BMNH, USNM).
Type locality: USA – Texas.

autocasis Meyrick, 1907d (*Prays*), Proc. Linn. Soc. N.S. Wales 32(1): 77. (BMNH: Lectotype ♀ [designated here]); described from 2 specimens (BMNH).
Type locality: Australia – New South Wales, Sydney.
Distribution: Australasian – Australia (New South Wales, West Australia).
Note: Both syntypes of *Prays autocasis* are in the BMNH. One of those was designated as lectotype and the other as paralectotype as shown by I. F. B. Common labels. This lectotype, however, has never been published. We stabilize the type series and preserve the identity of *autocasis* by formalizing the unpublished lectotypy here. The data labels read: "LECTO- | TYPE" [circular label with indigo margins], "Sydney | n.s.Wales | 26.4.79" [hand-written], "*autocasis* Meyr." [hand-written], "Meyrick Coll. | B.M. 1938–290.," "*Prays autocasis* | 1/2 Meyr. | E. Meyrick det. | in Meyrick Coll.," "LECTOTYPE ♀ | *Prays* | *autocasis* Meyr., | 1907. | selected by | I.F.B. Common, 1966."

beta Moriuti, 1977a (*Prays*), Fauna Jap. Ypon. s. l.: 121. (UOP: Holotype ♂); described from 3♂, 3♀ (UOP).
Type locality: Japan – Honshu, Niigata Pref., Etigo.
Distribution: Palearctic – Japan (Honshu), Russia (Amur, Primorsky), South Korea.
Hosts: Oleaceae – *Fraxinus chinensis* subsp. *rhynchophylla* (Hance) A.E. Murray, *F. japonica* Blume ex Koch (Sohn, 2007; Moriuti, 1977a).

calycias Meyrick, 1907d (*Prays*), Proc. Linn. Soc. N.S. Wales 32(1): 76. (BMNH: Lectotype ♂ [designated here]); described from 5 specimens (BMNH).
Type locality: Australia – Queensland, Brisbane.
Distribution: Australasian – Australia (New South Wales, Queensland).
Note: There are five syntypes of *Prays calycias* in the BMNH. One of them was labeled as lectotype and all others as paralectotypes, probably by I. F. B. Common in 1966. The lectotype designation, however, has never been published. To represent the unique species and stabilize the type series, we

designate the so labeled specimen as the lectotype. The type labels read: "Lecto- | type" [circular label with indigo margins], "Brisbane | Queensland | AST. 0.00" [hand-written], "*calycias* Meyr." [hand-written], "Meyrick Coll. | B.M. 1938-290," "*Prays calycias* | 1/5 Meyr. | E. Meyrick det. | in Meyrick Coll.," "LECTOTYPE | *Prays calycias* | Meyrick | teste J. Sohn, 2014."

chrysophyllae Silvestri, 1915 (*Prays*), Boll. Lab. Zool. Agr. Portici 9: 265. (MCSN: Syntypes); not given, but described from both sexes.
Type locality: Eritrea – Nefasit.
Distribution: Ethiopian – Eritrea.
Host: Oleaceae – *Olea europaea* subsp. *cuspidata* (Wall. ex G. Don) Cif. (Silvestri, 1915).

cingulata Yu & Li, 2004 (*Prays*), SHILAP Revta. lepid. 32(125): 15. (NUIC: Holotype ♂); described from 4♂, 3♀ (NUIC).
Type locality: China – Guizhou, Xishui.
Distribution: Oriental – China (Guizhou).

citri Millière, 1873 (*Acrolepia*), Pet. Nouv. Ent. 5: 310. (Type not found); described from 1 specimen.
Type locality: France – Corsica.
Distribution: Australasian – Fiji Islands, Guam, Samoa; Ethiopian – Mauritius, Seychelles, South Africa, Zimbabwe; Oriental – India, Philippines, Sri Lanka; Palearctic – Algeria, Canary Islands, Corsica, Crete, Dodecanese Islands, Egypt, France, Greece, Israel, Italy, Lebanon, Libya, Madeira Archipelago, Malta, Morocco, Portugal, Sardinia, Sicily, Spain, Tunisia, Turkey.
Hosts: Rutaceae – *Casimiroa edulis* La Llave, *Citrus* x *aurantium* L., *C. limon* (L) Osbeck, *C. sinensis* (L.) Osbeck (Surhone et al. 2010; Aguiar & Karsholt, 2006).

curulis Meyrick, 1914f (*Prays*), Jour. Bomb. Nat. Hist. Soc. 23(1): 125. (BMNH: Holotype ♀); described from 1 specimen.
Type locality: India – Assam, Khasi Hills.
Distribution: Oriental – India (Assam); Palearctic – Nepal.
Host: Symplocaceae – *Symplocos*? (Fletcher, 1928).

delta Moriuti, 1977a (*Prays*), Fauna Jap. Ypon. s. l.: 122. (UOP: Holotype ♂); described form 3♂, 2♀ (UOP).
Type locality: Japan – Honshu, Nagano Pref., Mt. Tatesinayama.
Distribution: Palearctic – China (Heilongjiang), Japan (Honshu), South Korea.

Hosts: Oleaceae – *Ligustrum tschonoskii* Decaisne, *Syringa persica* L. (Moriuti, 1977a; Sohn & Wu, 2011).

ducalis Meyrick, 1914f (*Prays*), Jour. Bomb. Nat. Hist. Soc. 23(1): 125. (BMNH: Holotype ♀); described from 1 specimen.
Type locality: Sri Lanka – Uva Prov., Namunukuli.
Distribution: Oriental – Sri Lanka.

endocarpa Meyrick, 1919 (*Prays*), Exot. Micr. 2(8): 234. (BMNH: Lectotype ♀ [designated here]); described from 1♂, 1♀ (BMNH).
Type locality: Indonesia – Java.
Distribution: Oriental – Indonesia (Java), Malaysia.
Host: Rutaceae – *Citrus* sp. (Meyrick, 1919).
Note: Diakonoff (1967) thought that Clarke (1965) had designated the lectotype of *endocarpa* from one of the two syntypes, a male and a female. Clarke (1965) found only the female specimen and reported it as the holotype. H. Stringer (see Diakonoff, 1967: 239) found the male syntype later and assumed it was a paratype. However, Meyrick did not designate a holotype, requiring revision of the type status for the species. To correct the type status and preserve the uniqueness of the species, we designate as lectotype the female specimen that Clarke (1965) regarded as the holotype. Note that one label states a female genitalia slide was made while on another of the labels, the sex is given as male in error. Pin data labels read: "LECTOTYPE" [circular label with red open circle at center], "Type | H.T." [circular label with red margins], "E.Java | [+in circle] Citrus | 1912 | Dr. Van Hall | 1918-123" [hand-written], "♀ genitalia on | slide 3-VI 1948 | J.F.G.C. 7564" [with black submarginal lines], "*PRAYS* | *ENDOCARPA* | Meyr. | Exot. Micr. 2. 234 (1919) || TYPE ♂" [with black borders], "LECTOTYPE | *Prays endocarpa* | Meyrick | teste J. Sohn, 2014."

endolemma Diakonoff, 1967 (*Prays*), USNM Bull. 257: 237. (USNM: Holotype ♂); described from 3♂, 3♀ (USNM).
Type locality: Philippines – Luzon Island, Los Baños.
Distribution: Oriental – Philippines (Batangas, Luzon).
Host: Rutaceae – galls in rind of lemon-fruits (*Citrus* sp.) (Diakonoff, 1967).

epsilon Moriuti, 1977a (*Prays*), Fauna Jap. Ypon. s. l.: 123. (UOP: Holotype ♂); described from 3♂, 1♀ (UOP).
Type locality: Japan – Hokkaido, Iburi, Tomakomai.
Distribution: Palearctic – Japan (Hokkaido).

erebitis Meyrick, 1914f (*Prays*), Jour. Bomb. Nat. Hist. Soc. 23(1): 124. (BMNH: Lectotype ♂ [designated by Clarke, 1965: 375]); described from 2 specimens (BMNH).
Type locality: India – Tamil Nadu, Ootacamund.
Distribution: Oriental – India (Tamil Nadu).

fraxinella Bjerkander, 1784 (*Tinea*), Kungl. Vetensk. Akad. Nya Handl. 5: 324. (Type not found); not given.
Type locality: not given.
Distribution: Nearctic – Canada (introduced: British Columbia); Palearctic – Albania, Austria, Belgium, Bulgaria, Czech Republic, Denmark, England, Estonia, Finland, France, Germany, Hungary, Ireland, Italy, Latvia, Lithuania, Luxembourg, Netherlands, Norway, Poland, Portugal, Romania, Russia (Amur, European Central, European Northwest, Kaliningrad, Primorsky), Slovakia, Spain, Sweden, Switzerland.
Host: Oleaceae – *Fraxinus excelsior* L. (Foggo, 1992).

caenobitella Hübner, [1813] 1816 (*Tinea*), Samml. Eur. Schmett. Vol. 8, pl. 45, Tineæ 5, Ignobiles B, fig. 309 [in Hübner, 1796–1838]. (Type not found); not given.
Type locality: not given.

circulella Fabricius, 1794 (*Tinea*), Ent. Syst. Suppl. 3(2): 301. (ZMUC: Lectotype ♀ [designated by Karsholt & Nielsen, 1976: 245]); not given.
Type locality: Germany – Kiliae [=Kiel].
coenobitella; Hübner, 1825 [1826] (*Prays*), Verz. bekannt. Schmett.: 413. [misspelling of *caenobitella*].

curtisella Donovan, 1793 (*Prays*), Nat. Hist. Brit. Ins. 2: 77. (Type lost); not given.
Type locality: England.

curtisii Haworth, 1828 (*Erminea*), Lep. Brit. 4: 514. [unjustified emendation of *curtisella*].

hortuella Stephens, 1834 (*Anacampsis*), Ill. Brit. Ent. Haust. 4: 205 (Type not found); not given.
Type locality: [England] – metropolitan district [London?].

maculella Fabricius, 1794 (*Tinea*), Ent. Syst. Suppl. 3(2): 306. (Type lost); not given.
Type locality: not given.

friesei Klimesh, 1992 (*Prays*), Entomofauna 13(4): 113. (NHMW: Holotype ♂); described from 9♂, 11♀ (NHMW).
Type locality: Canary Islands – Tenerife.
Distribution: Palearctic – Canary Islands, Madeira Archipelago.
Hosts: Oleaceae – *Jasminum barrelieri* Webb & Berthel., *Picconia excelsa* (Aiton) DC. (Klimesch, 1992; Aguiar & Karsholt, 2006).

fulvocanellus Walsingham, 1907c (*Prays*), Fauna Hawaiiensis 1(5): 652. (BMNH: Syntypes); described from 1♂, 1♀ (BMNH).
Type locality: Hawaii Islands – Kauai Island, Kaholuamano.
Distribution: Australasian – Hawaii (Kauai).
Hosts: probably Rutaceae – *Pelea*, *Fagara* (Zimmerman, 1978).
Note: Zimmerman (1978) erroneously referred to the two syntypes of *fulvocanellus* as holotype and paratype. Walsingham (1907c) did not specify the type series.

galapagosella Landry & Landry, 1998 (*Prays*), Trop. Lepid. 9(1): 31. (CNC: Holotype ♂); described from 19♂, 16♀ (CNC).
Type locality: Ecuador – Galápagos Islands, Santa Crúz Island, Finca Vilema.
Distribution: Neotropical – Ecuador (Galápagos Islands).

gamma Moriuti, 1977a (*Prays*), Fauna Jap. Ypon. s. l.: 122. (Issiki collection [not traced]: Holotype ♂); described from 1 specimen.
Type locality: Japan – Honshu, Hirosima Pref., Sandankyo.
Distribution: Palearctic – Japan (Honshu), South Korea.

ignota Clarke, 1986 (*Prays*), Smith. Contr. Zool. 416: 338. (USNM: Holotype ♀); described from 4♀ (USNM).
Type locality: French Polynesia – Marquesas Islands, Hiva Oa Island, Mt. Feani.
Distribution: Australasian – French Polynesia (Marquesas Islands).

inconspicua Yu & Li, 2004 (*Prays*), SHILAP Revta. lepid. 32(125): 16. (NUIC: Holotype ♂); described from 1 specimen.
Type locality: China – Henan, Shan Co.
Distribution: Palearctic – China (Henan).

inscripta Meyrick, 1907d (*Prays*), Proc. Linn. Soc. N.S. Wales 32(1): 75. (BMNH: Lectotype ♂ [designated here]); described from 10 specimens (BMNH).
Type locality: Australia – New South Wales, Sydney.
Distribution: Australasian – Australia (New South Wales).
Host: Rutaceae – *Phebalium* sp. (Common, 1990).

Note: There are ten syntypes of *Prays inscripta* in the BMNH. One of those was labeled as lectotype and all the others as paralectotypes, probably by I. F. B. Common. This lectotype, however, has never been published. We stabilize the type series of *inscripta* by designating the unpublished lectotype here. The pin data labels read: "LECTO- | TYPE" [circular label with indigo margins], "Sydney | N.S. Wales | 30. 8. 84" [hand-written], "*inscripta* Meyr." [hand-written], "Meyrick Coll. | B.N. 1938-290," "*Prays* | *inscripta* | 1/10 Meyr. | E. Meyrick det. | in Meyrick Coll.," "LECTOTYPE | *Prays inscripta* | Meyrick | teste J. Sohn, 2014."

iota Moriuti, 1977a (*Prays*), Fauna Jap. Ypon. s. l.: 124. (UOP: Holotype ♂); described from 7♂, 8♀ (UOP).
Type locality: Japan – Honshu, Nagano Pref., Tokugotoge.
Distribution: Palearctic – Japan (Hokkaido, Honshu).
Hosts: Adoxaceae – *Viburnum furcatum* Blume & Maxim., *V. wrightii* Miq. (Moriuti, 1977a; Oku, 2003).

kalligraphos Sohn & Wu, 2011 (*Prays*), Tijdschr. v. Ent. 154: 28. (IZCAS: Holotype ♂); described from 1 specimen.
Type locality: China – Sichuan, Dujiangyan, Mt. Qingcheng.
Distribution: Oriental – China (Sichuan).

kappa Moriuti, 1977a (*Prays*), Fauna Jap. Ypon. s. l.: 127. (UOP: Holotype ♂); described from 2♂, 1♀ (UOP).
Type locality: Japan – Hokkaido, Osima, Onuma.
Distribution: Palearctic – Japan (Hokkaido).

lambda Moriuti, 1977a (*Prays*), Fauna Jap. Ypon. s. l.: 127. (UOP: Holotype ♂); described from 32♂, 35♀ (UOP).
Type locality: Japan – Honshu, Osaka Pref., Sakai.
Distribution: Palearctic – Japan (Honshu, Kyushu).
Host: Adoxaceae – *Viburnum awabuki* Hort. Berol. ex K. Koch (Moriuti, 1977a).

liophaea Meyrick, 1927a (*Prays*), Exot. Micr. 3(12): 361. (TMP: Syntypes) described from 2 specimens (TMP).
Type locality: South Africa – Natal, Stellenbosch.
Distribution: Ethiopian – South Africa (Natal).

lobata Yu & Li, 2004 (*Prays*), SHILAP Revta. lepid. 32(125): 16. (NUIC: Holotype ♂); described from 1 specimen.
Type locality: China – Hubei, Xianfeng.
Distribution: Palearctic – China (Hubei).

nephelomima Meyrick, 1907d (*Prays*), Proc. Linn. Soc. N.S. Wales 32(1): 76. (BMNH: Syntypes); described from 5♂ (BMNH).
Type localities: Australia – New South Wales, Murrurundi & Sydney.
Distribution: Australasian – Australia (New South Wales), New Caledonia, New Zealand; Oceanic – Solomon Islands.
Host: Rutaceae – *Citrus* sp. (Common, 1990; Mille & Cazeres, 2007).

oleae Bernard, 1788 (*Phalena*), Mém. Hist. Nat. Prov. 2: 290. (Type not found); described from 'numerous' specimens of both sexes.
Type locality: [France], Provence.
Distribution: Ethiopian – Namibia, South Africa; Palearctic – Albania, Algeria, Austria, Corsica, Crete, England (introduced), France, Greece, Italy, Malta, Portugal, Russia (European parts), Sardinia, Sicily, Spain, [former] Yugoslavia.
Host: Oleaceae – *Olea europaea* L. (Bernard, 1788; Pelekassis, 1962; Mey, 2007).

moschettinella Costa, 1839 (*Oecophora*), Corrisp. Zool. 1: 118. (Type not found); not given.
Type locality: [Italy – ?Sicily]

oleaella; Costa, 1839 (*Tinea*), Corrisp. Zool. 1: 118. [misspelling of *oleella*].

oleella Fabricius, 1794 (*Tinea*), Ent. Syst. Suppl. 3(2): 308. (Type lost); not given.
Type locality: not given.

olivella Briganti, 1822 (*Tinea*), Atti R. Ist. Sci. Nat. Napoli 3: 132. (Type not found); not given.
Type locality: [Italy].

oliviella; Kaltenbach, 1874 (*Oecophora*), Pflanzenf. Ins.: 436. [misspelling of *olivella*, unnecessary emendation of *oleella*].

servillella Costa, 1836 (*Tinea*), Fau. Reg. Napoli. Lepid. 1836: Tinea: 7. (Type not found); not given.
Type locality: Italy – Puglia, Lecce Prov., Otranto.

servilliella; Costa, 1839 (*Tinea*), Corrisp. Zool. 1: 118. [misspelling of *servillella*].

oleaeoides Gibeaux, 1985b (*Prays*), Nouv. Revue Ent. (N.S.) 2: 280. (MNHN: Holotype ♀); described from 1 specimen.
Type locality: Madagascar – Centre, Boby Cirque .
Distribution: Ethiopian – Madagascar.

omicron Moriuti, 1977a (*Prays*), Fauna Jap. Ypon. s. l.: 130. (UOP: Holotype ♂); described from 20♂, 22♀ (UOP).
Type locality: Japan – Honshu, Siga Pref., Mt. Hirasan.
Distribution: Palearctic – Japan (Honshu, Shikoku).
Host: Adoxaceae – *Viburnum dilatatum* Thunb. (Moriuti, 1977a).

parilis Turner, 1923 (*Prays*), Trans. R. Soc. S. Austr. 47: 168. (ANIC: Syntypes); described from 3 specimens (ANIC).
Type locality: Australia – Queensland, Brisbane.
Distribution: Australasian – Australia (Queensland).
Host: Rutaceae – *Citrus* sp. (Common, 1990).

peperitis Meyrick, 1907b (*Prays*), Jour. Bomb. Nat. Hist. Soc. 17(4): 984. (BMNH: Lectotype ♂ [designated by Clarke, 1965: 375]); described from 7 specimens, 5 of which are present in the BMNH (Clarke, 1965).
Type locality: Sri Lanka – Central Prov., Maskeliya.
Distribution: Oriental – Sri Lanka.

peregrina Agassiz, 2007 (*Prays*), Nota lepid. 30(2): 408. (BMNH: Holotype ♀); described from 3♂, 3♀ (BMNH; Agassiz collection [private]).
Type locality: England – Middlesex, Greenford, Melville Avenue.
Distribution: Palearctic – England.

ruficeps Heinemann, 1854 (*Oecophora*), Bresl. Zeitschr. Ent. 8: 4. (Type not found); described from 1 specimen.
Type locality: Germany – Lower Saxony, Braunschweig.
Distribution: Palearctic – Austria, Denmark, Finland, Germany, England, Latvia, Lithuania, Norway, Poland, Russia (European Central, Kaliningrad), Sweden, Switzerland.

> *simplicella* Herrich-Schäffer, [1855] (*Swammerdamia*), Syst. Bearb. Schmett. Eur. 5: 280 [see Herrich-Schäffer, 1853–1855]. (Type not found); described from 1 specimen.
> Type locality: Germany – Lower Saxony, Braunschweig.

sparsipunctella Turati, 1924 (*Prays*), Atti Soc. Ital. Sci. Nat. 63: 159. (Type not found); [probably] described from 1 specimen.
Type locality: Lybia – Bengasi [=Benghazi]
Distribution: Palearctic – Lybia.

sublevatella Viette, 1957 (*Prays*), Mém. Inst. Sci. Madag. (E)8: 149. (MNHN: Holotype ♂); described from 1 specimen.
Type locality: Réunion Island – Bébour Forest.
Distribution: Ethiopian – Réunion Island.

temulentus Meyrick, 1910a (*Yponomeuta*), Rec. Ind. Mus. 5, Part 4(22): 228. (IMCK: Holotype ♂); described from 1 specimen.
Type locality: India – Theog, Simla Hills.
Distribution: Oriental – India (Theog).

tineiformis Sohn & Wu, 2011 (*Prays*), Tijdschr. v. Ent. 154: 30. (IZCAS: Holotype ♂); described from 1♂, 3♀ (IZCAS).
Type locality: China – Hainan, Ledong, Mt. Jianfengling.
Distribution: Oriental – China (Hainan).

tristis Gibeaux, 1985b (*Prays*), Nouv. Revue Ent. (N.S.) 3: 280. (MNHN: Holotype ♀); described from 1 specimen.
Type locality: Madagascar – Amoron'i Mania Region, Italaviana.
Distribution: Ethiopian – Madagascar.

tyrastis Meyrick, 1907d (*Prays*), Proc. Linn. Soc. N.S. Wales 32(1): 75. (BMNH: Syntypes); described from 3 specimens (BMNH).
Type Localities: Australia – Victoria, Melbourne; West Australia, Geraldton.
Distribution: Australasian – Australia (Victoria, West Australia).

xeroloxa Meyrick, 1935c (*Prays*), Exot. Micr. 4(19): 602. (BMNH: Lectotype ♂ [designated by Clarke, 1965: 375]); described from 4 specimens, 2 of which are present in the BMNH (Clarke, 1965).
Type locality: Indonesia – Java, Telawa.
Distribution: Oriental – Indonesia (Java).
Host: Rutaceae – *Harrisonia* sp. (Meyrick, 1935c).

SCYTHROPIIDAE Friese, 1966
Beitr. zur Ent. 16(3/4): 449.

SCYTHROPIA Hübner, 1816 [1825], Verz. bekannt. Schmett.: 414.
Type species: *Phalaena* (*Tinea*) *crataegella* Linnaeus, 1767, by subsequent designation by Stephens, (1834).

crataegella Linnaeus, 1767 (*Phalaena* (*Tinea*)), Syst. Nat. 12 1(2): 885, #353. (LSL: Lectotype ♂ [designated by Robinson & Nielsen, 1983: 207]); not given.
Type locality: not given.
Distribution: Palearctic – Albania, Austria, Belgium, Bulgaria, Corsica, Czech Republic, Denmark, England, Estonia, Finland, France, Germany, Greece, Hungary, Italy, Lithuania, Luxembourg, Netherlands, Norway, Poland, Romania, Russia (Kaliningrad, Middle Volzhsky, South Ural), Sicily, Slovakia, Spain, Sweden, Switzerland, [former] Yugoslavia.
Hosts: Rosaceae – *Cotoneaster* sp., *Crataegus monogyna* Jacq., *Crataegus* spp., *Malus sylvestris* Mill., *Prunus insititia* (L.), *P. spinosa* L., *Prunus* spp., *Pyrus* sp., *Sorbus aucuparia* L. (Weber, 1945; Werner, 1958; Carter, 1984; Agassiz, 1996; Parenti, 2000). Records from *Quercus* sp. (Fagaceae) and *Euonymus* sp. (Celastraceae) by Werner (1958) are suspicious.

cornella Fabricius, 1775 (*Tinea*), Sys. Ent.: 657. (ZMUC: Syntype); not given (1♀ in ZMUC).
Type locality: Germany – Saxony.

obscura Weber, 1945 (*Scythropia crataegella* f.), Mitt. schweiz. ent. Ges. 19: 394 (SFIT: Syntype); not given (1♂ in SFIT).
Type locality: Switzerland – Maroggia.

sparsella Denis & Schiffermüller, 1775, *Tinea*; Schmett. Wien: 141, #66. (Type lost); not given.
Type locality: not given.

YPONOMEUTIDAE Stephens, 1829.
Syst. Cat. Brit. Ins.: 139.

SARIDOSCELINAE Moriuti, 1977a.
Fauna Jap. Ypon. s. l.: 35.

EUCALANTICA Busck, 1904, Proc. USNM 27: 750.
Type species: *Calantica polita* Walsingham, 1881b by original designation.

costaricae Sohn & Nishida, 2011 (*Eucalantica*), Zookeys 118: 86. (INBIO: Holotype ♂); described from 6♂, 2♀ (EMEC, INBIO, UCR & USNM).
Type locality: Costa Rica – Cartago Prov., Macizo de la Muerte.

Distribution: Neotropical – Costa Rica (Cartago, San José).
Host: Ericaceae – possibly *Vaccinium floribundum* Kunth (Sohn & Nishida, 2011).

ehecatlella Sohn & Nishida, 2011 (*Eucalantica*), Zookeys 118: 87. (INBIO: Holotype ♂); described from 2♂ (INBIO).
Type locality: Costa Rica – Heredia Prov., Volcán Barva.
Distribution: Neotropical – Costa Rica (Heredia).

icarusella Sohn & Nishida, 2011 (*Eucalantica*), Zookeys 118: 88. (USNM: Holotype ♂); described from 7♂, 4♀ (BMNH, EMEC, UCR, USNM).
Type locality: Costa Rica – San José Prov., Cerro de la Muerte.
Distribution: Neotropical – Costa Rica (San José).

polita Walsingham, 1881b (*Calantica*), Proc. Zool. Soc. Lond. 1881(2): 302. (BMNH: Lectotype ♂ [designated by Sohn & Nishida, 2011: 82]); not given (1♂ in USNM; 2♂ in MCZ).
Type locality: USA – California, Lake Co.
Distribution: Nearctic – Canada (British Columbia), USA (California, Oregon, Washington).
Host: Ericaceae – *Vaccinium ovatum* Pursh (Powell & Opler, 2009); host records from *Rhododendron* (possibly *Rhododendron macrophyllum* D. Don ex G. Don) are questionable (Sohn & Nishida, 2011).

powelli Sohn, 2011 (*Eucalantica*), In: Sohn & Nishida, 2011. Zookeys 118: 90. (EMEC: Holotype ♂); described from 2♂, 4♀ (EMEC).
Type locality: Costa Rica – Cartago Prov., Cerro de la Muerte.
Distribution: Neotropical – Costa Rica (Cartago).

pumila Sohn, 2011 (*Eucalantica*), In: Sohn & Nishida, 2011. Zookeys 118: 90. (INBIO: Holotype ♂); described from 1 specimen.
Type locality: Costa Rica – Heredia Prov., Volcán Barva.
Distribution: Neotropical – Costa Rica (Heredia).

vaquero Sohn, 2011 (*Eucalantica*), In: Sohn & Nishida, 2011. Zookeys 118: 90. (USNM: Holotype ♂); described from 3♂, 3♀ (EMEC, USNM).
Type locality: USA – New Mexico, Pecos National Forest.
Distribution: Nearctic – USA (New Mexico, Arizona); Neotropical – Mexico (Tepalcates, Veracruz).

SARIDOSCELIS Meyrick, 1894, Trans. Ent. Soc. Lond., 1894: 28.
Type species: *Saridoscelis sphenias* Meyrick, 1894, by monotypy.

kodamai Moriuti, 1961 (*Saridoscelis*), Trans. Lep. Soc. Japan 11: 65. (UOP: Holotype ♂); described from 4♂, 4♀ (UOP).
Type locality: Japan – Honshu, Kii Peninsula, Mt. Natisan.
Distribution: Palearctic – Japan (Hokkaido, Honshu, Shikoku), Russia (South Kamchatka).
Hosts: Ericaceae – *Pieris japonica* (Thunb.) D. Don ex G. Don, *Leucothoe grayana* Maxim. var *oblongifolia* (Miq.) H. Hara (Moriuti, 1961, 1977a).

nudata Meyrick, 1913a (*Saridoscelis*), Exot. Micr. 1(5): 146. (BMNH: Holotype ♀); described from 1 specimen.
Type locality: India – Nilgiris, Pykara.
Distribution: Oriental – India (Nilgiris).
Note: Type locality is from the species description; the locality label on the pin reads Khasi Hills, India.

sphenias Meyrick, 1894 (*Saridoscelis*), Trans. Ent. Soc. Lond., 1894: 28. (BMNH: Holotype ♀); described from 1 specimen.
Type locality: Myanmar – Koni.
Distribution: Oriental – Burma (Myanmar), China (Jiangxi), India, Taiwan; Palearctic – Japan (Honshu).
Host: Ericaceae – *Vaccinium bracteatum* Thunb. (Moriuti, 1961).

issikii Moriuti, 1961 (*Saridoscelis*), Trans. Lep. Soc. Japan 11(4): 66. (UOP: Holotype ♂); described from 5♂, 7♀ (UOP).
Type locality: Japan – Honshu, Hyogo Pref., Nishinomiya.

synodias Meyrick, 1932a (*Saridoscelis*), Exot. Micr. 4(8): 228. (BMNH: Holotype ♀); described from 1 specimen.
Type locality: Japan – Honshu, Gifu Pref., Ikeno.
Distribution: Palearctic – Japan (Hokkaido, Honshu, Kyushu).
Note: Meyrick (1932a) reported the sex of the holotype as male, but it is actually female, based upon the presence of two acanthae on the hindwing.

YPONOMEUTINAE Stephens, 1829.
Syst. Cat. Brit. Ins.: 139.

CEDESTINAE Kloet & Hincks, 1945
Check List Brit. Ins.: 133.

HOFFMANNIINAE Spuler, 1910
Schmett. Eur. 2: 445.

HYPONOMEUTINAE Sodoffsky, 1837
Bull. Soc. Imp. Nat. Mosc. 10(6): 94.

ZELLERIINAE Turner, 1913
Proc. Linn. Soc. N.S. Wales 38(1): 193.

BANGHAASIA Friese, 1960, Beitr. zur Ent. 10(1/2): 115.
Type species: *Banghaasia ildefonsella* Friese, 1960, by original designation.

ildefonsella Friese, 1960 (*Banghaasia*), Beitr. zur Ent. 10(1/2): 116. (ZMB: Holotype ♂); described from 2♂, 1♀ (ZMB).
Type locality: Spain – Segovia Prov., San Ildefonso.
Distribution: Palearctic – Spain.

CEDESTIS Zeller, 1839, Isis Oken 32(3): 204.
Type species: *Argyresthia farinatella* Zeller, 1839, by subsequent designation by Fletcher, 1929, Mem. Dep. Agric. India (Ent.) 11: 206.

DESCEDISTIS; Inoue et al. 1982, Moths of Japan 2: 198. [misspelling of *Dyscedestis*].

DYSCEDESTIS Spuler, 1910, Schmett. Eur. 2: 449. A junior objective synonym of *Cedestis*.
Type species: *Argyresthia farinatella* Zeller, 1839, by monotypy.

EUCEDESTIS Paclt, 1951, Rev. Franc. Lep. 13: 127.
Type species: *Oecophora gysselinella* Duponchel, 1838 [1840], by original designation.

exiguata Moriuti, 1977a (*Cedestis*), Fauna Jap. Ypon. s. l.: 249. (UOP: Holotype ♂); described from 1 specimen.
Type locality: Japan – Shikoku, Ehime Pref., Otide.
Distribution: Palearctic – Japan (Honshu, Shikoku), South Korea.

gysseleniella Zeller, 1839 (*Cedestis*), Isis Oken 32(3): 205. (Types not found); not given.
Type locality: [Poland] – Gl[ogow].
Distribution: Palearctic – Austria, Belgium, Bulgaria, China (Shanxi), Czech Republic, Denmark, England, Estonia, Finland, France, Germany, Greece,

Hungary, Italy, Latvia, Lithuania, Netherlands, Norway, Poland, Romania, Russia (European Central, European Northeast, European Northwest, Irkutsk, Kaliningrad, Karelia, Middle Volzhsky, Murmansk, South Ural), Slovakia, Spain, Sweden, Switzerland.
Hosts: Pinaceae – *Abies alba* Mill., *A. fabri* (Mast.) Craib, *Pinus contorta* Douglas ex Loudon, *P. mugo* Turra, *P. sylvestris* L., *P. tabuliformis* Carrière. (Agassiz, 1996; Gershenson & Ulenberg, 1998; Hua, 2005); the host record from *Vaccinium myrtillus* L. (Ericaceae) in Gershenson & Ulenberg (1998): 62, is suspicious.
Note: *Cedestis gysseleniella* was a Kuhlwein manuscript name made nomenclaturally available by Zeller (1839) and by Duponchel (1838 [1840]). Zeller's authorship has priority due to the earlier publication year.

gysselinella Duponchel, 1838 [1840] (*Oecophora*), in Godart & Duponchel, Hist. Nat. Lép. France, 11: 461. (MNHN: Syntypes); not given.
Type locality: France.

leucopterostigmatis Sohn & Wu, 2010 (*Cedestis*), In: Sohn et al. 2010, J. Nat. Hist. 44(45/46): 2804. (IZAS: Holotype ♀); described from 1 specimen.
Type locality: China – Jiangxi, Mt. Lushan.
Distribution: Oriental – China (Jiangxi).

subfasciella Stephens, 1834 (*Telea*), Ill. Brit. Ent. Haust. 4: 247. (BMNH: Syntype); not given (1♀ in BMNH).
Type locality: England – Darenth Wood.
Distribution: Palearctic – Austria, Belgium, Czech Republic, Denmark, England, Estonia, Finland, France, Germany, Greece, Hungary, Ireland, Italy, Latvia, Lithuania, Netherlands, Norway, Poland, Portugal, Romania, Russia (European Central, European Northwest, Kaliningrad, Middle Volzhsky, Murmansk), Slovakia, Spain, Sweden, Switzerland, [former] Yugoslavia.
Hosts: Pinaceae – *Abies alba* Mill., *Pinus sylvestris* L., *P. mugo* Turra, *Pinus* spp. (Agassiz, 1996; Gershenson & Ulenberg, 1998).

farinatella Duponchel, 1838 [1840] (*Oecophora*), in Godart & Duponchel, Hist. Nat. Lép. France, 11: 473. (MNHN: Syntypes); not given.
Type locality: France.

farinella; Duponchel, 1838 [1840] (*Oecophora*), in Godart & Duponchel, Hist. Nat. Lép. France, 11: pl. 305 legend, fig. 11. [misspelling of *farinatella*].

EUHYPONOMEUTA Toll, 1941, Ztschr. Wien. Entomol. Ver. 26: 171.
Type species: *Tinea stannella* Thunberg, 1788, by monotypy, but cited by Toll [1941] as *Hyponomeuta stanellus,* an incorrect subsequent spelling.

rotunda Jin & Wang, 2009 (*Euhyponomeuta*), Oriental Insects 43: 271–273. (NUIC: Holotype ♂); described from 15♂, 14♀ (NUIC).
Type locality: China – Henan, Mt. Wangwu.
Distribution: Palearctic – China (Anhui, Henan, Shanxi, Tianjin).

secundus Moriuti, 1977a (*Euhyponomeuta*), Fauna Jap. Ypon. s. l.: 203. (UOP: Holotype ♂); described from 1 specimen.
Type locality: Japan – Hokkaido, Mt. Daisetusan.
Distribution: Palearctic – Japan (Hokkaido), Russia (Zabaykalsk).

stannella Thunberg, 1788 (*Tinea*), Mus. Nat. Acad. Upsal. 6: 80. (DEUU: Lectotype [sex unknown; designated by Karsholt & Nielsen, 1985: 456]); not given (2 specimens in DEUU).
Note: Thunberg briefly described *T. stannella* in 1788 in Mus. Nat. Acad. Upsal. 6: 80, then in 1794 described it again, giving more descriptive detail, in Mus. Nat. Acad. Upsal. 7: 83. This has led to confusion by some authors on the correct date of the species authorship.
Type locality: Sweden – Upplandia.
Distribution: Palearctic – Austria, Czech Republic, England, Finland, France, Germany, Hungary, Italy, Norway, Poland, Romania, Russia (European Northwest), Slovakia, Sweden, Switzerland.
Host: Crassulaceae – *Sedum telephium* L. (Friese, 1960).

rufimitrellus Zeller, 1839 (*Yponomeuta*), Isis Oken 32(3): 194. (Type not found); described from 10♂, 1♀.
Type locality: [Norway, Svalbard Archipelago] Spitzbergen.

stanellus; Toll, 1941 (*Euhyponomeuta*), Ztschr. Wien. Entomol. Ver. 26: 171 [misspelling of *stannella*].

stanniella; Herrich-Schäffer, 1854 [1855] (*Hyponomeuta*), Syst. Bearb. Schmett. Eur. 5: 91 [see Herrich-Schäffer, 1853–1855]. [misspelling of *stannella*].

EUHYPONOMEUTOIDES Gaj, 1954, Ent. Ber. (Amsterdam) 15: 11.
Type species: *Euhyponomeutoides albithoracellus* Gaj, 1954, by original designation.

NORDMANIANA Friese, 1960, Beitr. zur Ent. 10(1/2): 84.
Type species: *Zelleria ribesiella* de Joannis, 1900, by original designation.

albithoracellus Gaj, 1954 (*Euhyponomeutoides*), Ent. Ber. (Amsterdam) 15: 11. (Gaj collection, Krakow, Poland [not found]: Holotype ♂); described from 1 specimen.
Type locality: Poland – Nowy Sacz Co., Krynica.
Distribution: Palearctic – Austria, Estonia, Finland, France, Germany, Italy, Latvia, Lithuania, Norway, Poland, Romania (?), Russia (Karelia, Primorsky, Sakhalin), Slovakia, Sweden, Switzerland.
Hosts: Grossulariaceae – *Ribes nigrum* L., *R. uva-crispa* L.; Saxifragaceae – Saxifragaceae spp. (Friese, 1960).

rufella Tengström, 1848 (*Tinea*), Not. Sällsk. Faun. Fenn. Förh. 1: 111. (ZMH: Syntypes); not given (4♂ in ZMH). A junior homonym of *Tinea rufella* Panzer, 1804 [current status unknown].
Type locality: Finland – Uleåborg [=Oulu].

gracilariella Busck, 1904 (*Zelleria*), Proc. USNM 27: 753. (USNM: Lectotype ♂ [designated by Sohn & Lewis, 2012: 8]); not given (4♂, 2♀, 3 other specimens of unknown sex in USNM).
Type locality: Canada – British Columbia, Kaslo.
Distribution: Nearctic – Canada (Alberta, British Columbia, Nova Scotia, Ontario, Saskatchewan), USA (California).
Hosts: Grossulariaceae – *Ribes cereum* Douglas, *R. lacustre* (Pers.) Poir., *R. menziesii* Pursh, *R. sanguineum* Pursh, *R. viscosissimum* Pursh (Busck, 1904; Miller & Hammond, 2003; Powell & Opler, 2009).

ribesella Busck, 1904 (*Zelleria*), Proc. USNM 27: 754. (USNM: Syntype); not given (1♂ in USNM).
Type locality: Canada – British Columbia, Kaslo.

namikoae Moriuti, 1977a (*Euhyponomeutoides*), Fauna Jap. Ypon. s. l.: 215. (UOP: Holotype ♂); described from 1♂, 3♀ (UOP).
Type locality: Japan – Honshu, Nagano Pref., Sigakogen.
Distribution: Palearctic – Japan (Honshu).

petrias Meyrick, 1907a (*Zelleria*), Jour. Bomb. Nat. Hist. Soc. 17(3): 746. (BMNH: Lectotype ♂ [designated by Clarke, 1965: 412]); described from 2 specimens (BMNH).

Type locality: India – Punjab, Simla.
Distribution: Oriental – India (Punjab).

ribesiella de Joannis, 1900 (*Zelleria*), Bull. Soc. Ent. Fr.: 391. (MNHN: Syntypes); described from 29♂, 36♀ (MNHN).
Type locality: France – La Louvesc, Ardèche.
Distribution: Palearctic – Austria, Belgium, Finland, France, Germany, Norway, Poland, Russia (European Northwest, European South Taiga), Slovakia, Sweden, Switzerland.
Hosts: Grossulariaceae – *Ribes alpinum* L., *R. nigrum* L., *R. uva-crispa* L. (de Joannis, 1900; Nordman, 1926)

> *alpherakiella* Krulikowski, 1909 (*Gracilaria*), Rev. russe Ent. 8(3/4): 275. (ZMKU: Syntypes); described from 3 specimens (ZMKU).
> Type locality: Russia – Kirov, Urzhum.

spadix Sohn & Wu, 2010 (*Euhyponomeutoides*), In: Sohn et al., J. Nat. Hist. 44(45/46): 2806. (IZAS: Holotype ♀); described from 10♀ (IZAS).
Type locality: China – Sichuan, Wulong.
Distribution: Oriental – China (Sichuan).

trachydelta Meyrick, 1931b (*Hyponomeuta*), Exot. Micr. 4(6): 172. (BMNH: Holotype ♂); described from 1 specimen.
Type locality: Japan – Hokkaido, Mt. Teinesan.
Distribution: Oriental – China (Jiangxi); Palearctic – Japan (Hokkaido, Honshu, Kyushu), Russia (Amur, Primorsky, Zabaykalsk), South Korea.
Hosts: Celastraceae – *Euonymus alatus* (Thunb.) Siebold (Moriuti, 1977a). Note: A host record on *Euonymus hamiltonianus* ssp. *sieboldianus* (Blume) H. Hara by Issiki (1957) is possibly erroneous (see Moriuti, 1977a: 214).

> *lushanensis* Gozmany, 1960 (*Paradoxus*), Ann. Hist.-Nat. Mus. Nation. Hung. 52: 423. (TMB: Holotype ♀); described from 1 specimen.
> Type locality: China – Jiangxi, Lushan Dist.

EUMONOPYTA Moriuti, 1977a, Fauna Jap. Ypon. s. l.: 201.
Type species: *Eumonopyta unicornis* Moriuti, 1977a, by original designation.

unicornis Moriuti, 1977a (*Eumonopyta*), Fauna Jap. Ypon. s. l.: 201. (USNM: Holotype ♂); described from 1 specimen.

Type locality: Japan – Honshu, Tokyo.
Distribution: Palearctic – Japan (Honshu).

KESSLERIA Nowicki, 1864, Microlep. Spec. Nov.: 12.
Type species: *Kessleria zimmermanni* Nowicki, 1864, by monotypy.

> *CIRCOSTOLA* Meyrick, 1889, Trans. N. Z. Inst. 21: 163.
> Type species: *Circostola copidota* Meyrick, 1889, by monotypy.
>
> *HOFMANNIA* Wocke, 1876, in Heinemann & Wocke, Schmett. Deutsch. 2(2)2: 644.
> Type species: *Zelleria saxifragae* Stainton, 1868, subsequently designated by Meyrick, 1915b, Trans. N. Z. Inst. 47: 227.

albanica Friese, 1960 (*Kessleria*), Beitr. zur Ent. 10(1/2): 68. (LNK: Holotype ♂); described from 1 specimen.
Type locality: Albania – Nika.
Distribution: Palearctic – Albania, [former] Yugoslavia.

albescens Rebel, 1899 (*Hofmannia*), Verh. Zool.-Bot. Ges. Wien 49: 177. (NHMW: Lectotype ♂ [designated by Friese, 1960: 44]); described from 3 specimens (NHMW).
Type locality: Italy – Bozen [=Bolzano].
Distribution: Palearctic – Austria, Italy, Switzerland.

albomaculata Huemer & Tarmann, 1991 (*Kessleria*), Mitt. Münch. Ent. Ges. 81: 42. (MNCN: Holotype ♂); described from 2♂ (MNCN).
Type locality: France – Hautes-Pyrénées, Cauterets.
Distribution: Palearctic – France.

alpicella Stainton, 1851 (*Tinea*), Suppl. Cat. Brit. Tin. & Pteroph.: 17. (BMNH: Lectotype ♂ [designated by Huemer & Tarmann, 1991: 17]); not given (2 specimens in BMNH).
Type locality: [Austria?].
Distribution: Palearctic – Albania, Austria, Bulgaria, Czech Republic, France (?), Germany, Italy, Poland, Romania, Slovakia, Switzerland, [former] Yugoslavia.
Hosts: Saxifragaceae – *Saxifraga bryoides* L., *S. paniculata* Mill., *S. rotundifolia* L. (Huemer & Tarmann, 1991).
Note: The unpublished name *Oecophora alpicella* of Fisher von Röslerstamm was made available by Stainton (1851), who described the species from

specimens sent to him from the Mann collection that were collected by F. V. Röslerstamm. Stainton included the species under the genus *Tinea*, then stated "perhaps hardly a true *Tinea*," but by placing it in *Tinea*, created the original combination of *Tinea alpicella*. Later, Herrich-Schäffer ([1853], Syst. Bearb. Schmett. Eur. 5, fig. 359 [see Herrich-Schäffer, 1853–1855]) illustrated one of the specimens. This record, however, was not binomial and thus is not valid.

alpicella Herrich-Schäffer, [1853] (*Swammerdamia*), Syst. Bearb. Schmett. Eur. 5: 282 [see Herrich-Schäffer, 1853–1855]. (Type(s) not found); described from 4 specimens.
Type locality: Austria – Schneeberg.
Note: The original type series included males and females. However, Stainton (1868) discovered the females were a separate species, which he described as *Zelleria saxifragae*. Additionally, a female illustrated and reported by Herrich-Schäffer ([1856], Neue Schmett.: fig. 45 [see Herrich-Schäffer, 1856–1861]) as *S. alpicella* was also *Z. saxifragae* (discussion in Stainton, 1868).

alternans Staudinger, 1870 [1871] (*Swammerdamia*), Berlin. Ent. Ztschr. 14: 291. (ZMB: Lectotype ♂ [designated by Friese, 1960: 75]); described from 4♂ (DEIB, ZMB, ZSM).
Type locality: Switzerland – Graubünden, Ober-Engadin.
Distribution: Palearctic – France, Italy, Russia (Zabaykalsk), Switzerland.
Hosts: Saxifragaceae – *Saxifraga oppositifolia* L., *S. paniculata* Mill. (Huemer & Tarmann, 1991).

alternana; Kirby, 1870, Zool. Rec. 7: 423. [misspelling of *alternans*].

brachypterella Huemer & Tarmann, 1991 (*Kessleria*), Mitt. Münch. Ent. Ges. 81: 36. (BMNH: Holotype ♂); described from 3♂, 2♀ (BMNH).
Type locality: France – Hautes-Pyrénées, Pic du Midi de Bigorre.
Distribution: Palearctic – France, Spain.

brachyptera; Vives Moreno, 1992 (*Kessleria*), Cat. Sis. Sin. Lep.: 73. [misspelling of *brachypterella*].

brevicornuta Huemer & Tarmann, 1991 (*Kessleria*), Mitt. Münch. Ent. Ges. 81: 34. (Arenberger collection [private], Wien, Austria: Holotype ♂); described from 2♂, 1♀ (Arenberger collection & NHMW).
Type locality: Spain – Sierra de Gredos, Garganta de las Pozas.
Distribution: Palearctic – Spain.

burmanni Huemer & Tarmann, 1991 (*Kessleria*), Mitt. Münch. Ent. Ges. 81: 28. (TLMF: Holotype ♂); described from 84♂, 59♀ (Burmann collection, DEIB, NHMW, TLMF, Whitebread collection, ZMB).
Type locality: Austria – Tyrol, Nordkette.
Distribution: Palearctic – Austria, Switzerland, [former] Yugoslavia.
Hosts: Saxifragaceae – *Saxifraga macropetala* (A. Kern. Ex Engl.), *S. caesia* L., *S. oppositifolia* L. (Huemer & Tarmann, 1991).

caflischiella Frey, 1880 (*Swammerdamia*), Lep. Schweiz: 344. (BMNH: Syntype); not given (1♂ in BMNH).
Type locality: Switzerland – Gamsen.
Distribution: Palearctic – Austria, Italy, Russia (South Ural), Switzerland.
Host: Saxifragaceae – *Saxifraga moschata* Wulfen (Huemer & Tarmann, 1991).

calculosa Meyrick, 1914f (*Hyponomeuta*), Jour. Bomb. Nat. Hist. Soc. 23(1): 127. (BMNH: Lectotype ♂ [designated by Clarke, 1965: 323]); described from 14 specimens (BMNH).
Type locality: India – Assam, Khasi Hills.
Distribution: Oriental – China (Zhejiang), India (Assam).

copidota Meyrick, 1889 (*Circostola*), Trans. N. Z. Inst. 21: 163. (BMNH: Lectotype ♂ [designated by Moriuti, 1977a: 208]); not given [more than 2 specimens].
Type Locality: New Zealand – Nelson.
Distribution: Australasian – New Zealand.

corusca Meyrick, 1914f (*Hyponomeuta*), Jour. Bomb. Nat. Hist. Soc. 23(1): 127. (BMNH: Holotype ♀); described from 1 specimen.
Type locality: India – Assam, Khasi Hills.
Distribution: Oriental – India (Assam).

conisca; Sharp, 1914, Zool. Rec. (7) 51: 260 [misspelling of *corusca*].

diabolica Huemer & Tarmann, 1991 (*Kessleria*), Mitt. Münch. Ent. Ges. 81: 33. (ZMUC: Holotype ♂); described from 1 specimen.
Type locality: Spain – Avila, Sierra de Gredos.
Distribution: Palearctic – Portugal, Spain.

fasciapennella Stainton, 1849c (*Zelleria*), Syst. Cat. Brit. Tin. & Pteroph. 8: 22. (BMNH: Lectotype ♂ [designated by Huemer & Tarmann, 1991: 55]); described from 2 specimens.

Type locality: England – Scotland, Pentland.
Distribution: Palearctic – Austria, England (Scotland), Estonia, Finland, Latvia, Lithuania, Norway, Poland, Portugal (?), Russia (European Northwest, Irkutsk, Karelia, Komi), Sweden, Switzerland.
Host: Celastraceae – *Parnassia palustris* L. (Huemer & Tarmann, 1991).

longipenella Friese, 1960 (*Kessleria*), Beitr. zur Ent. 10(1/2): 83. (DEIB: Holotype ♂); described from 1 specimen.
Type locality: Russia – Karelia Republic, Petrosawodsk.

hauderi Huemer & Tarmann, 1991 (*Kessleria*), Mitt. Münch. Ent. Ges. 81: 31. (TLMF: Holotype ♂); described from 49♂, 4♀ (Burmann collection, DEIB, Klimesch collection, MCSN, NHMS, NHMW, OLML, TLMF, ZMAK, ZMB, ZSM).
Type locality: Austria – Steiermark, Eisenerzer Reichenstein.
Distribution: Palearctic – Austria.
Host: Saxifragaceae – *Saxifraga oppositifolia* L. (Huemer & Tarmann, 1991).

helvetica Huemer & Tarmann, 1991 (*Kessleria*), Mitt. Münch. Ent. Ges. 81: 46. (TLMF: Holotype ♂); described from 10♂, 3♀ (NHMS, TLMF, Whitebread collection, ZMB).
Type locality: Switzerland – Valais, Zermatt.
Distribution: Palearctic – France, Switzerland.

inexpectata Huemer & Tarmann, 1991 (*Kessleria*), Mitt. Münch. Ent. Ges. 81: 45. (TLMF: Holotype ♂); described from 11♂, 2♀ (TLMF).
Type locality: France – Provence-Alpes-Côte d'Azur, Alpes Maritimes, Punta Marguareis.
Distribution: Palearctic – France.

insubrica Huemer & Tarmann, 1993 (*Kessleria*), Veröff. Tirol. Landes. Ferdin. 73: 41. (TLMF: Holotype ♀); described from 8♀ (TLMF).
Type locality: Italy – Bergamo, Alpi Orobie, Val d'Arera.
Distribution: Palearctic – Italy.
Host: Saxifragaceae – *Saxifraga caesia* L. (Huemer & Tarmann, 1993).

insulella Moriuti, 1977a (*Kessleria*), Fauna Jap. Ypon. s. l.: 209. (ELKU: Holotype ♂); described from 3♂, 1♀ (ELKU).
Type locality: Taiwan – Taichung, Hassenzan [=Pahsienshan].
Distribution: Oriental – Japan (Ryukyu Islands), Taiwan; Palearctic – Japan (Kyushu).

klimeschi Huemer & Tarmann, 1991 (*Kessleria*), Mitt. Münch. Ent. Ges. 81: 47. (TLMF: Holotype ♂); described from 13♂, 14♀ (Arenberger collection, Klimesch collection, MCSN, NHMW, TLMF).
Type locality: Italy – Udine, Montasio Malga Pecol.
Distribution: Palearctic – Italy, [former] Yugoslavia.
Hosts: Saxifragaceae – *Saxifraga incrustata* Vest, *S. paniculata* Mill. (Huemer & Tarmann, 1991)

leucotoma Meyrick, 1935a (*Hyponomeuta*), in Caradja & Meyrick: Mat. Microlep.-Fauna Chin. Prov. Kiangsu, Chekiang, Hunan: 88. (MGAB: Holotype ♀); described from 1 specimen.
Type locality: China – Hunan, Mt. Hoeng-Shan.
Distribution: Oriental – China (Hunan).
Note: Meyrick (1935a: 89) erroneously gave the type locality as Lungtan in his description of *leucotoma*. See Friese (1962: 318) for details.

macedonica Huemer & Tarmann, 1991 (*Kessleria*), Mitt. Münch. Ent. Ges. 81: 26. (ZSM: Holotype ♂); described from 1 specimen.
Type locality: Serbia – Shar Planina, Crni Vrh.
Distribution: Palearctic – Serbia.

martinella Walker, 1863 (*Tinea*), List Lepid. Ins. BM 28: 472. (BMNH: Syntype); not given (1♂ specimen in BMNH).
Type locality: Canada – Ontario, Hudson Bay, St. Martin's Falls.
Distribution: Nearctic – Canada (Ontario).

> *afflictella* Walker, 1864b (*Tinea*), List Lepid. Ins. BM 30: 1002 (Addenda: 472). (BMNH: Holotype ♂ [erroneously stated as ♀ in the original description; the holotype has male genitalia]); described from 1 specimen.
> Type locality: Canada – Ontario, Hudson Bay, St. Martin's Falls.

mixta Huemer & Tarmann, 1991 (*Kessleria*), Mitt. Münch. Ent. Ges. 81: 18. (NHMW: Holotype ♂); described from 1 specimen.
Type locality: Albania – Korab.
Distribution: Palearctic – Albania.

munda Meyrick, 1921b (*Hyponomeuta*), Exot. Micr. 2(14): 438. (BMNH: Holotype ♂); described from 1 specimen.
Type locality: India – Mumbai, Kharaghoda.
Distribution: Oriental – India (Mumbai, Punjab).

neuguinea Moriuti, 1981 (*Kessleria*), Trans. Lep. Soc. Japan 32(1/2): 83. (UOP: Holotype ♂); described from 1 specimen.
Type locality: New Guinea – Wareng.
Distribution: Australasian – New Guinea.

nivescens Burmann, 1980 (*Kessleria*), Nach. der Bayer. Ent. 29(6): 105. (Burmann collection [private], Innsbruck, Austria: Holotype ♂); described from 5♂ (Burmann collection, DEIB, IPEG).
Type locality: Italy – Verona, Monte Baldo, Telegrafo.
Distribution: Palearctic – Austria, Italy, Slovenia.
Hosts: Saxifragaceae – *Saxifraga caesia* L., *S. tombeaensis* Boiss. ex Engl. (Huemer & Tarmann, 1991).

nivosa Meyrick, 1938a (*Zelleria*), In: Caradja & Meyrick, Dt. Ent. Z. Iris 52(1): 16. (BMNH: Lectotype ♂ [designated here]); described from 2 specimens.
Type locality: China – Yunnan, Likiang [=Lijiang], Yülingshanmass [possibly Yulongxueshan].
Distribution: Oriental – China (Yunnan).
Note: Meyrick (1938a) did not specify the type series of *nivosa*. It was described from two specimens, only one of which is still in the BMNH collection. Clarke (1965) reported it as "type" but did not specify further. We correct that error here by designating it as the lectotype. Labels read: male, "Lecto- | type" [a circular label with indigo borders], "Type" [circular label with red margins], "Likiang | China | H. 0.7.34" [hand-writtten], "*nivosa* Meyr." [hand-writtten], "*Zelleria* | *nivosa* | 1/2 Meyr. | E. Meyrick det. | in Meyrick Coll.," "♂ genitalia on | slide 24.VI 1948 | J.F.G.C. 7709" [rectangular label with black lined margins], "LECTOTYPE | *Zelleria nivosa* | Zeller | teste J. Sohn, 2014."

nuvosa; Hua, 2005 (*Zelleria*), List Chin. Ins. 3: 11. [misspelling of *nivosa*].

parnassiae Braun, 1940 (*Zelleria*), Trans. Am. Ent. Soc. 66: 277. (ANSP: Holotype ♂); described from 12♂,12♀ (ANSP, CNC, MCAS, USNM).
Type locality: USA – Washington, Olympic Mountains, Hoh Lake.
Distribution: Nearctic – Canada (Alberta), USA (Washington).
Host: Celastraceae – *Parnassia fimbriata* K.D. Koenig. (Braun, 1940).

petrobiella Zeller, 1868 (*Scythropia*), Verh. Zool.-Bot. Ges. Wien 18: 607. (BMNH: Lectotype ♂ [designated by Huemer & Tarmann, 1991: 38]); not given, mentioned as "♂♀" (1♂, 1♀ in BMNH).
Type locality: [former] Yugoslavia.

Distribution: Palearctic – Austria, Italy, Slovenia.
Host: Saxifragaceae – *Saxifraga caesia* L. (Huemer & Tarmann, 1991).

pseudosericella Moriuti, 1977a (*Kessleria*), Fauna Jap. Ypon. s. l.: 210. (UOP: Holotype ♂); described from 2♂, 3♀ (UOP, USNM).
Type locality: Japan – Honshu, Nagano Pref., Sigakogen.
Distribution: Palearctic – Japan (Honshu).

pyrenaea Friese, 1960 (*Kessleria*), Beitr. zur Ent. 10(1/2): 76. (NHMW: Holotype ♂); described from 1 specimen.
Type locality: France – Catalan Pyrénées, Mount Canigou.
Distribution: Palearctic – France.

saxifragae Stainton, 1868 (*Zelleria*), Trans. Ent. Soc. Lond. 1868: 139. (BMNH: Syntype); not given (1 specimen in BMNH).
Type locality: Germany – Kaiserberg, near Oberaudorf.
Distribution: Palearctic – Austria, England, France, Germany, Ireland, Israel, Italy, Poland, Portugal, Romania, Russia (Zabaykalsk), Slovakia, Spain, Switzerland, [former] Yugoslavia.
Hosts: Saxifragaceae – *Saxifraga azoides* L., *S. hirsuta* L., *S. oppositifolia* L., *S. paniculata* Mill., *S. spathularis* Brot. (Huemer & Tarmann, 1991).

wehrlii Huemer & Tarmann, 1991 (*Kessleria*), Mitt. Münch. Ent. Ges. 81: 22. (TLMF: Holotype ♂); described from 16♂, 1♀ (DEIB, NHMS, TLMF).
Type locality: France – Alpes Maritimes, Mount Gélas Massif, Mount Colomb.
Distribution: Palearctic – France.

zimmermanni Nowicki, 1864 (*Kessleria*), Microlep. Spec. Nov.: 13. (DEIB: Lectotype ♂ [designated by Huemer & Tarmann, 1991: 40]); not given.
Type locality: Poland – Tatra Mts.
Distribution: Palearctic – Poland, Slovakia.
Hosts: Saxifragaceae – *Saxifraga caesia* L., *S. macropetala A.* Kerner, *S. paniculata* Mill. (Burmann, 1956, 1973; Baraniak, 1988).

> *tatrica* Friese, 1960 (*Kessleria*), Beitr. zur Ent. 10(1/2): 71. (NHMW: Holotype ♂); described from 1 specimen.
> Type locality: Poland – Tatra Mts.

KLAUSIUS Moriuti, 1977a, Fauna Jap. Ypon. s. l.: 231.
Type species: *Klausius minor* Moriuti, 1977a, by original designation.

angustus Jin, Wang & Li, 2010 (*Klausius*), Ent. News 121(1): 5. (NUIC: Holotype ♂); described from 10♂, 3♀ (NUIC).
Type locality: China – Guangdong, Mt. Dadong.
Distribution: Oriental – China (Fujian, Guangdong, Hunan); Palearctic – China (Gansu).

latus Jin, Wang & Li, 2010 (*Klausius*), Ent. News 121(1): 2. (NUIC: Holotype ♂); described from 5♂, 3♀ (NUIC).
Type locality: China – Guizhou, Mt. Fanjing.
Distribution: Oriental – China (Guizhou); Palearctic – China (Tianjin).

major Moriuti, 1977a (*Klausius*), Fauna Jap. Ypon. s. l.: 233. (UOP: Holotype ♂); described from 10♂ (UOP).
Type locality: Japan – Kyushu, Kagosima Pref., Mt. Kirisimayama.
Distribution: Palearctic – Japan (Honshu, Kyushu).

minor Moriuti, 1977a (*Klausius*), Fauna Jap. Ypon. s. l.: 232. (UOP: Holotype ♂); described from 1♂, 1♀ (UOP).
Type locality: Japan – Honshu, Kyoto, Mt. Hieisan.
Distribution: Palearctic – Japan (Honshu), South Korea.

LAMPRESTHIA Moriuti, 1977a, Fauna Jap. Ypon. s. l.: 233.
Type species: *Lampresthia lucella* Moriuti, 1977a, by original designation.

lucella Moriuti, 1977a (*Lampresthia*), Fauna Jap. Ypon. s. l.: 234. (UOP: Holotype ♂); described from 3♂, 4♀ (EIHU, UOP).
Type locality: Japan – Honshu, Nagano Pref., Sigakogen.
Distribution: Palearctic – Japan (Hokkaido, Honshu).

METANOMEUTA Meyrick, 1935a, in Caradja & Meyrick: Mat. Microlep.-Fauna Chin. Prov. Kiangsu, Chekiang, Hunan: 87.
Type species: *Metanomeuta fulvicrinis* Meyrick, 1935a, by monotypy.

> *MATANOMEUTA*; Moriuti, 1977a, Fauna Jap. Ypon. s. l.: 11. [misspelling of *Metanomeuta*]

fulvicrinis Meyrick, 1935a (*Metanomeuta*), in Caradja & Meyrick: Mat. Microlep.-Fauna Chin. Prov. Kiangsu, Chekiang, Hunan: 87. (MGAB: Lectotype ♂ [designated by Friese, 1962: 323]); described from 9 specimens (BMNH, MGAB).

Type locality: China – Zhejiang, Mt. Tienmushan.
Distribution: Palearctic – China (Anhui, Fujian, Guangxi, Guizhou, Henan, Hubei, Hunan, Jiangxi, Sichuan, Zhejiang), Japan (Honshu, Kyushu, Shikoku), South Korea.
Note: Prior to Clarke (1965), Friese (1962) had designated a lectotype of *fulvicrinis*, as reported by Popescu-Gorj (1992), invalidating Clarke's designation. Friese (1962) found all 9 type specimens from the MGAB and BMNH collections, but only six remain available, according to Popescu-Gorj (1992).

zonoceros Meyrick, 1935a (*Metanomeuta*), in Caradja & Meyrick: Mat. Microlep.-Fauna Chin. Prov. Kiangsu, Chekiang, Hunan: 88. (MGAB: Lectotype ♂ [designated by Friese, 1962: 323]); described from 3 specimens (BMNH, MGAB).
Type locality: China – Hunan, Mt. Hoengshan.
Note: Prior to Clarke (1965), Friese (1962) designated the lectotype of *zonoceros*, as reported by Popescu-Gorj (1992), invalidating Clarke's designation.

spinisparsula Jin & Wang, 2008 (*Metanomeuta*), Acta Zootaxon. Sin. 33(1): 55. (NUIC: Holotype ♂); described from 1♂, 1♀ (NUIC).
Type locality: China – Hunan, Shimen Co.
Distribution: Oriental – China (Hunan).

yuexiensis Jin & Wang, 2008 (*Metanomeuta*), Acta Zootaxon. Sin. 33(1): 54. (NUIC: Holotype ♂); described from 9♂, 1♀ (NUIC).
Type locality: China – Anhui, Yuexi Co.
Distribution: Palearctic – China (Anhui).

NIPHONYMPHA Meyrick, 1914d, Exot. Micr. 1(6): 174. An objective replacement name for *Calantica* Zeller, 1847a.
Type species: *Calantica albella* Zeller, 1847a, by subsequent designation (for *Calantica* Zeller, 1847a) by Busck, 1912, Proc. Ent. Soc. Wash. 14: 85.

CALANTICA Zeller, 1847a, Isis Oken 40(11): 811. A junior homonym of *Calantica* Grey, 1825 (Crustacea).
Type species: *Calantica albella* Zeller, 1847a, by subsequent designation by Busck, 1912, Proc. Ent. Soc. Wash. 14: 85.
Note: Zeller attributed the authorship of *Calantica* to Heyden. However, H. v Heyden had not published the name, making Zeller the author.

TRISOPHISTA Meyrick, 1924, Exot. Micr. 3(4): 117.
Type species: *Trisophista doctissima* Meyrick, 1924, by monotypy.

argentea Busck, 1912 (*Calantica*), Proc. Ent. Soc. Wash. 14: 86. (USNM: Lectotype ♂ [designated here]); not given (1♂ in BMNH; 1♂ in USNM).
Type locality: Mexico – Orizaba.
Distribution: Neotropical – Mexico (Orizaba).
Note: The lectotype of *argentea* Busck, 1912 is designated here to eliminate future confusion; it is based on the male syntype in the USNM collection. The pin data labels read: "LECTOTYPE [red letters] | *Calantica* | *argentea* Busck" [rectangular label with red margins], "Orizaba | Mex[ico]," "R Muller | Collector," "Oct | 10," "Cotype | No. 14527 | U.S.N.M.," "2653" [red, rectangular label], "2653" [hand written], "Calantica | argentea | Cotype Busck" [handwritten on rectangular label with black-lined margins].

dealbatella Zeller, 1847a (*Calantica*), Isis Oken 40(11): 811. (BMNH: Syntype); not given (1 ♂ in BMNH).
Type locality: Germany – Hesse, Taunus.
Distribution: Palearctic – Austria, Corsica, Czech Republic, France, Germany, Hungary, Italy, Romania, Sicily, Slovakia, Switzerland, Turkey, [former] Yugoslavia.
Hosts: Fagaceae – *Quercus robur* L., *Quercus* sp. (Zeller, 1847a; Friese, 1960).

> *albella* Zeller, 1847a (*Calantica*), Isis Oken 40(11): 812. (BMNH: Syntype); not given (1 ♂ in BMNH).
> Type locality: Germany – Hesse, Taunus.
> Note: Zeller (1847a) described specimens he had in hand as *Calantica dealbatella* on page 811, then stated at the end of the description on page 812 that the species just described might be the same as *albella* Heyden *in litt.*, but he was not sure since he had not seen a specimen of the latter, and did not want it to be confused with his *dealbatella*. The synonymy proved to be true when Friese (1960: 39) wrote that he had Zeller's specimens as well as others collected from the type locality that proved to be conspecific. Then, Friese (1960: 38–39), chose to ignore the validly described *dealbatella* and called the type species *Calantica albella* Heyden *in litt.* (unpublished), instead of *dealbatella* Zeller, 1847a, without explanation or defense for his choice of name. *Calantica albella* Heyden was never published, which makes it unavailable. *Calantica dealbatella* by Zeller (1847a) is therefore the correct name for the species and type of the genus. Leraut (1980) followed the classification of Friese (1960), which has lead to the perpetuation of the error by some subsequent authors.

doctissima Meyrick, 1924 (*Trisophista*), Exot. Micr. 3(4): 118. (BMNH: Lectotype ♀ [designated by Clarke, 1965: 396]); described from 4 specimens (BMNH).

Type locality: Rwanda – Lake Tshohoa.
Distribution: Ethiopian – Democratic Republic of the Congo, Kenya, Rwanda.

longispina Yu & Li, 2002 (*Niphonympha*), Reichenbachia 34(45): 376. (NUIC: Holotype ♂); described from 1 specimen.
Type locality: China – Guizhou, Suoluo.
Distribution: Oriental – China (Guizhou).

pauli Viette, 1967 (*Trisophista*), Bull. Soc. Ent. Fr. 72: 291. (MNHN: Holotype ♂); described from 1 specimen.
Type locality: Madagascar – Centre, vicinity of Ambohimahasoa, Tsarafidy.
Distribution: Ethiopian – Madagascar.

varivera Yu & Li, 2002 (*Niphonympha*), Reichenbachia 34(45): 375. (NUIC: Holotype ♂); described from 12♂, 14♀ (NUIC).
Type locality: China – Henan, Xixia, Huangshi'an.
Distribution: Oriental – China (Guizhou, Henan); Palearctic – China (Gansu, Shaanxi).

vera Moriuti, 1963a (*Niphonympha*), Kontyû 31(3): 215. (UOP: Holotype ♂); described from 11♂, 5♀ (UOP).
Type locality: Japan – Shikoku, Ehime Pref., Iyo Prov., Zyozyusya.
Distribution: Palearctic – Japan (Honshu, Kyushu, Shikoku), South Korea.

wuzhishana Jin & Li, 2012 (*Niphonympha*), Entomotaxonomia 34(2): 256. (NUIC: Holotype ♂); described from 8♂ (NUIC).
Type locality: China – Hainan, Mt. Wuzhi.
Distribution: Oriental – China (Hainan).

OCNEROSTOMA Zeller, 1847b, Linn. Ent. 2: 298.
Type species: *Ocnerostoma piniariella* Zeller, 1847b, by monotypy.

> *OCNEROSTEMA*; Heydenreich, 1851, Lepid. Euro. Cat. Method.: 88. [misspelling of *Ocnerostoma*].

friesei Svensson, 1966 (*Ocnerostoma*), Opusc. Ent. 31(3): 196. (MZLU: Holotype ♂); described from 12 ♂, 11♀ (Johansson collection [private]; MZLU).
Type locality: Sweden – Skåne, Simris.
Distribution: Palearctic – Austria, Belgium, Czech Republic, Denmark, England, Estonia, Finland, France, Germany, Ireland, Israel, Japan (Honshu), Latvia,

Lithuania, Netherlands, Norway, Poland, Romania (?), Russia (Karelia, Middle Volzhsky, Murmansk, South Khabarovsk, Zabaykalsk), Slovakia, South Korea, Spain, Sweden, Switzerland.
Hosts: Pinaceae – *Pinus cembra* L., *P. densiflora* Siebold & Zucc., *P. mugo* Turra, *P. sylvestris* L. (Svensson, 1966; Moriuti, 1977a; Kulfan & Holecová, 2010).

transbaicalensis Pozmogova, 1969 (*Ocnerostoma fresei* ssp.), Trudy Burya. Inst. Estest. Nauk 7: 53 (BSCM: Holotype [sex not given]); not given.
Type locality: Russia – Buryatsky, Dodogolsky.

piniariella Zeller, 1847b (*Ocnerostoma*), Linn. Ent. 2: 299. (BMNH: Syntypes); not given (1♀ in BMNH).
Type *locality:* [Germany – Berlin].
Distribution: Ethiopian – Aldabra; Nearctic – Canada (British Columbia, Ontario, Québec), USA (California, Maine, Montana); Oriental – China (Zhejiang); Palearctic – Austria, Belgium, Bulgaria, Czech Republic, Denmark, England, Estonia, Finland, France, Germany, Ireland, Italy, Latvia, Luxembourg, Netherlands, Norway, Poland, Romania, Russia (European Central, Kaliningrad, Middle Volzhsky), Slovakia, Spain, Sweden, Switzerland.
Hosts: Pinaceae – *Pinus cembra* L., *P. contorta* Douglas ex Loudon, *P. monticola* Douglas ex Loudon, *P. mugo* Turra, *P. radiata* D. Don, *P. resinosa* Aiton, *P. sylvestris* L. (Freeman, 1960; Prentice, 1965; Agassiz, 1996; Kulfan & Holecová, 2010); host records for *Abies alba* Mill., *Picea abies* (L.) H. Karst. (Pinaceae) and *Juniperus* sp. (Cupressaceae) by Friese (1960), and López (2007) are doubtful (Pitkin et al. 2012).

copiosella Frey, 1856 (*Ocnerostoma*) [in part], Tin. Pter. Schweiz: 197. (BMNH: Syntype); not given (1♂ in BMNH).
Type locality: Switzerland – Engadin, St. Moritz.

galactitella Eversmann, 1844 (*Oecophora*), Lep. Volgo-Ural.: 595. (ZMAS: Syntypes); described from 2 specimens.
Type locality: Russia – Orenburg.

strobivorum Freeman, 1960 (*Ocnerostoma*), Can. Ent. 92, Supp. 16: 27. (CNC: Holotype ♂); described from 10♂, 3♀ (CNC).
Type locality: Canada – Ontario, Constance Bay.
Distribution: Nearctic – Canada (Newfoundland, Ontario, Québec), USA (Maine, Washington).
Host: Pinaceae – *Pinus strobus* L. (Freeman, 1960).

OPSICLINES Meyrick, 1907d, Proc. Linn. Soc. N.S. Wales 32(1): 68.
Type species: *Opsiclines leucomorpha* Lower, 1900, by monotypy.

leucomorpha Lower, 1900 (*Zelleria*), Proc. Linn. Soc. N.S. Wales 25: 422. (SAMA: Holotype ♀); described from 1 specimen.
Type locality: Australia – South Australia, Adelaide, Parkside.
Distribution: Australasian – Australia (South Australia).

> *leucophasa* Lower, 1908 (*Elachista*), Trans. R. Soc. S. Austr. 32: 118. (SAMA: Holotype ♀); described from 1 specimen.
> Type locality: Australia – South Australia, Semaphore.

ORENCOSTOMA Moriuti, 1971b, Kontyû 39(3): 253.
Type species: *Orencostoma bicornigerum* Moriuti, 1971b, by original designation.

bicornigerum Moriuti, 1971b (*Orencostoma*), Kontyû 39(3): 255. (BMNH: Holotype ♂); described from 1♂, 1♀ (BMNH).
Type locality: India – Kashmir, Gulmarg.
Distribution: Oriental – India (Kashmir).

PARADOXUS Stainton, 1869, Ent. M. Mag. 6: 42.
Type species: *Paradoxus osyridellus* Stainton, 1869, by monotypy.
Note: *Paradoxus* and *P. osyridellus* were described by Millière 1869: Ann. Soc. Linn. Lyon 3(23): 42. However, Stainton (1869) had already made the manuscript names nomenclaturally available (for discussion, see Sattler & Tremewan, 1973).

deformis Meyrick, 1914f (*Paradoxus*), Jour. Bomb. Nat. Hist. Soc. 23(1): 124. (BMNH: Holotype ♂); described from 1 specimen.
Type locality: Sri Lanka – Sabaragamuwa Prov., Kegalle.
Distribution: Oriental – Sri Lanka.

fulgens Christoph, 1888 (*Tinea*), Hor. Soc. Ent. Ross. 22: 312. (ZMB: Syntypes); not given (1♂, 1♀ in ZMB).
Type locality: Russia – Caucasus, Borshom.
Distribution: Palearctic – Georgia, Russia (West Caucasus).

> *caucasica* Friese, 1960 (*Kessleria*), Beitr. zur Ent. 10(1/2): 77. (ZMB: Holotype ♂); described from 3♂, 1♀ (ZMB).
> Type locality: Russia – Caucasus, Kirala.

osyridellus Stainton, 1869 (*Paradoxus*), Ent. M. Mag. 6: 42. (Type not found); described from 1 specimen.
Type locality: France – Cannes.
Distribution: Palearctic – France, Greece, Portugal, Sardinia, Spain, Turkey, [former] Yugoslavia.
Host: Santalaceae – *Osyris alba* L. (Millière, 1869).

> *oxyridella*; Vives Moreno, 1992 (*Kessleria*), Cat. Sis. Sin. Lep.: 73. [misspelling of *osyridellus*].

restrictellus Chrétien, 1915 (*Paradoxus*), Ann. Soc. Ent. Fr. 84: 311. (MNHN: Type not found, possibly lost (J. Minet, pers. comm. to JAL)); not given.
Type locality: Tunisia – Gafsa.
Distribution: Palearctic – Tunisia.

PARAHYPONOMEUTA Toll, 1941, Ztschr. Wien. Entomol. Ver. 26: 172.
Type species: *Butalis egregiella* Duponchel, 1838 [1840], by monotypy.

bakeri Walsingham, 1894b (*Hyponomeuta*), Trans. Ent. Soc. Lond. 1894(4): 542. (BMNH: Syntypes); described from 5 specimens (BMNH).
Type locality: Madeira – San Antonio da Serra.
Distribution: Palearctic – Madeira Archipelago.
Hosts: Ericaceae – *Erica arborea* L., *E. scoparia* L. (Aguiar & Karsholt, 2006).
Note: Aguiar & Karsholt (2006) resurrected this species which had been treated as a synonym of *egregiella* by Friese (1960: 64).

egregiella Duponchel, 1838 [1840] (*Butalis*), in Godart & Duponchel, Hist. Nat. Lép. France, 11: 345. (MNHN: Holotype ♀); described from 1 specimen.
Type locality: France – Avesnes.
Distribution: Palearctic – Corsica, France, Italy, Portugal, Sardinia, Spain, Switzerland, [former] Yugoslavia.
Hosts: Ericaceae – *Erica arborea* L., *E. scoparia* L. (Schütze, 1931).

> *obsoleta* Gibeaux, 1982 (*Kessleria egregiella* f.), Alexanor 12 (8): 371 (MNHN: Holotype ♀); described from 2♂, 3♀ (MNHN).
> Type locality: France – Alpes-Maritimes, Cannes.

PARASWAMMERDAMIA Friese, 1960, Beitr. zur Ent. 10(1/2): 101.
Type species: *Swammerdamia lapponica* Petersen, 1932, by original designation.

albicapitella Scharfenberg, 1805 (*Phalaena Tinea*), In: Bechstein & Scharfenberg, Voll. Natur. der Sch. Forst. 3: 803. (Type not found); not given.
Type locality: [Germany].
Distribution: Nearctic – Canada (introduced: British Columbia [see Landry et al. 2013.]); Palearctic – Albania, Austria, Belgium, Bulgaria, China, Corsica, Czech Republic, Denmark, England, Estonia, Finland, France, Germany, Greece, Ireland, Italy, Netherlands, Norway, Poland, Romania, Russia (Primorsky, Sakhalin), Sardinia, Slovakia, Spain, Sweden, Switzerland, [former] Yugoslavia.
Hosts: Rosaceae – *Crataegus* spp., *Prunus spinosa* L., *Pyrus* sp.? [Friese, 1960; Agassiz, 1996; White, 2000].

atricapitella; White, 2000 (*Paraswammerdamia*), Ent. Rec. J. Var. 112: 42. [misspelling of *albicapitella*].

caesiella Hübner, [1813] 1816 (*Tinea*), Samml. Eur. Schmett. Vol. 8, pl. 53, Tineæ 2, Pyralidiformes B, fig. 360 [not 260] [in Hübner, 1796–1838]. (Type not found); not given. A junior homonym of *Tinea caesiella* Hübner, 1796 [unnecessary replacement name].

spiniella Hübner, [1809] (*Tinea*), Gesch. Europ. Schmett., Larvae Lepidoptera 8, Tineæ 3, Tortriciformes B, figs. 1a, b [in Hübner, 1793–1842]. (Type not found); not given.
Type locality: [Europe].

conspersella Tengström, 1848 (*Tinea*), Not. Sällsk. Faun. Fenn. Förh. 1: 112. (ZMH: Syntype); not given (1♀ in ZMH).
Type locality: Finland – Kuusamo.
Distribution: Nearctic – Canada (Québec); Palearctic – Denmark, Estonia, Finland, France, Israel, Latvia, Lithuania, Netherlands, Norway, Russia (European Northwest, Kaliningrad, Murmansk, South Ural, Zabaykalsk), Sweden, Switzerland, Ukraine.
Host: Ericaceae – *Empetrum nigrum* L. (Friese, 1960).

moënsis Strand, 1920 (*Swammerdamia conspersella* ab.), Archiv Naturgesch. (A) 85(4): 59. (ZMO: Syntypes); described from 2 specimens (ZMO).
Type locality: Norway – Groenlien [=Gronlia].

disemanta Meyrick, 1933 (*Hyponomeuta*), Exot. Micr. 4(12): 373. (BMNH: Holotype ♀); described from 1 specimen.

Type locality: Sierra Leone – Old Calabar [=Njala].
Distribution: Ethiopian – Sierra Leone.
Note: Gershenson & Ulenberg (1998) stated the type depository as LNK in error. The holotype is in the BMNH collection as Clarke (1965: 324) reported.

iranella Friese, 1960 (*Paraswammerdamia*), Beitr. zur Ent. 10(1/2): 104. (LNK: Holotype ♂); described from 1♂, 1♀ (LNK, ZSM).
Type locality: Iran – Elbursgebirge, Keredi.
Distribution: Palearctic – Iran.
Hosts: Rosaceae – *Cotoneaster microcarpus* (Rehder & E.H. Wilson) Flinck & B. Hylmö, *Prunus cerasifera* Ehrh. (Gershenson & Ulenberg, 1998).

lapponica Peterson, 1932 (*Swammerdamia*), Archiv Naturgesch. (N. F.) 1: 210. (BMNH: Holotype ♂); described from 3 specimens (BMNH; Svensson collection [private]).
Type locality: Finland – Lapland, Pello.
Distribution: Nearctic – Canada (Québec); Palearctic – Finland, Norway, Russia (Karelia, Murmansk, Zabaykalsk), Sweden.
Host: Betulaceae – *Betula nana* L. (Kutenkova, 1980).

monticolella Moriuti, 1977a (*Paraswammerdamia*), Fauna Jap. Ypon. s. l.: 247. (UOP: Holotype ♂); described from 3♂ (UOP).
Type Locality: Japan – Hokkaido, Mt. Daisetusan.
Distribution: Palearctic – Japan (Hokkaido).

nebulella Goeze, 1783 (*Phalaena Tinea*), Entom. Beyt. zu des Ritter: 162. (Type not found); not given.
Distribution: Palearctic – Albania, Austria, Belgium, Bulgaria, Czech Republic, Denmark, England, Estonia, Finland, France, Germany, Greece, Hungary, Ireland, Italy, Latvia, Lithuania, Luxembourg, Netherlands, Norway, Poland, Portugal, Romania, Russia (European Northwest, Kaliningrad, West Caucasus), Sardinia, Slovakia, Spain, Sweden, Switzerland, [former] Yugoslavia.
Hosts: Rosaceae – *Cotoneaster horizontalis* Decne, *Crataegus monogyna* Jacq., *Crataegus* spp., *Rosa* spp., *Sorbus aucuparia* L. (Agassiz, 1996; Gershenson & Ulenberg, 1998).

lutarea Haworth, 1828 (*Erminea*), Lep. Brit. 4: 515. (BMNH: Syntypes); not given (1♂ in BMNH).
Type Locality: Great Britain – Darn Wood.

oxyacanthella Duponchel, 1842 [1843] (*Tinea*), In: Godart & Duponchel, Hist. Nat. Lép. France, Suppl. 4: 205. (Type lost (J. Minet, pers. comm. to JAL)); not given.
Type locality: France.

oxyocanthella; Gómez-Bustillo, 1979 (*Paraswammerdamia*), SHILAP Revta. lepid. 7(28): 243. [misspelling of *oxyacanthella*].

ornichella Friese, 1960 (*Paraswammerdamia*), Beitr. zur Ent. 10(1/2): 110. (ZMB: Holotype ♂); described from 9♂, 8♀ (BMNH, ZMB).
Type locality: Russia – Sarepta.
Distribution: Palearctic – Russia (Volga-Donskoy, South Ural), Ukraine.
Host: Rosaceae – *Prunus tenella* Batsch. (Gershenson, 1986).

praetincta Meyrick, 1914f (*Hyponomeuta*), Jour. Bomb. Nat. Hist. Soc. 23(1): 126. (BMNH: Lectotype ♂ [designated by Clarke, 1965: 328]); described from 4 specimens (BMNH).
Type locality: India – Assam, Khasi Hills.
Distribution: Oriental – India (Assam).

ruthiella Steuer, 1993 (*Paraswammerdamia*), Rudolstädter Nat. Schr. 5: 15. (ZMB: Holotype ♀); described from 1 specimen.
Type locality: Germany – Thuringia, Schwarzatal near Bad Blankenburg.
Distribution: Palearctic – Germany.

PSEUDOSWAMMERDAMIA Friese, 1960, Beitr. zur Ent. 10(1/2): 91.
Type species: *Phalaena Tinea combinella* Hübner, 1786, by original designation.

combinella Hübner, 1786 (*Phalaena Tinea*), Beitr. Gesch. Schmett. 1(1): 5. (Type not found); not given.
Type locality: Germany – Bavaria, Augsburg.
Distribution: Palearctic – Austria, Belgium, Bulgaria, Czech Republic, Denmark, England, France, Germany, Hungary, Ireland, Italy, Luxembourg, Netherlands, Norway, Poland, Romania, Russia (European Central, Kaliningrad, Middle Volzhsky, Volga-Donskoy), Sardinia, Sicily, Slovakia, Spain, Sweden, Switzerland, Turkey, Turkmenia, [former] Yugoslavia.
Hosts: Rosaceae – *Prunus domestica* L., *P. spinosa* L. (Friese, 1960).

apicella Donovan, 1793 (*Phalaena*), Nat. Hist. Brit. Ins. 2: 54. (Type not found); described from 1 specimen.
Type locality: England – Epping Forest.

auro-finitella Duponchel, 1842 [1843] (*Elachista*), in Godart & Duponchel, Hist. Nat. Lép. France, Supp. 4: 516. (MNHN: Syntype); not given (1♀ in MNHN).
Type locality: France.

compta Haworth, 1828 (*Erminea*), Lep. Brit. 4: 515. Unjustified ememdation of *comptella*.

comptella Hübner, 1796 (*Tinea*), Samml. Eur. Schmett. Vol. 8, pl. 13, Tineæ 3, Tortriciformes B, fig. 89 [in Hübner, 1796–1838]. (Type not found); not given.
Type locality: not given.

PTILOTEINA Gershenson & Ulenberg, 1998, Kon. Ned. Akad. Weten. Verh. Afdel. Nat., Tweede Reeks 99: 90.
Type species: *Yponomeuta melanaster* Meyrick, 1907a, by original designation.

melanaster Meyrick, 1907a (*Yponomeuta*), Jour. Bomb. Nat. Hist. Soc. 17(3): 746. (BMNH: Lectotype ♂ [designated by Clarke, 1965: 331]); described from 2 specimens (BMNH).
Type locality: Sri Lanka – North Western Prov., Puttalam.
Distribution: Oriental – India (Karnataka), Philippines (Manila, Luzon), Sri Lanka.
Host: Rubiaceae – *Hedyotis wightii* (Hook f.) Sivar., Biju & P. Mathew (Meyrick, 1931a).

delicata Schultze, 1908 (*Psecadia*), Philipp. Jour. Sci., Ser. A 3: 36. (ECMP: Syntype); not given (1♂ in ECMP).
Type locality: Philippines – Manila.

nigricola Meyrick, 1912 (*Hyponomeuta*), Ann. S. Afr. Mus. 10: 66. (SAMC: Syntypes); described from 4 specimens.
Type Localities: South Africa – Transvaal, Barberton; Zululand [=Natal], Mfongosi.
Distribution: Ethiopian – South Africa (Natal, Transvaal).

SWAMMERDAMIA Hübner, 1825 [1826], Verz. Bekannt. Schmett.: 425.
Type species: *Swammerdamia heroldella* Hübner, 1825 [1826], by subsequent designation by Walsingham & Durrant, 1909, Ent. M. Mag. Second Series 20: 155.

MACAROPHANTA Meyrick, 1932c, Exot. Micr. 4(11): 339.
Type species: *Macarophanta aulosema* Meyrick, 1932c, by monotypy.

MACROPHANTA; Moriuti, 1977a, Fauna Jap. Ypon. s. l.: 238. [misspelling of *Macarophanta*].

SWAMMERDAMMIA; Heinemann, 1870, Schmett. Deutsch. 2(1)1: 104. [misspelling of *Swammerdamia*].

aulosema Meyrick, 1932c (*Macarophanta*), Exot. Micr. 4(11): 339. (BMNH: Syntypes); described from 15 specimens (BMNH).
Type locality: India – Kashmir, Killanmarg.
Distribution: Oriental – India (Kashmir).

beirnei Doğanlar, 1979 (*Swammerdamia*), Can. Ent. 111: 967. (CNC: Holotype ♂); described from 3♂, 3♀ (CNC).
Type locality: Canada – British Columbia, Burnaby.
Distribution: Nearctic – Canada (British Columbia).
Host: Rosaceae – *Malus domestica* Borkh. (Doğanlar, 1979).

buraetella Junnilainen, 2001 (*Swammerdamia*), Ent. Fennica 12(2): 116. (ZMH: Holotype ♂); described from 82♂, 18♀ (ZMH).
Type locality: Russia – Buryatia, Ivolginsk-Taphar.
Distribution: Palearctic – Russia (Buryatia).

caesiella Hübner, [1796] (*Tinea*), Samml. Eur. Schmett. Vol. 8, pl. 25, Tineæ 5, Ignobiles B, fig. 172 [in Hübner, 1796–1838]. (Type not found); not given.
Type locality: not given.
Distribution: Nearctic – Canada (British Columbia, Newfoundland, Nova Scotia, Ontario, Québec), USA (Alaska, Connecticut, Illinois, Maine, Massachusetts, Minnesota, New Hampshire, Pennsylvania, West Virginia). Palearctic – Austria, Belgium, Bulgaria, China (Beijing, Hebei, Shanxi), Czech Republic, Denmark, England, Estonia, Finland, France, Germany, Greece, Hungary, Ireland, Italy, Japan (Hokkaido, Honshu), Latvia, Netherlands, Norway, Poland, Romania, Russia, Sardinia, Sicily, Slovakia, South Korea, Spain, Sweden, Switzerland, [former] Yugoslavia.
Hosts: Betulaceae – *Alnus glutinosa* (L.) Gaertn., *Betula papyrifera* Marshall, *B. pendula* Roth., *Betula* spp.; Fagaceae – *Castanea dentata* (Marshall) Borkh. (Duckworth, 1965; Prentice, 1965; Gershenson, 1974; Moriuti, 1977a; Agassiz, 1996; Bengtsson & Johansson, 2011a).

caesia Haworth, 1828 (*Erminea*), Lep. Brit. 4: 516. [unjustified emendation of *caesiella*].

casiella; Byun et al. 2009 (*Swammerdamia*), Checklist Microlep. Kor.: 134. [misspelling of *caesiella*].

castaneae Busck, 1914 (*Swammerdamia*), Proc. Ent. Soc. Wash. 16: 148. (USNM: Lectotype ♀ [designated by Sohn & Lewis, 2012: 5]); not given (1♂, 1♀ in USNM).
Type locality: USA – Pennsylvania, Huntingdon Co., Charter Oak.

cuprescens Braun, 1918 (*Swammerdamia*), Can. Ent. 50: 231. (ANSP: Holotype ♂); described from 4 specimens.
Type locality: Canada – British Columbia, Field.

griseocapitella Stainton, 1851 (*Tinea*), Suppl. Cat. Brit. Tin. & Pteroph.: 3. (BMNH: Syntypes); not given (1♂ in BMNH).
Type locality: England – Stirlingshire, Torwood.

helordella; Gómez-Bustillo, 1979 (*Swammerdamia*), SHILAP Revta. lepid. 7(28): 242. [misspelling of *heroldella*].

heroldella Hübner, 1825 [1826] (*Swammerdamia*), Verz. bekannt. Schmett.: 425. [unnecessary replacement name for *caesiella* Hübner, 1796].

nubeculella Tengström, 1848 (*Tinea*), Not. Sällsk. Faun. Fenn. Förh. 1: 112. (ZMH: Lectotype ♂ [designated here]); not given (1♂, 1♀ in ZMH).
Type Locality: Finland – Uleåborg.
Note: The type series of *Tinea nubeculella* is composed of a male and a female. Peterson (1932) suggested they are conspecific with *Swammerdamia heroldella* (=*caesiella*). The male specimen is *S. caesiella*, but the female is *Paraswammerdamia conspersella* (Lauri Kaila, pers. comm. to JCS). We designate the male specimen as lectotype of *nubeculella* to avoid further confusion. The pin data labels read: "Uleåborg," "W. Nylander" "Mus Zool. H:fors | Spec. type No 7048 | Tinea nubeculella Tngstr.," "ohne abdomen" [=without abdomen] [hand-written].

caudinigra Fan, 2007 (*Swammerdamia*), Acta Zootaxon. Sin. 32(3): 557. (NUIC: Holotype ♂); described from 4♂, 2♀ (NUIC).
Type locality: China – Guizhou, Daozhen, Dashahe.
Distribution: Oriental – China (Guizhou).

compunctella Herrich-Schäffer, [1855] (*Tinea*), Syst. Bearb. Schmett. Eur. 5: 281 [see Herrich-Schäffer, 1853–1855]. (BMNH: Syntypes); not given (1♂ in BMNH).

Type locality: not given [Germany?].
Distribution: Palearctic – Croatia, Czech Republic, England, Germany, Italy, Ireland, Kazakhstan, Poland, Russia (European Central, European Northwest, Kaliningrad, Karelia, Middle Volzhsky, Murmansk, Volga-Donskoy, Zabaykalsk), Spain, Switzerland.
Hosts: Rosaceae – *Crataegus* sp., *Sorbus aucuparia* L. (Friese, 1960; Agassiz, 1996).

conpunctella; Agenjo, 1960 (*Swammerdamia*), Bol. Serv. Plagas Forestales 3(5): 98. [misspelling of *compunctella*].

nebolusella; Friese, 1960, Beitr. zur Ent. 10(1/2): 99. [misspelling of *nebulosella*].

nebulosella Stainton, 1870 (*Swammerdamia*), Nat. Hist. Tin. 11: 44. (Type not found); described from 1 specimen.
Type locality: Switzerland – Zürich.

glaucella Junnilainen, 2001 (*Swammerdamia*), Ent. Fennica 12(2): 113. (Junnilainen collection [private]: Holotype ♂); described from 167♂, 33♀ (J. Junnilainen, J.-P. Kaitila, T. & K. Nupponen collections [private], ZMH).
Type locality: Russia – Cheliabinsk Dist., Ural Mts., Miass.
Distribution: Palearctic – Kazakhstan, Mongolia, Russia (Buryatia, Tuva, Ural), South Korea.

passerella Zetterstedt, [1840] (*Oecophora*), Insecta Lapponica Descripta: 1009 [see Zetterstedt, 1838–1840]. (MZLU: Holotype ♂); described from 1 specimen.
Type locality: Sweden – 'Lapponia, Umensi [=Lycksele Lappmark]'.
Distribution: Palearctic – England, Estonia, Finland, Norway, Russia (European Northwest, Murmansk), Sweden.
Host: Betulaceae – *Betula nana* L. (Agassiz, 1981).

nanivora Stainton, 1871 (*Swammerdamia*), Ent. Ann. 1871: 96. (BMNH: Holotype ♀): described from 1 specimen.
Type locality: England – Strathglass, Inverness-shire.

variegata Tengström, 1869 (*Swammerdamia*); Not. Sällsk. Faun. Fenn. Förh. 10: 364. (ZMH: Lectotype ♀ [designated here]); described from 3 specimens (ZMH).
Type locality: Russia – Republic of Karelia, Petrosavodsk.

Note: Tengström (1869) described *Swammerdamia variegata* based on one specimen from Tivdi and two from Petrosavodsk without designating a holotype. He gave a male sign '♂' at the end of his description but in fact only one specimen (from Petrosavodsk) is a male and the other two specimens in the type series are females. According to Lauri Kaila, 2013 (pers. comm. to JCS), the specimens from Petrosavodsk are two species, *Paraswammerdamia conspersella* [the male] and *Swammerdamia passerella* [the female]. *S. variegata* was treated as a junior synonym of *S. passerella* by Agassiz (1981). To reaffirm his synonymy and avoid further confusion, the lectotype of *S. variegata* is designated here, based on the female from Petrosavodsk. The labels of the lectotype read: "Petrosav[odsk]," "Günther" "Coll. Tengstr," "Mus Zool. H:fors. | Spec. type No 7139 | *Swammerdamia variegata* Tngstr."

pyrella de Villers, 1789 (*Phalaena Tinea*), Caroli Linn. Ent. 2: 515. (Type not found); not given.
Type locality: [probably Sweden].
Distribution: Nearctic – Canada (British Columbia), USA (Michigan, Washington); Palearctic – Austria, Belgium, Bulgaria, China (Henan), Czech Republic, Denmark, England, Estonia, Finland, France, Germany, Hungary, Iran, Ireland, Italy, Japan (Hokkaido, Honshu), Latvia, Lithuania, Luxembourg, Netherlands, Poland, Portugal (?), Romania, Russia (European Northwest, Kaliningrad, Karelia, Murmansk, Primorsky, Sakhalin, South Kamchatka, Volga-Donskoy, Zabaykalsk), Slovakia, Spain, Sweden, Switzerland, [former] Yugoslavia.
Hosts: Rosaceae – *Crataegus* spp., *Cydonia obdonnia* Mill., *Malus domestica* Borkh., *M. sylvestris* Mill., *Prunus avium* (L.) L., *P. cerasus* L., *P. domestica* L., *P. salicina* Lidl., *P. ssiori* F. Schmidt, *Pyrus communis* L. (Moriuti, 1977a; Gershenson & Ulenberg, 1998); host records (see Gershenson & Ulenberg, 1998) from *Betula nana* L. and *B. pendula* Roth. (Betulaceae) need verification.

cerasiella Hübner, 1816 (*Tinea*), Samml. Eur. Schmett. Vol. 8, pl. 48, Tineæ 2, Pyralidiformes B, fig. 332 [in Hübner, 1796–1838]. (Type not found); not given.
Type locality: not given.

pelicaria pyri Retzius, 1783 (*Phalaena*), In: De Geer, Gen. Spec. Ins.: 54 (Type not found); not given. Nomen nudum [not binominal] (ICZN, 1985: article 11c).
Type locality: not given.

sapporella Matsumura, 1931 (*Argyresthia*), 6000 Ill. Ins. Jap.: 1096, no. 2264. (EIHU: Lectotype ♂ [designated by Moriuti, 1977a: 241]); not given. Type locality: Japan.

sedella Moriuti, 1977a (*Swammerdamia*), Fauna Jap. Ypon. s. l.: 244. (UOP: Holotype ♂); described from 14♂, 14♀ (UOP).
Type locality: Japan – Honshu, Osaka Pref., Mt. Inunakisan.
Distribution: Palearctic – Japan (Hokkaido, Honshu, Shikoku).
Host: Crassulaceae – *Sedum makinoi* Maxim. (Moriuti, 1977a).

villiersi Gibeaux, 1984 (*Swammerdamia*), Revue Fr. Ent. (N. S.) 6(2): 83. (MNHN: Holotype ♂); described from 10♂, 12♀ (MNHN).
Type locality: Madagascar – Centre, Imaitso Forest, Anjavidilava.
Distribution: Ethiopian – Madagascar.

zhengi Li & Fan, 2007 (*Swammerdamia*), Acta Zootaxon. Sin. 32(3): 557. (NUIC: Holotype ♂); described from 6♂, 4♀ (NUIC).
Type locality: China – Gensu, Yuzhong Co., Mt. Xinglong.
Distribution: Palearctic – China (Gansu, Hebei, Qinghai, Shaanxi).

zhangi; Sohn & Wu, 2010 (*Swammerdamia*), In: Sohn et al., J. Nat. Hist. 44(45/46): 2809. [misspelling of *zhengi*].

TEINOPTILA Sauber, 1902, In: Semper, Die Schmett. Philipp. Inseln. Die Nachtfelter. Heterocera. 2(6): 701.
Type species: *Teinoptila interruptella* Sauber, 1902, by monotypy.

CHOUTINEA Huang, 1982, Entomotaxonomia 4 [numbered 6 in error] (4): 269. Type species: *Choutinea shaanxiensis* Huang, 1982, by monotypy.

antistatica Meyrick, 1931a (*Hyponomeuta*), Exot. Micr. 4(3): 87. (BMNH: Lectotype ♀ [designated here]); described from 2 specimens (BMNH).
Type locality: Indonesia – Java, Merbaboe [=Mt. Merbabu].
Distribution: Oriental – China (Guizhou, Yunnan), Indonesia (Java).
Hosts: Celastraceae – *Euonymus japonicus* Thunb., *Maytenus hookeri* Loes. (Meyrick, 1931a; Kuo, 1981).
Note: Clarke (1965: 320), selected a specimen from the type series of 2 specimens, and reported it as the "type," meaning holotype. However, Meyrick did not differentiate between specimens in the type series, making them syntypes. We correct the type status by designating the female Clarke selected as the

lectotype. The pin labels read: Merbaboe, 1,000 m, VI 16 | W. Roepke | type | Slide No. 7527. JCS added two labels reading "Lecto- | type" [circular label with indigo margins], and "LECTOTYPE | *Teinoptila* | *antistatica* M. | teste J. Sohn, 2014."

bolidias Meyrick, 1913a (*Hyponomeuta*), Exot. Micr. 1(5): 137. (BMNH: Holotype ♀); described from 1 specimen.
Type locality: Thailand – Saraburi Prov., Muoklek.
Distribution: Oriental – China (Hunan, Yunnan, Zhejiang), Malaysia (West), Thailand (Saraburi); Palearctic – China (Gansu, Hubei, Shaanxi).

> *shaanxiensis* Huang, 1982 (*Choutinea*), Entomotaxonomia 6(4): 269. (NAFU: Holotype ♂); described from 19 ♂, 19 ♀ (NAFU).
> Type locality: China – Shaanxi, Ziyang.

brunnescens Moore, 1879 [1888] (*Hyponomeuta*), In: Hewitson & Moore, Descr. Lep. Atk.: 282. (ZMB: Syntypes); not given (1♂ in ZMB).
Type locality: India – West Bengal, Darjeeling.
Distribution: Oriental – Bhutan, China (Yunnan), India (Assam, West Bengal); Palearctic – Nepal.

calcarata Meyrick, 1924 (*Hyponomeuta*), Exot. Micr. 3(4): 117. (BMNH: Lectotype ♂ [designated here]); described from 4 specimens (♂, ♀) (BMNH).
Type locality: Bermuda Islands – Highlands, Bailey's Bay.
Distribution: Nearctic – Bermuda Islands, USA (Florida).
Note: Note: Clarke (1965: 320), selected a specimen from the type series of 4, declaring it the "type," meaning holotype, but Meyrick (1924) included 4 specimens without specification of a holotype. We correct the type status by designating the male which Clarke selected as the type to be the lectotype. The pin data labels read: Bermuda Agric. Station | The Highlands, Bailey's Bay | 1924 | L. Ogilvie | At light | Type | Slide No. 7528. JCS added two labels: "Lecto- | type" [circular label with indigo margins], and "LECTOTYPE | *Yponomeuta* | *calcarata* M. | teste J. Sohn, 2014."

clavata Jin, Wang, & Li, 2009 (*Teinoptila*), Zootaxa 2249: 49. (NUIC: Holotype ♂): described from 3♂ (NUIC).
Type locality: China – Guangxi, Yongfu Co.
Distribution: Oriental – China (Guangxi).

corpuscularis Meyrick, 1907c (*Yponomeuta*), Jour. Bomb. Nat. Hist. Soc. 18(1): 153. (BMNH: Holotype ♀); described from 1 specimen.

Type locality: India – Tamil Nadu, Nilgiri Hills.
Distribution: Oriental – India (Tamil Nadu), Sri Lanka (Matala).

guttella Moriuti, 1977a (*Teinoptila*), Fauna Jap. Ypon. s. l.: 206. (UOP: Holotype ♂); described from 2♂, 1♀ (UOP).
Type locality: Taiwan – Chibon.
Distribution: Oriental – Japan (Ryukyu Islands), Taiwan.
Host: Celastraceae – *Maytenus diversifolia* (Maxim.) Ding Hou (Sugi & Tominaga, 2002).

ingens Gershenson & Ulenberg, 1998 (*Teinoptila*), Kon. Ned. Akad. Weten. Verh. Afdel. Nat., Tweede Reeks 99: 100. (MRAC: Holotype ♂); described from 1 specimen.
Type locality: Burundi – Kitega [=Gitega].
Distribution: Ethiopian – Burundi.

interruptella Sauber, 1902 (*Teinoptila*), In: Semper, Die Schmett. Philipp. Inseln. Die Nachtfelter. Heterocera. 2(6): 701. (NMS: Lectotype ♂ [designated by Diakonoff, 1967: 236]); described from 10 specimens (NMS).
Type locality: Philippines – Luzon, Guimaras.
Distribution: Australasian – Australia (Queensland), New Guinea; Oriental – Indonesia (Tukangbesi Island), Philippines (Luzon).

puncticornis Walsingham, 1891b (*Hyponomeuta*), Trans. Ent. Soc. Lond.: 90. (BMNH: Lectotype [designated here]); described from 1♂, 1♀ (BMNH).
Type locality: Mozambique – Maputo Bay [formerly Delagoa Bay].
Distribution: Ethiopian – Mozambique, Senegal, South Africa (Transvaal).
Note: Walsingham (1891b) labeled both specimens as "type" without further distinction. For stabilization of the species, we designate the male as lectotype. The pin data labels read: "Lecto- | type" [circular label with indigo margins], "Type" [circular label with red margins], "Delagoa Bay | E. AFRICA | Druce No 95. 1884" [hand-written], "Walsingham | Collection. | 1910-427," "*Hyponomeuta* | *puncticornis* | Walsm. Pr. Ent. Soc. | TYPE ♂" [hand-written] || "LECTOTYPE | *Teinoptila* | *puncticornis* W. | teste J. Sohn, 2014."

THECOBATHRA Meyrick, 1922a, Exot. Micr. 2(18): 553.
Type species: *Thecobathra acropercna* Meyrick, 1922a, by monotypy.

PSEUDOCALANTICA Friese, 1960, Beitr. zur Ent. 10(1/2): 36.
Type species: *Niphonympha anas* Stringer, 1930, by original designation.

acrivalvata Fan, Jin, & Li, 2008 (*Thecobathra*), Zootaxa 1821: 19. (NUIC: Holotype ♂); described from 1 specimen.
Type locality: China – Guizhou, Mt. Fanjing.
Distribution: Oriental – China (Guizhou).

acropercna Meyrick, 1922a (*Thecobathra*), Exot. Micr. 2(18): 553. (BMNH: Lectotype ♀ [designated by Clarke, 1965:395]); described from 2 specimens (BMNH).
Type locality: India – Assam, Cherrapunji.
Distribution: Oriental – India (Assam).

albana Liu, 1980 (*Thecobathra*), Entomotaxonomia 2(1): 36. (IZAS: Holotype ♂); described from 3♂, 3♀ (IZAS).
Type locality: China – Hainan, Jianfengling.
Distribution: Oriental – China (Hainan, Jiangxi).

anas Stringer, 1930 (*Niphonympha*), Ann. Mag. Nat. Hist. Series 10 6: 420. (BMNH: Holotype ♂); described from 13 specimens (8♂, 3♀ in BMNH).
Type locality: Japan.
Distribution: Oriental – China (Guizhou, Hunan, Jiangxi, Sichuan, Yunnan, Zhejiang); Palearctic – China (Anhui, Hubei), Japan (Honshu, Kyushu, Shikoku), South Korea.
Hosts: Fagaceae – *Castanopsis cuspidata* (Thunb.) Schottky, *Quercus glauca* Thunb., *Q. serrata* Murray (Moriuti, 1977a).

> *unimaculata* Matsumura, 1931 (*Scythropiodes*), 6000 Ill. Ins. Jap.: 1099. (EIHU: Syntypes); not given (1♂ in EIHU).
> Type locality: Japan.

argophanes Meyrick, 1907a (*Pyrozela*), Jour. Bomb. Nat. Hist. Soc. 17(3): 747. (BMNH: Lectotype ♀ [designated by Moriuti, 1971a: 249]); described from 5 specimens (BMNH).
Type locality: India – Assam, Khasi Hills.
Distribution: Oriental – China (Yunnan, Zhejiang), India (Assam).

> *argophenes*; Fan, Jin, & Li, 2008, Zootaxa 1821: 22 (checklist). [misspelling of *argophanes*].

badagongshana Fan, Jin, & Li, 2008 (*Thecobathra*), Zootaxa 1821: 20. (NUIC: Holotype ♂); described from 1♂, 1♀ (NUIC).
Type locality: China – Hunan, Sangzhi Co., Mt. Badagong.
Distribution: Oriental – China (Hunan).

basilobata Fan, Jin, & Li, 2008 (*Thecobathra*), Zootaxa 1821: 15. (NSU: Holotype ♂); described from 3♂ (NSU).
Type locality: Taiwan – Taitung Co., Mt. Baxian.
Distribution: Oriental – Taiwan.

bidentata Liu, 1980 (*Thecobathra*), Entomotaxonomia 2(1): 36. (IZAS: Holotype ♂); described from 1 specimen.
Type locality: China – Jiangxi, Yudu Xian, Qilushan.
Distribution: Oriental – China (Jiangxi).

casta Meyrick, 1907a (*Pyrozela*), Jour. Bomb. Nat. Hist. Soc. 17(3): 747. (BMNH: Lectotype ♂ [designated by Moriuti, 1971a: 237]); described from 2 specimens (BMNH).
Type locality: Bhutan.
Distribution: Oriental – Bhutan, India (Assam).

chiona Liu, 1983 (*Thecobathra*), Acta Zootaxon. Sin. 8: 81. (IZAS: Holotype ♂); described from 5♂, 1♀ (IZAS).
Type locality: China – Sichuan, Dukou City.
Distribution: Oriental – China (Sichuan).

delias Meyrick, 1913a (*Calantica*), Exot. Micr. 1(5): 148. (BMNH: Lectotype ♂ [designated by Moriuti, 1971a: 245]); described from 13 specimens (BMNH).
Type locality: India – Assam, Khasi Hills.
Distribution: Oriental – China (Guangdong, Hunan, Jiangsu, Sichuan, Yunnan, Zhejiang), India (Assam); Palearctic – China (Xizang).

dilechria Bradley, 1982 (*Thecobathra*), J. Nat. Hist. 16: 371. (BMNH: Holotype ♂); described from 6♂, 26♀ (BMNH).
Type locality: Australia – Norfolk Island, Mt. Pitt Rd.
Distribution: Australasian – Australia (Norfolk Island).

eta Moriuti, 1963a (*Pseudocalantica*), Kontyû 31(3): 218. (UOP: Holotype ♂); described from 23♂, 24♀ (UOP).
Type locality: Japan – Kyushu, Kagoshima Pref., Mt. Kirisimayama.
Distribution: Oriental – China (Zhejiang); Palearctic – Japan (Honshu, Kyushu, Shikoku).

flavida Yu & Li, 2001 (*Thecobathra*), SHILAP Revta. lepid. 29(114): 118. (NUIC: Holotype ♂); described from 1 specimen.

Type locality: China – Hubei, Wufeng Co.
Distribution: Palearctic – China (Hubei).

flovida; Yu & Li, 2001 (*Thecobathra*), SHILAP Revta. lepid. 29(114): 115 (abstract). [misspelling of *flavida*]

kappa Moriuti, 1963a (*Pseudocalantica*), Kontyû 31(3): 220. (USNM: Holotype ♂); described from 2♂ (USNM).
Type locality: Taiwan – Mt. Tattaka [=Sungkang].
Distribution: Oriental – China (Sichuan), Taiwan.

kurokoi Moriuti, 1982a (*Thecobathra*), Trans. Lep. Soc. Japan 33(1/2): 61. (UOP: Holotype ♂); described from 1 specimen.
Type locality: Thailand – Nakorn Nayok, Kao Yai (National Park).
Distribution: Oriental – Thailand (Nakorn Nayok).

lambda Moriuti, 1963a (*Pseudocalantica*), Kontyû 31(3): 222. (Issiki collection [not traced]: Holotype ♂); described from 8♂, 3♀ (Issiki collection).
Type locality: Taiwan – Kansirei [=Kuantzuling].
Distribution: Oriental – China (Hunan, Jiangxi, Sichuan, Zhejiang), Taiwan, Thailand (Nakorn Nayok, Kansirei).
Host: Altingiaceae – *Liquidambar formosana* Hance (Moriuti, 1971a).

latibasis Fan, Jin & Li, 2008 (*Thecobathra*), Zootaxa 1821: 18. (NSU: Holotype: ♂); described from 1 specimen.
Type locality: China – Guangxi, Jinxiu Co.
Distribution: Oriental – China (Guangxi).

longisaccata Fan, Jin & Li, 2008 (*Thecobathra*), Zootaxa 1821: 17. (NSU: Holotype ♂); described from 1 specimen.
Type locality: China – Yunnan, Ruili.
Distribution: Oriental – China (Yunnan).

microsignata Liu, 1980 (*Thecobathra*), Entomotaxonomia 2(1): 37. (IZAS: Holotype ♀); described from 3♀ (IZAS).
Type locality: China – Yunnan, Yulongshan.
Distribution: Oriental – China (Yunnan).

nakaoi Moriuti, 1965a (*Pseudocalantica*), Kontyû 33(1): 9. (UOP: Holotype ♂); described from 1 specimen.

Type locality: Nepal – Taplejung Dist., Taplejung.
Distribution: Palearctic – Nepal.

ovata Liu, 1980 (*Thecobathra*) Entomotaxonomia 2(1): 35. (IZAS: Holotype ♂); described from 2♂, 1♀ (IZAS).
Type locality: China – Jiangxi, Dayu Xian.
Distribution: Oriental – China (Jiangxi).

paranas Fan, Jin & Li, 2008 (*Thecobathra*), Zootaxa 1821: 20. (NSU: Holotype ♂); described from 1 specimen.
Type locality: China – Yunnan, Ruili.
Distribution: Oriental – China (Yunnan).

partinuda Fan, Jin & Li, 2008 (*Thecobathra*), Zootaxa 1821: 17. (NSU: Holotype ♂); described from 7♂ (NSU).
Type locality: Taiwan – Taitung Co., Mt. Baxian.
Distribution: Oriental – Taiwan.

sororiata Moriuti, 1971a (*Thecobathra*), Kontyû 39(3): 240. (BMNH: Holotype ♀); described from 1 specimen.
Type locality: China – Zhejiang, Tienmushan.
Distribution: Oriental – China (Hunan, Jiangxi, Sichuan, Zhejiang); Palearctic – China (Anhui, Henan, Jiangsu).

> *nivalis* Moriuti, 1971a (*Thecobathra*), Kontyû 39(3): 241. (BMNH: Holotype ♂); described from 1 specimen.
> Type locality: China – [Sichuan], Omai.

tetragona Liu, 1983 (*Thecobathra*), Acta Zootaxon. Sin. 8: 80. (IZAS: Holotype ♀); described from 7♂, 12♀ (IZAS).
Type locality: China – Sichuan, Wenchuan Co.
Distribution: Oriental – China (Sichuan).

yasudai Moriuti, 1965a (*Pseudocalantica*), Kontyû 33(1): 7. (UOP: Holotype ♂); described from 2♂ (UOP).
Type locality: Nepal – Taplejung Dist., Taplejung.
Distribution: Oriental – China (Yunnan), India (Sikkim); Palearctic – Nepal.

yunnana Liu, 1984 (*Thecobathra*), Acta Zootaxon. Sin. 9: 324. (IZAS: Holotype ♂); described from 1 specimen.

Type locality: China – Yunnan, Lianghe Co.
Distribution: Oriental – China (Yunnan).

XYROSARIS Meyrick, 1907d, Proc. Linn. Soc. N.S. Wales 32(1): 71.
Type species: *Xyrosaris dryopa* Meyrick, 1907d, by monotypy.

> *XYROSARIA*; Kearfott, 1903, Jour. N. Y. Ent. Soc. 11: 150. [misspelling of *Xyrosaris*].

acroxutha Turner, 1923 (*Xyrosaris*), Trans. R. Soc. S. Austr. 47: 168. (ANIC: Holotype ♂); described from 1 specimen.
Type locality: Australia – Queensland, "National Park."
Distribution: Australasian – Australia (Queensland).

campsiptila Meyrick, 1914f (*Xyrosaris*), Jour. Bomb. Nat. Hist. Soc. 23(1): 124. (BMNH: Lectotype ♂ [designated by Clarke, 1965: 404]); described from 3 specimens (BMNH).
Type locality: Sri Lanka – North Western Prov., Puttalam.
Distribution: Oriental – India (Karnataka), Sri Lanka.

dryopa Meyrick, 1907d (*Xyrosaris*), Proc. Linn. Soc. N.S. Wales 32(1): 72. (BMNH: Holotype ♂); described from 1 specimen.
Type locality: Australia – Queensland, Brisbane.
Distribution: Australasian – Australia (Queensland).

lichneuta Meyrick, 1918a (*Xyrosaris*), Exot. Micr. 2(6): 188. (BMNH: Lectotype ♂ [designated by Clarke, 1965: 404]); described from 2 specimens (BMNH).
Type locality: India – Assam, Khasi Hills.
Distribution: Oriental – China (Hunan, Zhejiang), India (Assam), Japan (Ryukyu Isls.), Taiwan; Palearctic – China (Hebei, Henan, Hubei, Jiangsu), Japan (Hokkaido, Honshu, Kyushu, Shikoku), Russia (Primorsky), South Korea.
Hosts: Celastraceae – *Celastrus orbiculatus* Thunb., *Euonymus alatus* (Thunb.) Siebold, *E. alatus* f. *ciliatodentatus* (Franchet & Savatier) Hiyama, *E. alatus* f. *striatus* (Thunb.) Kitag., *E. hamiltonianus* ssp. *sieboldianus* (Blume) H. Hara, *E. oxyphyllus* Miq., *Euonymus* sp. (Friese, 1962; Moriuti, 1977a).

> *melanopsamma* Meyrick, 1931b (*Xyrosaris*), Exot. Micr. 4(6): 171. (BMNH: Lectotype ♀ [designated by Clarke, 1965: 404]); described from 2 specimens (BMNH).
> Type locality: Japan – Honshu, Gifu Pref., Ikeno.

lirinopa Meyrick, 1922a (*Xyrosaris*), Exot. Micr. 2(18): 551. (MNHN: Holotype ♀); described from 1 specimen.
Type locality: China – Shanghai.
Distribution: Oriental – China (Shanghai).

maligna Meyrick, 1907b (*Xyrosaris*), Jour. Bomb. Nat. Hist. Soc. 17(4): 984. (BMNH: Lectotype ♀ [designated by Clarke, 1965: 404]); described from 6 specimens (BMNH).
Type locality: Sri Lanka – North Western Prov., Puttalam.
Distribution: Oriental – India (Karnataka), Sri Lanka.

mnesicentra Meyrick, 1913a (*Xyrosaris*), Exot. Micr. 1(5): 136. (BMNH: Lectotype ♂ [designated by Clarke, 1965: 407]); described from 4 specimens (BMHN).
Type locality: Guyana – Bartica.
Distribution: Neotropical – Guyana.

obtorta Meyrick, 1923a (*Xyrosaris*), Trans. Ent. Soc. Lond. 3/4: 554. (BMNH: Lectotype ♂ [designated here]); described from 2 specimens (BMNH).
Type locality: Mauritius – Rodriguez Island.
Distribution: Ethiopian – Mauritius.
Note: Only a male type specimen of *Xyrosaris obtorta* Meyrick is still available in the BMNH. This specimen bears a type label indicating "H.T." (holotype). However, Meyrick (1923a) did not designate a holotype in the type series. To preserve the species identity, we designate this only remaining type specimen as the lectotype. The pin data labels read: "Lecto- | type" [circular label with indigo margins], "Type | H.T." [circular label with red margins], "Rodriguez Id. | viii–xi 1918/ | H. P. Thomasset | & H. J. Snell Coll.," "173" [hand-written], "pres. by Cambridge | University Museum | B.M. 1924-238," "*XYROSARIS* | *OBTORTA*, Meyr. | Tr. Ent.Soc. Lond. 1923 p.554 || TYPE" [rectangular label with broad, black margins], || "LECTOTYPE | *Xyrosaris obtorta* | Meyrick | teste J. Sohn, 2014."

secreta Meyrick, 1912 (*Xyrosaris*), Ann. S. Afr. Mus. 10: 67. (SAMC: Holotype ♀); described from 1 specimen.
Type locality: South Africa – Natal, Zululand, Mfongosi.
Distribution: Ethiopian – South Africa (Natal).

YPONOMEUTA Latreille, 1796, Préc. Caract. gén. Ins.: 146.
Type species: *Phalaena evonymella* Linnaeus, 1758, by subsequent designation by Latreille, 1802, Hist. Nat. Crust. Ins. 3: 417.

> *COENIPHANTES*; Hübner, 1822, Syst.-alph. Verz. Europ. Schmett.: 67. [misspelling of *Coenyphantes*].

COENYPHANTES Hübner, 1822, Syst.-alph. Verz. Europ. Schmett.: 65. A junior objective synonym.
Type species: *Phalaena evonymella* Linnaeus, 1758, by subsequent designation by Bradley, 1966b, Ent. Gaz. 17: 219.

ERMINEA Haworth, 1811, Lep. Brit. 3: 512. A junior objective synonym.
Type species: *Phalaena evonymella* Linnaeus, 1758 by subsequent designation by Stephens, 1834, Ill. Brit. Ent. Haust. 4: 244.

HYPHANTES Hübner, 1806 [1873], Tentamen: 2. Suppressed (ICZN Opinion 97).
Type species: *Phalaena evonymella* Linnaeus, 1758, by monotypy.

HYPOMENEUTA; Zeller, 1846, Ber. Lepidopt. Tausch.: 165. [misspelling of *Hyponomeuta*].

HYPONOMENTA; Turner, 1898 Trans. R. Soc. S. Austr. 22: 201. [misspelling of *Hyponomeuta*].

HYPONOMEUTA Billberg, 1820, Enumer. insect. Museo G. J. Billberg: 91. [unjustified emendation of *Yponomeuta*].

NYGMIA Hübner, 1825 [1826], Verz. bekannt. Schmett.: 412. A a junior objective synonym, a junior homonym of *Nygmia* Hübner, 1816 [1820] (*Lymantriidae*).
Type species: *Phalaena evonymella* Linnaeus, 1758, by subsequent designation by Fletcher, 1928 Cat. Indian Insects 17: 8.

YPOMENEUTA; Eversmann, 1844, Lep. Volgo-Ural.: 564. [misspelling of *Yponomeuta*].

acronops Becker, 2013 (*Yponomeuta*), SHILAP Revta. Lepid. 41(163): 307. (USNM: Holotype ♂; Paratypes 20 ♂, 18 ♀ in BMNH, IES, USNM, VOB); described from 39 specimens.
Type locality: Cuba – Santiago, Sierra Maestra, Pico, 1500 meters.
Distribution: Neotropical – Cuba.

africanus Stainton, 1860 (*Hyponomeuta*), Trans. Ent. Soc. Lond., New Series 5(6): 222. (BMNH: Holotype [sex unknown, not traced]); described from 1 specimen.

Type locality: South Africa – Knysna.
Distribution: Ethiopian – South Africa (Cape Prov.).
Note: Gershenson & Ulenberg (1998) reported that the BMNH is the type depository of *Hyponomeuta africanus* Stainton. The junior author (JCS) searched the BMNH collection twice (2009 & 2011) but could not find the type. The publication date of vol. 5(6) is given at the bottom of page 217 as July 1860, rather than 1862, the year previously credited to be the year of publication by other workers. Also, in the issue of the Transactions of the Entomological Society of London New Series Volume 5 for 1858–1861, on page 152 of the Proceedings for 1858–1859 [1861], the article by Stainton describing *Hyponomeuta africanus* is also listed among those as having been published in 1860.

albonigratus Gershenson, 1972 (*Yponomeuta*), Ent. Obozr. 51(3): 635. (ZMAS: Holotype ♂); described from 34♂, 40♀ (NASU, ZMAS).
Type locality: Tadzhikistan – Kkoja-Obigram.
Distribution: Palearctic – Israel, Kirgizistan, Tadzhikistan, Uzbekistan.
Host: Salicaceae – *Salix oxycarpa* Andersson. (Gershenson & Ulenberg, 1998).
Note: To identify the holotype of *Y. albonigratus*, Gershenson (1972) stated the type locality as Tajikistan and that it bears the labels, "Holotypus" and "Genitali prep. No. 11542 ♂." This incomplete statement was insufficient regarding recommendation 73c of the ICZN. The holotype data were not published until later by Gershenson & Kozhevnikova, 2004.

alienellus Walker, 1863 (*Hyponomeuta*), List Lepid. Ins. BM 28: 532. (BMNH: Holotype ♂); described from 1 specimen.
Type locality: Sri Lanka.
Distribution: Oriental – Sri Lanka.

anatolicus Stringer, 1930 (*Hyponomeuta*), Ann. Mag. Nat. Hist. 10(6): 419. (BMNH: Holotype ♀); described from 3 specimens (BMNH).
Type locality: Japan – Honshu, Toyama Pref., Fushiki.
Distribution: Palearctic – Japan (Honshu, Kyushu), Russia (Primorsky, South Kamchatka), South Korea.
Host: Celastraceae – *Euonymus* sp. (Moriuti, 1977a).

athyris Meyrick, 1928b (*Hyponomeuta*), Exot. Micr. 3(13): 415. (BMNH: Holotype ♀); described from 1 specimen.
Type locality: South Africa – Natal, Weenen.
Distribution: Ethiopian – South Africa (Natal).

bipunctella Matsumura, 1931 (*Hyponomeuta*), 6000 Ill. Ins. Jap.: 1097. (EIHU: Lectotype ♂ [designated by Moriuti, 1977a: 194]); not given (11 specimens in EIHU).
Type locality: Japan.
Distribution: Palearctic – Japan (Hokkaido, Honshu, Shikoku), Russia (Primorsky).
Host: Celastraceae – *Euonymus fortunei* (Trucz.) Hand.-Maz. (Moriuti, 1977a).

cagnagella Hübner, [1813] (*Tinea*), Samml. Eur. Schmett. Vol. 8, pl. 58, Tineæ 2, Pyralidiformes B, figs 391, 392 (in Hübner, 1796–1838). (Type not found); not given.
Type locality: Europe.
Distribution: Nearctic – Canada (introduced: New Brunswick, Nova Scotia, Ontario, Québec), USA (introduced: Massachusetts, Michigan, New York, Wisconsin); Palearctic – Albania, Austria, Belgium, Bulgaria, Corsica, Crete, Czech Republic, Denmark, England, Estonia, Finland, France, Germany, Greece, Hungary, Ireland, Israel, Italy, Latvia, Lithuania, Luxembourg, Netherlands, Norway, Poland, Romania, Russia (European Central, European Northwest, European South Taiga, Kaliningrad), Sardinia, Sicily, Slovakia, Spain, Sweden, Switzerland, Turkey, [former] Yugoslavia.
Hosts: Celastraceae – *Euonymus alatus* (Thunb.) Siebold, *E. europaea* L., *E. japonicus* Thunb., *E. kiautschovicus* Loes., *E. verrucosus* Scop. (Servadei, 1930; Agassiz, 1996; Gershenson & Ulenberg, 1998; Malinoski & Clement, 2009).

cagnatella; Hübner, 1825 [1826] (*Nygmia*), Verz. bekannt. Schmett.: 412. [misspelling of *cagnagella*].

evominella; Gómez-Bustillo, 1979 (*Yponomeuta*), SHILAP Revta. lepid. 7(28): 242. [misspelling of *evonymella* (sensu Scopoli, 1763)].

evonymella sensu Scopoli, 1763 (Phalaena), Ent. Carn.: 246. [misidentification].

evonymi Zeller, 1844 (*Hyponomeuta*), Isis Oken 37: 223. [unnecessary replacement name for *Tinea cagnagella*].

euonymi; Dallas, 1866, Zool. Rec. 3: 494. [misspelling of *evonymi*].

catharotis Meyrick, 1935a (*Hyponomeuta*), in Caradja & Meyrick: Mat. Microlep.-Fauna Chin. Prov. Kiangsu, Chekiang, Hunan: 88. (MGAB: Lectotype ♂

[designated by Friese, 1962: 307]); stated as "many specimens" (7♂, 1♀ in MGAB; 1♂ in BMNH).
Type locality: China – Jiangsu, Nanking.
Distribution: Oriental – China (Hunan); Palearctic – China (Jiangsu), Russia (Amur, Primorsky, South Khabarovsk).
Host: Celastraceae – *Euonymus alatus* (Thunb.) Siebold (Gershenson & Ulenberg, 1998).
Note: Friese (1962) designated the lectotype of *catharotis* from the MGAB collection, prior to Clarke (1965) as reported by Popescu-Gorj (1992). Therefore, Clarke's type designation is invalid. Gershenson & Ulenberg (1998) erroneously cited Clarke's lectotype as valid.

chalcocoma Meyrick, 1938a (*Hyponomeuta*), in Caradja & Meyrick, Dt. Ent. Z. Iris 52(1): 17. (BMNH: Lectotype ♂ [designated by Clarke, 1965: 323]); described from 5 specimens (BMNH).
Type locality: China – Yunnan, Likiang [=Lijiang], Yülingshanmass [possibly Yulongxueshan].
Distribution: Oriental – China (Yunnan).

cinefacta Meyrick, 1935a (*Hyponomeuta*), in Caradja & Meyrick, Mat. Microlep.-Fauna Chin. Prov. Kiangsu, Chekiang, Hunan: 89, erratum. (MGAB: Lectotype ♂ [designated by Friese, 1962: 312]); described from 10 specimens (5♂ in MGAB; 1♂ in BMNH).
Type locality: China – Zhejiang, Mt. Tienmushan.
Distribution: Oriental – China (Zhejiang); Palearctic – China (Gansu, Hebei, Henan, Jiangsu, Liaoning, Shaanxi), Russia (Primorsky).
Host: Celastraceae – *Euonymus* sp. (Gershenson & Ulenberg, 1998).
Note: Friese (1962) designated the lectotype of *cinefacta* from the MGAB collection, prior to Clarke (1965). Therefore, Clarke's type designation is invalid. Gershenson & Ulenberg (1998) erroneously cited Clarke's lectotype as valid. See also Popescu-Gorj (1992).

> *cinefcata*; Meyrick, 1935a, in Caradja & Meyrick: Mat. Microlep.-Fauna Chin. Prov. Kiangsu, Chekiang, Hunan: 89. [misspelling of *cinefacta*].

diffluellus Heinemann, 1870 (*Hyponomeuta*), Schmett. Deutsch. 2(1)1: 111. (ZMAS: Lectotype ♂); described from 2♂ (ZMAS).
Type locality: Poland – Wroclaw.
Distribution: Palearctic – Poland, Russia (Stavropol Krai).
Host: Celastraceae – *Euonymus europaea* L. (Heinemann, 1870).

Note: Heinemann (1870) attributed the authorship of *diffluellus* to Wocke but this is invalid because Wocke never published on the species.

enneacentra Meyrick, 1925c (*Hyponomeuta*), Treubia 6(3–4): 431. (BMNH: Lectotype ♀ [designated here]); described from 3 specimens (BMNH).
Type locality: Indonesia – Kei [=Kai] Islands.
Distribution: Oriental – Indonesia (Kai Islands).
Note: Of the three syntypes, only one is still in the BMNH collection. To avoid possible confusion, we designate it as the lectotype of *enneacentra*. The pin data labels read: "Lecto- | type" [circular label with indigo margins], "SYN- | TYPE" [circular label with a pale blue margin], "Kei I[slan]d | New Guinea | HCS 0.22" [hand-written] "*enneacentra* Meyr." [hand- written] "Meyrick Coll. | B.M. 1938-290," "*Hyponomeuta* | *enneacentra* | 1/1 Meyr. | E. Meyrick det. | in Meyrick Coll.," "LECTOTYPE | *Yponomeuta* | *enneacentra* M. | teste J. Sohn, 2014."

eurinellus Zagulajev, 1969 (*Yponomeuta*), Ent. Obozr. 48(1): 195. (ZMAS: Holotype ♂); described from 1♂, 4♀ (ZMAS).
Type locality: Russia – Primorsky Krai, near Vladivostok.
Distribution: Palearctic – China (North), Japan (Hokkaido, Honshu, Kyushu, Shikoku), Russia (Amur, Primorsky, Sakhalin, South Kamchatka, South Khabarovsk), South Korea.
Hosts: Celastraceae – *Euonymus macropterus* Rupr., *E. oxyphyllus* Miq. (Gershenson & Ulenberg, 1998).

> *solitariellus* Moriuti, 1977a (*Yponomeuta*), Fauna Jap. Ypon. s. l.: 183. (UOP: Holotype ♂); described from 42 specimens (UOP, ZMAS).
> Type locality: Japan: Honshu, Wakayama Pref., Mount Kozindake.

evonymella Linnaeus, 1758 (*Phalaena Tinea*), Syst. Nat. 10(1): 534. (LSL: Lectotype ♂ [designated by Robinson & Nielsen, 1983: 209]); not given.
Type locality: not given.
Distribution: Oriental – India (?); Palearctic – Austria, Belgium, Bulgaria, China (Inner Mongolia, Jiangsu, Jilin, Sachalin), Czech Republic, Denmark, England, Estonia, Finland, France, Germany, Greece, Hungary, Ireland, Italy, Latvia, Lithuania, Luxembourg, Netherlands, Norway, Poland, Romania, Russia, Sicily, Slovakia, South Korea, Spain, Sweden, Switzerland, Turkey, [former] Yugoslavia.
Hosts: Rosaceae – *Prunus cerasus* L., *P. domestica* L., *P. padus* L., *P. ssiori* F. Schmidt, *Sorbus aucuparia* L. (Gershenson & Ulenberg, 1998).

eronymella; Pathania & Rose, 2003 (*Phaloena* [sic]), J. Nat. Conserv. 15(2): 529, 530. [misspelling of *evonymella*].

euonymella; Fabricius, 1781 (*Tinea*), Species Insectorum 2: 290. [misspelling of *evonymella*].

evonymè Haworth, 1811 (*Erminea*), Lep. Brit. 3: 512. [unjustified emendation of *evonymella*].

padi Zeller, 1844 (*Hyponomeuta*), Isis Oken 37: 225. [unnecessary replacement name for *Phalaena Tinea evonymella* Linnaeus].

falkovitshi Gershenson & Ulenberg, 1998 (*Yponomeuta*), Kon. Ned. Akad. Weten. Verh. Afdel. Nat., Tweede Reeks 99: 125. (ZMAS: Holotype ♀); described from 1 specimen.
Type locality: Russia – Primorsky Krai, Tigroviy.
Distribution: Palearctic – Russia (Primorsky).

favillacea Meyrick, 1922c (*Hyponomeuta*), Zool. Meded. Leiden 7: 88. (RMNH: Holotype ♀); described from 1 specimen.
Type locality: Indonesia – Java, Buitenzorg.
Distribution: Oriental – Indonesia (Java).

fumigatus Zeller, 1852 [1854] (*Hyponomeuta*), K.V. Akad. Handl. 1852: 101. (NRM: Holotype ♀); described from 1 specimen.
Type locality: South Africa – Natal, Caffraria.
Distribution: Ethiopian – Democratic Republic of the Congo, Kenya, South Africa (Knysna, Natal), Tanzania.

funesta Meyrick, 1914d (*Hyponomeuta*), Exot. Micr. 1(6): 173.
Note: Meyrick (1914d) proposed this replacement name for the species cited as *Hyponomeuta fumigatus* Zeller by Stainton (1860 [nec 1862]) because he thought it represented a distinct species. Gershenson & Ulenberg (1998), however, found that *Hyponomeuta funesta* Meyrick, 1914d is a variant of *Y. fumigatus* Zeller.

glaphyropis Meyrick, 1908c (*Hyponomeuta*), Proc. Zool. Soc. Lond. 1908 (46–48): 736. (BMNH: Neotype ♀ [designated here]); described from 6 specimens (BMNH).
Type locality: South Africa – Natal, Weenen.

Note: Gershenson & Ulenberg (1998) listed three specimens that Meyrick had identified as *Hyponomeuta glaphyropis*. Of these, two of them have data labels which correspond with the original description. It is very likely these are part of the syntypes designated by Meyrick (1908c) although they do not have a "Type" label. Unfortunately, Meyrick did not mention where the type specimens were deposited and there is no trace of those specimens. To avoid further confusion, we designate a ♀ as neotype of *Hyponomeuta glaphyropis* Meyrick from one of the two possible syntypes. The label data reads: female, "Weenen | Natal | R. 12.93," "*Hyponomeuta* | *glaphyropis* | 9/5 Meyr. | E. Meyrick det. | in Meyrick coll.," "BM ♀ | Genitalia slide | No. 28387," "NEOTYPE | *Yponomeuta* | *glaphyropis* M. | teste J. Sohn, 2014."

gershensoni Sinev, 2008 (*Yponomeuta*), Cat. Lep. Russ.: 324. A replacement name for *zagulajevi* Gershenson, 1977.
Distribution: Palearctic – China (Beijing), Russia (Amur, Primorsky).
Host: Celastraceae – *Euonymus alatus* (Thunb.) Siebold (Gershenson & Ulenberg, 1998).

zagulajevi Gershenson, 1977 (*Yponomeuta*), Ent. Obozr. 56(1): 150 (ZMAS: Holotype ♀); described from 2♀ (ZMAS). A junior homonym of *Yponomeuta zagulajevi* Efremov, 1976.
Type locality: Russia – Primorsky Krai, Okeanskaya.

gigas Rebel, 1892 (*Hyponomeuta*), Ann. Nat. Hofmus. Wien 7: 271. (NHMW: Syntypes): not given, mentioned only as "♂" and "♀" (NHMW).
Type locality: Canary Islands – Grand Canary Island.
Distribution: Palearctic – Canary Islands.
Hosts: Salicaceae – *Populus alba* L., *Salix canariensis* C. Sm. ex Link (Gershenson & Ulenberg, 1998).

innotata Walsingham, 1907b (*Hyponomeuta gigas* var.), Proc. Zool. Soc. Lond. 1907: 976. (BMNH: Syntypes); described from 36 specimens (BMNH).
Type Localities: Canary Islands – Grand Canary Island & Tenerife Island.

griseatus Moriuti, 1977a (*Yponomeuta*), Fauna Jap. Ypon. s. l.: 196. (UOP: Holotype ♂); described from 2♂, 2♀ (UOP).
Type locality: Japan – Ryukyu Archipelago, Amami Islands, Amamiosima Island.

Distribution: Oriental – China (Jiangxi, Shanghai, Zhejiang), Japan (Ryukyu Isls.), Taiwan; Palearctic – China (Anhui, Henan, Shaanxi, Shandong), South Korea.
Host: Celastraceae – *Euonymus japonicus* Thunb. (Liu & Huang, 1996).

griseomaculatus Gershenson, 1969 (*Yponomeuta*), Ent. Obozr. 48(1): 199. (ZMAS: Holotype ♀); described from 10♀ (ZMAS).
Type locality: Russia – Primorsky Krai, Yakovlevka.
Distribution: Palearctic – Russia (Primorsky).

grossipunctella Guenée, 1879 (*Hyponomeuta*), Ann. Soc. Ent. Fr. (Série 5) 9(18–19): 282. (Type not found); not given.
Type locality: Australia.
Distribution: Australasian – Australia.

> *grandipunctella*; Kirby, 1879 (*Hyponomeuta*), Zool. Rec. (Ins.) 16: 189 [misspelling of *grossipunctella*].

hemileuca Meyrick, 1932c (*Hyponomeuta*), Exot. Micr. 4(11): 346. (BMNH: Lectotype ♂ [designated here]); described from 8 specimens (BMNH).
Type locality: Uganda – Busana.
Distribution: Ethiopian – Democratic Republic of the Congo, Uganda.
Note: Meyrick (1932c) did not designate a holotype from the type series. Clarke (1965) found only one of the syntypes and reported it as the "Type," meaning holotype. We designate the specimen as the lectotype to correct the status and eliminate possible confusion. The data labels read: "Lecto- | type" [circular label with indigo margins], "Type" [circular label with red margins], "Uganda | Busana [hand-written] | 21. iv. 1932 | G. L. R. Hancock," "Brit. Mus. | 1933–430.," "*Hyponomeuta* | *hemileuca* | TYPE n.sp. | E. Meyrick det." [top three lines hand-written], "♂ genitalia on | slide 29.V 1948 | J.F.G.C. 7530," "LECTOTYPE | *Yponomeuta* | *hemileuca* M. | teste J. Sohn, 2014."

horologa Meyrick, 1935b (*Hyponomeuta*), Exot. Micr. 4(18): 556. (MRAC: Holotype ♂); described from 1 specimen.
Type locality: Zambia – Kivu, Lulenga.
Distribution: Ethiopian – Zambia.
Note: Gershenson & Ulenberg (1998) unnecessarily designated a lectotype for *Hyponomeuta horologa* Meyrick, 1935b because it was described from only one specimen, automatically making it the holotype.

irrorella Hübner, [1796] (*Tinea*), Samml. Eur. Schmett. Tin.: 44. (Type not found); not given.
Type locality: not given [Europe, from publication title].
Distribution: Palearctic – Austria, Belgium, Bulgaria, Corsica, Czech Republic, Denmark, England, Finland, France, Germany, Hungary, Italy, Latvia, Lithuania, Luxembourg, Netherlands, Poland, Romania, Russia (European Northwest, Lower Volga, Middle Volzhsky, Volga-Donskoy), Slovakia, Sweden, Switzerland, [former] Yugoslavia.
Hosts: Celastraceae – *Euonymus europaea* L., *E. verrucosus* Scop., *E. japonicus* Thunb. (Wakely, 1959; Gershenson & Ulenberg, 1998); records from *Prunus spinosa* L. and *P. domestica* L. (Rosaceae) by Friese (1960) are possibly erroneous.

huebneri Koçak, 1982 (*Yponomeuta*), Priamus 2(3): 108. [unnecessary replacement name for *irrorella* Hübner, 1796].

irrorea Haworth, 1811 (*Erminea*), Lep. Brit. 3: 512. [unjustified emendation of *irrorella* Hübner, 1796].
Note: Page 512 was the last page of Lep. Brit. 3 (in 1811). It was reprinted in 1828 as the first page of Lep. Brit. 4.

internellus Walker, 1863 (*Hyponomeuta*), List Lepid. Ins. BM 28: 533. (BMNH: Holotype ♂); described from 1 specimen.
Type locality: Australia – New South Wales, Sydney.
Distribution: Australasian – Australia (New South Wales, Queensland).
Host: Celastraceae – *Cassine australis* (Vent.) Kuntze. (Common, 1990).

javanellus Gershenson & Ulenberg, 1998 (*Yponomeuta*), Kon. Ned. Akad. Weten. Verh. Afdel. Nat., Tweede Reeks 99: 135. (RMNH: Holotype ♀); described from 1 specimen.
Type locality: Indonesia – Java, Nongkodjadjar.
Distribution: Oriental – Indonesia (Java).

kanaiella Matsumura, 1931 (*Hyponomeuta*), 6000 Ill. Ins. Jap.: 1097. (EIHU: Lectotype ♂ [designated by Moriuti, 1977a: 180]); not given (2♂ in EIHU).
Type locality: Japan – Honshu, Nagano Pref., Kamisuwa/Sinano.
Distribution: Oriental – China (Zhejiang); Palearctic – China (Hebei, Heilongjiang, Henan, Jilin, Shaanxi), Japan (Hokkaido, Honshu), South Korea.
Hosts: Celastraceae – *Euonymus alata* (Thunb.) Siebold, *E. alata* f. *ciliatodentatus* (Franch. & Sav.) Hiyama (Moriuti, 1977a).

kostjuki Gershenson, 1985 (*Yponomeuta*), Ent. Obozr. 64(3): 580. (NASU: Holotype ♀); described from 1 specimen.
Type locality: Russia – Kuril Islands, Kunashir Island.
Distribution: Palearctic – Russia (South Kamchatka).
Note: Gershenson & Ulenberg (1998) mistakenly reported the depository of the holotype of *kostjuki* Gershenson as ZMAS. Subsequently, this was corrected when Gershenson & Kozhenvnikova (2005) gave the correct type depository, but erroneously reported the sex of the specimen as male.

leucophaeus Gershenson & Ulenberg, 1998 (*Yponomeuta*), Kon. Ned. Akad. Weten. Verh. Afdel. Nat., Tweede Reeks 99: 136. (RMNH: Holotype ♀); described from 1 specimen.
Type locality: Indonesia – Java, Nongkodjadjar.
Distribution: Oriental – Indonesia (Java).

liberalis Meyrick, 1913a (*Hyponomeuta*), Exot. Micr. 1(5): 138. (BMNH: Holotype ♀); described from 1 specimen.
Type locality: Australia – Queensland, Mackay.
Distribution: Australasian – Australia (Queensland).

madagascariensis Gershenson, 2003 (*Yponomeuta*), Vestnik Zoologii 37(3): 94. (NASU: Holotype ♀); described from 1 specimen.
Type locality: Madagascar – Central Plateau, Périnet Park.
Distribution: Ethiopian – Madagascar.
Note: Gershenson & Kozhevnikova (2005) incorrectly reported the sex of the holotype of *madagascariensis* Gershenson as male.

mahalebella Guenée, 1845b (*Hyponomeuta*), Eur. Microlep. Ind. Meth.: 105. (Type lost [J. Minet, pers. comm. to JAL]); not given.
Type locality: France – vicinity of Paris.
Distribution: Palearctic – France, Germany, Italy, Sardinia, Sicily, Ukraine.
Hosts: Rosaceae – *Prunus avium* (L.) L., *P. cerasus* L., *P. mahaleb* L., *P. spinosa* L. (Gershenson, 1974).

malgassaella Viette, 1954 (*Parahyponomeuta*), Ann. Soc. Ent. Fr. 123: 82. (MNHN: Holotype ♂); described from 1♂, 5♀ (MNHN).
Type locality: Madagascar – Centre-South, Bekily.
Distribution: Ethiopian – Madagascar.

malinellus Zeller, 1838 (*Yponomeuta*), Isis Oken 31: 670. (Type not found); not given [mentioned as “in abundance from apple trees”].

Type locality: Poland – Glogów.
Distribution: Nearctic – Canada (introduced: British Columbia, Nova Scotia, Ontario), USA (introduced: Oregon, Washington); Palearctic – Albania, Austria, Azerbaijan, Belgium, Bulgaria, Corsica, Crete, Czech Republic, Denmark, England, Estonia, Finland, France, Germany, Greece, Hungary, Ireland, Italy, Japan, Latvia, Lithuania, Luxembourg, Netherlands, Norway, Poland, Romania, Russia (Caucasus, European Central, European Northeast, European Northwest, European South Taiga, Middle Volzhsky, Ural, Volga-Donskoy), Sardinia, Sicily, Slovakia, Spain, Sweden, Switzerland, Turkey, [former] Yugoslavia.
Hosts: Rosaceae – *Malus praecox* (Pall.) Borkh., *M. sylvestris* (L.) Mill., *Malus* spp. (Unruh et al. 1993; Agassiz, 1996; Gershenson & Ulenberg, 1998).

anomalella Dufrane, 1960 (*Hyponomeuta malinellus* f.), Bull. Inst. Roy. Sci. nat. Belg. 36 (29): 12. (IRSN: Holotype ♂); described from 1 specimen.
Type locality: France – Calvados, Mont-St-Michel.

malivorella Guenée, 1845a (*Yponomeuta padella* var.?); Ann. Soc. Ent. Fr. 2(3): 342. (Type lost (J. Minet, pers. comm. to JAL)); not given.
Type locality: France – "central."

mayumivorella Matsumura, 1931 (*Hyponomeuta*), 6000 Ill. Ins. Jap.: 1098. (EIHU: Lectotype ♀ [designated by Moriuti, 1977a: 192]); not given (8 specimens in EIHU).
Type locality: Japan – Hokkaido, Sapporo.
Distribution: Palearctic – Japan (Hokkaido, Honshu).
Hosts: Celastraceae – *Euonymus fortunei* var. *radicans* (Miq.) Rehder, *E. hamiltonianus* ssp. *sieboldianus* (Blume) H. Hara (Gershenson & Ulenberg, 1998).

meguronis Matsumura, 1931 (*Hyponomeuta*), 6000 Ill. Ins. Jap.: 1098. (EIHU: Lectotype ♀ [designated by Moriuti, 1977a: 189]); not given (1♂, 1♀ in EIHU).
Type locality: Japan – Honshu, Tokyo.
Distribution: Palearctic – Japan (Honshu, Kyushu, Shikoku).
Hosts: Celastraceae – *Euonymus fortunei* (Turcz.) Hand.-Maz., *E. japonicus* Thunb. (Moriuti, 1977a).

hexabola Meyrick, 1935c (*Hyponomeuta*), Exot. Micr. 4(19): 602. (BMNH: Holotype ♂); described from 1 specimen.
Type locality: Japan – Kyushu, Kumamoto Pref., Amakusa.

menkeni Gershenson & Ulenberg, 1998 (*Yponomeuta*), Kon. Ned. Akad. Weten. Verh. Afdel. Nat., Tweede Reeks 99: 142. (ZMAN: Holotype ♂); described from 7♂ (ZMAN).
Type locality: Japan – Honshu, Aomori Pref., Kumoinotaki.
Distribution: Palearctic – Japan (Honshu).
Hosts: Celastraceae – *Euonymus alatus* (Thunb.) Siebold, *E. hamiltonianus* ssp. *sieboldianus* (Blume) H. Hara (Gershenson & Ulenberg, 1998).

meraculus Bradley, 1962 (*Yponomeuta*), Bull. Br. Mus. Nat. Hist. (Ent) 12(5): 266. (BMNH: Holotype ♂); described from 7♂, 14♀ (BMNH).
Type locality: New Hebrides – Aneityum, Red Crest.
Distribution: Australasian – New Hebrides.

meridionalis Gershenson, 1972 (*Yponomeuta*), Ent. Obozr. 51(3): 635. (ZMAS: Holotype ♂); described from 9♂, 19♀ (ZMAS).
Type locality: Tadzhikistan – Varzobgorge.
Distribution: Palearctic – Israel, Tadzhikistan.
Hosts: Rosaceae – *Crataegus korolkowii* Regel ex Zabel, *C. songarica* K. Koch., *C. turkestanica* Pojark. (Gershenson & Ulenberg, 1998).
Note: To identify the holotype of *Y. meridionalis* Gershenson (1972) only mentioned that it bears the labels "Holotypus" and "Genitali prep. No. 11537 ♂." This incomplete statement was insufficient regarding recommendation 73c of the ICZN. The complete holotype data were later given by Gershenson & Kozhevnikova (2004), who erroneously reported the sex of the specimen as female.

minipunctatus Gershenson & Ulenberg, 1998 (*Yponomeuta*), Kon. Ned. Akad. Weten. Verh. Afdel. Nat., Tweede Reeks 99: 144. (BMNH: Holotype ♂); described from 1 specimen.
Type locality: India – Punjab, Khyra Gully.
Distribution: Oriental – India (Punjab).

mintenna Povel, 1985 (*Eumonopyta*), Proc. K. Ned. Akad. Wet. C. 88(4): 446. (RMNH: Holotype ♂); described from 1♂, 1♀ (RMNH).
Type locality: Indonesia – Java, Merbabu.
Distribution: Oriental – Indonesia (Java).
Host: Celastraceae – *Euonymus japonicus* Thunb. (Povel, 1985).

minuellus Walker, 1863 (*Hyponomeuta*), List Lepid. Ins. BM 28: 533. (BMNH: Holotype ♀); described from 5 specimens of which the four paratypes were later determined to be a different species (see Moriuti, 1972: 140).

Type locality: Nepal.
Distribution: Oriental – India (Assam, Himalaya, Kalapani, Kaschmir, Punjab); Palearctic – Nepal.

millepunctatellus Warren, 1888 (*Hyponomeuta*), Proc. Zool. Soc. Lond. 1888(21–23): 338. (BMNH: Holotype ♀); described from 2 specimens (BMNH).
Type locality: India – Kalapani.

numerosa Meyrick, 1921b (*Hyponomeuta*), Exot. Micr. 2(14): 438. (BMNH: Holotype ♀); described from 1 specimen.
Type locality: India – Punjab, Murree.

montanatus Moriuti, 1977a (*Yponomeuta*), Fauna Jap. Ypon. s. l.: 190. (UOP: Holotype ♀); described from 1 specimen.
Type locality: Japan – Honshu, Nagano Pref., Sigakogen.
Distribution: Palearctic – Japan (Honshu), South Korea.
Host: Celastraceae – *Euonymus oxyphyllus* Miq. (Moriuti, 1977a).

morbillosus Zeller, 1877 (*Hyponomeuta*), Hor. Soc. Ent. Ross. 13: 222. (ZMB: Syntypes); described from 3♂ (ZMB).
Type locality: Tanzania – Zanzibar.
Distribution: Ethiopian – Kenya, Tanzania.

multipunctella Clemens, 1861a (*Hyponomeuta*), Proc. Acad. Nat. Sci. Phil. 12: 8. (ANSP: Syntypes); not given (1♂ in ANSP).
Type locality: USA – Pennsylvania, Northampton Co., Easton.
Distribution: Nearctic – Canada (Ontario), USA (Alabama, Delaware, District of Columbia, Florida, Georgia, Illinois, Indiana, Kentucky, Maryland, Massachusetts, Michigan, Mississippi, Missouri, New Jersey, New York, North Carolina, Ohio, Pennsylvania, South Carolina, Tennessee, Virginia, West Virginia).
Hosts: Celastraceae – *Euonymus americanus* L., *E. americanus* var. *obovatus* (Nutt.) Dippel., *E. atropurpureus* Jacq. (Gaumer, 1878; Dyar, 1900a; MacKay, 1972).

euonymella Chambers, 1872 (*Hyponomeuta*), Can. Ent. 4: 42. (Type not found); not given.
Type locality: USA – Kentucky, Covington.

evonymellus; Chambers, 1878 (*Hyponomeuta*), Bull. U. S. Geol. Surv. 4: 110. [misspelling of *euonymella* Chambers].

leucothorax Meyrick, 1913a (*Hyponomeuta*), Exot. Micr. 1(5): 138. (BMNH: Syntypes); described from 2 specimens (BMNH).
Type locality: USA – Ohio, Cleveland.

orbimaculella Chambers, 1872 (*Hyponomeuta*), Can. Ent. 4: 88. [unnecessary replacement name for *euonymella* Chambers].

orbinaculella; Riley, 1891 (*Hyponomeuta*), in Smith, List Lep. Bor. Am., No. 5157: 97. [misspelling of *orbimaculella*].

ordinatellus Walker, 1863 (*Hyponomeuta*), List Lepid. Ins. BM 28: 530. (BMNH: Syntypes); described from at least 2 specimens [both sexes cited in the species description] (1♀ in BMNH).
Type locality: USA.

semialba Meyrick, 1913a (*Hyponomeuta*), Exot. Micr. 1(5): 139. (BMNH: Syntypes); described from 2 specimens (BMNH).
Type locality: Canada – Ontario, Toronto.

wakarusa Gaumer, 1878 (*Hyponomeuta*), Obs. Nat. 2(1): 3 (Type not found); described from 35 specimens.
Type locality: USA – Kansas, Wakarusa Valley.

myriosema Turner, 1898 (*Hyponomenta*), Trans. R. Soc. S. Austr. 22: 200. (ANIC: Lectotype ♂ [designated by Turner, 1923: 169]); described from 3 specimens (ANIC).
Type locality: Australia – Queensland, Brisbane.
Distribution: Australasian – Australia (New South Wales, Queensland).
Note: Later, Turner (1923) wrote: "My original material consisted of three examples, of which two are *paurodes,* only the type being *myriosemus.*" Hence, the only true *myriosema* becomes the lectotype. (the ICZN: article 74.5)

nigrifimbriata Christoph, 1882 (*Hyponomeuta*), Bull. Soc. Imp. Nat. Mosc. 57: 11. (ZMB: Holotype ♀); described from 1 specimen.
Type locality: Russia – Primorsky Krai, Askold [Island].
Distribution: Palearctic – Russia (Primorsky, South Khabarovsk).
Note: The sex of the type specimen was erroneously reported as male in the original description.

octocentra Meyrick, 1921a (*Hyponomeuta*), Ann. Transv. Mus. 8: 117. (TMP: Holotype ♂); described from 1 specimen.

Type locality: Zimbabwe – Emangeni.
Distribution: Ethiopian – Zimbabwe.

effeta Meyrick, 1924 (*Hyponomeuta*), Exot. Micr. 3(4): 117. (TMP: Holotype ♂); described from 1 specimen.
Type locality: Zimbabwe – Bulawayo.

ocypora Meyrick, 1932d (*Hyponomeuta*), Trans. Ent. Soc. Lond. 80(1): 115. (BMNH: Holotype ♂); described from 1 specimen.
Type locality: Ethiopia [=Abyssinia] – Djem-Djem Forest.
Distribution: Ethiopian – Ethiopia.

orientalis Zagulajev 1969 (*Yponomeuta*), Ent. Obozr. 48(1): 193. (ZMAS: Holotype ♂); described from 60♂, 100♀ (ZMAS).
Type locality: Russia – Sakhalin Island.
Distribution: Palearctic – China (North), Japan (Hokkaido, Honshu), Russia (Amur, Primorsky, Sakhalin, South Khabarovsk), South Korea.
Hosts: Rosaceae – *Malus domestica* Borkh., *M. halliana* Koehne, *M.* x *micromalus* Makino, *M. pallasiana* Juz., *M. sieboldii* (Regel) Rehder, *Malus* spp. (Gershenson & Ulenberg, 1998).

okuellus Arduino & Bullini, 1994 [1995] (*Yponomeuta*), Boll. Ist. Ent. Univ. Bologna 49: 145. (Bullini collection [private, Rome, Italy]: Holotype ♂); described from 2♂, 3♀ (Bullini collection).
Type locality: Japan – Honshu, Iwate Pref., Morioka.
Note: The article appears in the 1995 volume, not 1994, and begins on page 145, not 131 as referenced by others.

osakae Moriuti, 1977a (*Yponomeuta*), Fauna Jap. Ypon. s. l.: 181. (UOP: Holotype ♂); described from 6♂ (UOP).
Type locality: Japan – Honshu, Osaka Pref., Minoo.
Distribution: Palearctic – Japan (Honshu, Shikoku).
Host: Celastraceae – *Euonymus hamiltoniaus* ssp. *sieboldianus* (Blume) H. Hara (Moriuti, 1977a).

padella Linnaeus, 1758 (*Phalaena Tinea*), Syst. Nat. 10(1): 535. (LSL: Syntypes); not given (2 specimens in LSL).
Type locality: not given.
Distribution: Nearctic – Canada (introduced: British Columbia, Ontario), USA (introduced: New York, Washington); Palearctic – Albania, Austria, Belgium,

Bulgaria, Corsica, Crete, Czech Republic, Denmark, England, Estonia, Finland, France, Germany, Greece, Hungary, Ireland, Italy, Japan, Latvia, Lithuania, Luxembourg, Malta, Netherlands, Norway, Poland, Portugal (?), Romania, Russia (Kaliningrad, European Central, European Northwest, European South Taiga, Middle Volzhsky, South Ural, Sredneobskaya, Volga-Donskoy, West Caucasus), Sardinia, Slovakia, Spain, Sweden, Switzerland, Turkey, [former] Yugoslavia.
Hosts: Rosaceae – *Amelanchier lamarkii* F. G. Schroed., *Cotoneaster melanocarpus* G. Lodd., *C. nummularioides* Pojark., *C. nummularius* Fisch & C. A. Mey., *Crataegus monogyna* Jacq., *Cr. laevigata* (Poir.) DC, *Cr. rhipidophylla* Gand., *Cydonia oblonga* Mill., *Malus domestica* Borkh., *Malus praecox* (Pall.) Borkh., *M. sylvestris* (L.) Mill., *Mespilus germanica* L., *Prunus amygdalus* Stokes., *P. avium* (L.) L., *P. bucharica* (Korsh.) B. Fedtsch. & Rehder, *P. cerasifera* Ehrn., *P. cerasus* L., *P. domestica* L., *P. spinosa* L., *Prunus* spp., *P. stepposa* Kotov, *Pyrus bucharica* Litv., *Py. communis* L., *Sorbus aucuparia* L., *Sorbus torminalis* (L.) Crantz (Servadei, 1930; Friese, 1960; Gershenson & Ulenberg, 1998; Robinson et al. 2010); host records from non rosaceous plants such as Fagaceae, Oleaceae, Pinaceae, Rhamnaceae and Salicaceae (see Friese, 1960) are possibly erroneous.
Note: This species is often hardly distinguishable from a few related species, referred to as the *Y. padella*-complex. The host records listed above therefore may pertain to species other than *padella*.

> *alba* Dufrane, 1960 (*Hyponomeuta padellus* f.), Bull. Inst. Roy. Sci. nat. Belg. 36 (29): 12. (IRSN: Holotype ♂); described from 1♂, 3♀ (IRSN).
> Type locality: Belgium – Hainaut Prov., Frameries.
>
> *grisea* Dufrane, 1960 (*Hyponomeuta padellus* f.), Bull. Inst. Roy. Sci. nat. Belg. 36 (29): 11. (IRSN: Holotype ♂); described from 4 specimens (IRSN).
> Type locality: Belgium – Hainaut Prov., Mons.
>
> *pedallus*; Liu & Huang, 1996 (*Yponomeuta*), Forest Pests and Diseases (3): 3. [misspelling of *padellus*].
> *variabilis* Zeller, 1844 (*Hyponomeuta*), Isis Oken 37: 214. [unnecessary replacement name for *padellus*].

paradoxus Gershenson, 1979 (*Yponomeuta*), Trudy vses. ent. Obshch. 61: 103. (NASU: Holotype ♀); described from 2♀ (NASU).
Type locality: Armenia – Khosrov National Park.
Distribution: Palearctic – Armenia.

parvipunctus Gershenson & Ulenberg, 1998 (*Yponomeuta*), Kon. Ned. Akad. Weten. Verh. Afdel. Nat., Tweede Reeks 99: 151. (MRAC: Holotype ♀); described from 1 specimen.
Type locality: Democratic Republic of the Congo – Lubumbashi [=Elisabethville].
Distribution: Ethiopian – Democratic Republic of the Congo.

pauciflore Efremov, 1969 (*Yponomeuta*), Biol. sbornik: 26. (ZMAS: Holotype ♂); described from 1♂, 1♀ (ZMAS).
Type locality: Russia – Amur Oblast, Arkhar.
Distribution: Palearctic – Russia (Amur, Primorsky).
Host: Celastraceae – *Euonymus pauciflorus* Maxim. (Efremov, 1969).

paurodes Meyrick, 1907d (*Hyponomeuta*), Proc. Linn. Soc. N.S. Wales 32(1): 150. (BMNH: Lectotype ♂ [designated here]); described from 5 specimens (BMNH).
Type locality: Australia – Queensland, Townsville.
Distribution: Australasian – Australia (New South Wales, Queensland).
Hosts: Celastraceae – *Cassine ausralis* (Vent.) Kuntze, *Cassine* sp. (Common, 1990; Gershenson & Ulenberg, 1998).
Note: Gershenson & Ulenberg (1998) reported that the lectotype of *paurodes* was in the ANIC collection by mistake. The lectotypy, however, had not been published, and no specimen was specifically identified with accompanying data as the type. All specimens of the type series of *paurodes* are actually in the BMNH collection. To eliminate any future confusion, we here designate the lectotype of *paurodes*. The data labels read: "LECTO- | TYPE" [circular label with indigo margins], "Townville | Queensland | FPD. 0.01" [hand-written], "*paurodes* Meyr." [hand-writing], "Meyrick Coll. | B.M. 1938–290.," "*Hyponomeuta* | *paurodes* | 1/5 Meyr. | E. Meyrick det. | in Meyrick Coll." [top three lines hand-written], "LECTOTYPE | *Yponomeuta* | *paurodes* M. | teste J. Sohn, 2014"

> *paurocentera* Turner, 1898 (*Hyponomenta*), Trans. R. Soc. S. Austr. 22: 200. Nomen nudum.
> Note: This name was published as a Meyrick manuscript name, without an included description.

plumbella Denis & Schiffermüller, 1775 (*Tinea*), Schmett. Wien: 139. (Type lost); not given.
Type locality: Austria, surroundings of Vienna.
Distribution: Nearctic – USA (introduced: Massachusetts); Palearctic – Albania, Austria, Belgium, Bulgaria, Czech Republic, Denmark, England, Estonia,

Finland, France, Germany, Greece, Hungary, Ireland, Israel, Italy, Latvia, Lithuania, Luxembourg, Netherlands, Poland, Romania, Russia (European Central, European Northwest, European South Taiga, Kaliningrad, Middle Volzhsky, Ural, Volga-Donskoy, West Caucasus), Sicily, Slovakia, Spain, Sweden, Switzerland, Turkey, [former] Yugoslavia.
Hosts: Celastraceae – *Euonymus europaea* L., *E. verrucosus* Scop. (Gershenson & Ulenberg, 1998); host records not from the Celastraceae, such as Aceraceae, Rhamnaceae, and Rosaceae (see Friese, 1960) are possibly erroneous.

lentiginella Schrank, 1796 [1802] (*Tinea*), Samml. Nat. Phys. Aufs.: 220. (Type not found); not given.
Type locality: not given.

plumbea Haworth, 1828 (*Erminea*), Lep. Brit. 4: 513. [unjustified emendation of *plumbella*].

roscidella Hübner, 1793 (*Phalaena Tinea*), Samml. Vögel u. Schmett.: 16. (Type not found); not given.
Type locality: not given.

polysticta Butler, 1879a (*Hyponomeuta*), Ill. Het. Coll. Brit. Mus. 3: 81. (BMNH: Holotype ♀); [probably] described from 1 specimen.
Type locality: Japan – Honshu, Kanagawa Pref., Yokohama.
Distribution: Oriental – China (Jiangxi, Sichuan); Palearctic – China (Fujian, Shaanxi), Japan (Hokkaido, Honshu, Kyushu, Shikoku), Russia (Amur, Primorsky, Sakhalin, South Kamchatka, South Khabarovsk), South Korea.
Hosts: Celastraceae – *Euonymus maackii* Rupr., *E. oxyphyllus* Miq., *E. hamiltonianus* ssp. *sieboldianus* (Blume) H. Hara (Gershenson & Ulenberg, 1998).

tyrodes Meyrick, 1913a (*Hyponomeuta*), Exot. Micr. 1(5): 138. (BMNH: Holotype ♂); described from 1 specimen.
Type locality: Japan.

polystigmellus Felder & Felder, 1862 (*Hyponomeuta*), Wien. Ent. Monatsschr. 6: 40. (BMNH: Holotype ♂); not given.
Type locality: China – Zhejiang, Ning-Po.
Distribution: Oriental – China (Jiangxi, Sichuan, Zhejiang), Taiwan; Palearctic – Japan (Hokkaido, Honshu, Kyushu), Russia (Amur, Primorsky), South Korea.
Hosts: Celastraceae – *Euonymus alatus* (Thunb.) Siebold, *E. hamiltonianus* ssp. *sieboldianus* (Blume) H. Hara (Gershenson & Ulenberg, 1998).

Note: Felder & Felder (1862) ended the description with a male symbol (♂) and did not record the sexes or number of specimens in the type series. Moriuti (1977a) reported a holotype (male) and two paratypes (females) for this species. Since no statement indicates the presence of female specimens in the original description, the type status of the two "paratypes" mentioned by Moriuti is disputable.

pseudostrigillatus Gershenson & Ulenberg, 1998 (*Yponomeuta*), Kon. Ned. Akad. Weten. Verh. Afdel. Nat., Tweede Reeks 99: 157. (MRAC: Holotype ♂); described from 7♂, 1♀ (MRAC).
Type locality: Cameroon – Southwest, Buea.
Distribution: Ethiopian – Cameroon, Democratic Republic of the Congo.

pustulellus Walker, 1863 (*Hyponomeuta*), List Lepid. Ins. BM 28: 533. (BMNH: Lectotype ♀ [designated here]); described from 2 specimens (1♀ in BMNH).
Type locality: Australia – Queensland, Townsville.
Distribution: Australasian – Australia (New South Wales, Queensland).
Hosts: Celastraceae – *Cassine australis* (Vent.) Kuntze., *Cassine* sp. (Common, 1990; Gershenson & Ulenberg, 1998).
Note: The only available syntype of *pustulellus* Walker bears a 'holotype' label, but Walker did not designate a holotype. To avoid confusion if the other type specimen is found, and to stabilize the species, we designate the specimen in the BMNH as the lectotype. The type labels read: "Lecto- | type" [circular label with indigo margins], "Type | H.T" [circular label with red margin], "New | South | Wales" [hand-written on cream-colored, circular label], "*Hyponomeuta* | *pustulellus* | Wlkr. | Cat. Lep. BM. 28. p533 (1863) || TYPE ♀" [rectangular label with thick, black lined margins], "LECTOTYPE | *Yponomeuta* | *pustulellus* W. | teste J. Sohn, 2014."

refrigeratus Meyrick, 1931b (*Hyponomeuta*), Exot. Micr. 4(6): 172. (BMNH: Holotype ♂); described from 1 specimen.
Type locality: Japan – Hokkaido, Sapporo.
Distribution: Palearctic – Japan (Hokkaido, Honshu), Russia (Amur), South Korea.
Host: Celastraceae – *Euonymus maackii* Rupr. (Efremov, 1976).

> *zagulajevi* Efremov, 1976 (*Yponomeuta*), Zhivotny mir Sovetskogo Dalnego Vostoka, Blagoveshchensk 1: 108. (ZMAS: Holotype ♂); described from 10 specimens (3♂, 1♀ in ZMAS).
> Type locality: Russia, Amur region, West Blagoveshchensk.

rhamnellus Gershenson, 1974 (*Yponomeuta*), Fauna Ukraini 15(6): 64. (NASU: Holotype ♂); described from 6♂, 4♀ (NASU; ZMAS).
Type locality: Ukraine – Donetsk Oblast, Novoazovskii Dist., Khomutovsky Steppe.
Distribution: Palearctic – Estonia, Ukraine.
Host: Rhamnaceae – *Rhamnus cathartica* L. (Gershenson, 1974).

> *rhamniellus*; Forstén et al. 1998 (*Yponomeuta*), Memor. Soc. Fau. Flo. Fenn. 74: 90. [misspelling of *rhamnellus*].

rorrella Hübner, [1796] (*Tinea*), Samml. Eur. Schmett. Vol. 8, pl. 34, Tineæ 2, Pyralidise B, fig. 234 [in Hübner, 1796–1838]. (Type not found); not given.
Type locality: Europe [based on book title].
Distribution: Palearctic – Austria, Belgium, Bulgaria, China (Hebei, Ningxia), Czech Republic, Denmark, England, Finland, France, Germany, Greece, Hungary, Italy, Latvia, Lithuania, Luxembourg, Netherlands, Poland, Portugal, Romania, Russia (European Northwest, Kaliningrad, Lower Volga, Middle Volzhsky, Volga-Donskoy, West Caucasus), Sicily, Spain, Sweden, Switzerland.
Hosts: Salicaceae – *Populus alba* L., *Salix alba* L., *S. babylonica* L., *S. caprea* L., *S. cinerea* L., *S. dasyclados* Wimm., *S. elegans* Besser, *S. fragilis* L., *S. incana* Schrank, *S. viminalis* L. (Roberti, 1952; Gershenson & Ulenberg, 1998; Zaki & Masoodi, 1997; Küppers, 2008).

> *helicella* Freyer, 1842 (*Tinea*), Neu. Beitr. Schmett. 4: 154. (Type not found); not given.
> Type locality: Germany – Lech River.

> *rorea* Haworth, 1828 (*Erminea*), Lep. Brit. 4: 513. [unjustified emendation of *rorrella* Hübner, 1796].

> *rorella*; Hübner, 1825 [1826] (*Nygmia*), Verz. bekannt. Schmett.: 412. [misspelling of *rorrellus*].

sacculata Pathania & Rose, 2003 (*Yponomeuta*), J. Nat. Conserv. 15(2): 530. (DZPU: Holotype ♂); described from 13♂, 4♀ (DZPU).
Type locality: India – Himachal Pradesh, Sirmour, Renuka Lake.
Distribution: Oriental – India (Himachal Pradesh).

sedellus Treitschke, 1832 (*Yponomeuta*), Schmett. Europa 9(1): 223. (TMB: Syntypes); not given (2 specimens [sex indeterminable] in TMB).
Type locality: Germany – Meißen Mts.

Distribution: Oriental – China (Shanghai); Palearctic – Austria, Belgium, Bulgaria, China (Anhui), Czech Republic, Denmark, England, Estonia, Finland, France, Germany, Hungary, Israel, Italy, Japan (Hokkaido, Honshu, Shikoku), Latvia, Lithuania, Luxembourg, Netherlands, Norway, Poland, Romania, Russia, Sicily, Slovakia, South Korea, Spain, Sweden, Switzerland, [former] Yugoslavia.
Hosts: Crassulaceae – *Sedum album* L., *S. erythrostictum* Miq., *S. maximum* (L.) Suter, *S. spectabile* Boreau, *S. taqueti* Pragener, *S. telephium* L., (Servadei, 1930; Gershenson & Ulenberg, 1998); records of *Euonymus* (Celastraceae) by Werner (1958) are possibly erroneous.

apicalis Matsumura, 1931 (*Hyponomeuta*), 6000 Ill. Ins. Jap.: 1097. (EIHU: Syntypes); not given (1♀ in EIHU).
Type locality: Japan.

elementalis; Vives Moreno, 1992 (*Yponomeuta*), Cat. Sis. Sin. Lep.: 73. [misspelling of *elementaris*].

elementaris Meyrick, 1931b (*Hyponomeuta*), Exot. Micr. 4(6): 172. (BMNH: Holotype ♂); described from 1 specimen.
Type locality: Japan – Gifu.

viginti punctata Retzius, 1783 (*Phalaena*), In: De Geer, Gen. Spec. Ins.: 52. (Type not found). Nomen nudum [not binominal] (ICZN, 1985: article 11c).

vigintipunctella Servadei, 1930 (*Hyponomeuta*), Boll. Lab. Ent. R. Ist. sup. agr. Bologna 8: 254. [unnecessary emendation of *viginti punctata*].

sistrophora Meyrick, 1909c (*Hyponomeuta*), Ann. S. Afr. Mus. 5(7): 358. (SAMC: Holotype ♂); described from 2 specimens (SAMC).
Type locality: South Africa – Natal, Lower Umkomaas.
Distribution: Ethiopian – South Africa (Natal).

sociatus Moriuti, 1972 (*Yponomeuta*), Kontyû 40(3): 153. (UOP: Holotype ♂); described from 18♂, 12♀ (BMNH; UOP).
Type locality: Japan – Honshu, Aomori Pref., Hirosaki.
Distribution: Oriental – China (Jiangxi, Sichuan); Palearctic – Japan (Hokkaido, Honshu, Kyushu), South Korea.
Hosts: Celastraceae – *Celastrus orbiculatus* Thunb., *Euonymus macropterus* Rupr. (Moriuti, 1977a; Gershenson & Ulenberg, 1998).

spodocrossa Meyrick, 1935c (*Hyponomeuta*), Exot. Micr. 4(19): 602. (BMNH: Lectotype ♀ [designated by Clarke, 1965: 332]); described from 2 specimens (BMNH).
Type locality: Japan – Honshu, Miyagi Pref., Yumoto.
Distribution: Palearctic – China (Henan, Shaanxi), Japan (Hokkaido, Honshu).
Hosts: Celastraceae – *Euonymus hamiltonianus* ssp. *sieboldianus* (Blume) H. Hara, *E. macropterus* Rupr. (Moriuti, 1977a).

stenodoxa Meyrick, 1931a (*Hyponomeuta*), Exot. Micr. 4(3): 87. (BMNH: Holotype ♀); described from 1 specimen.
Type locality: India – Mumbai, Mahableshwar.
Distribution: Oriental – India (Mumbai).

strigillatus Zeller, 1852 [1854] (*Hyponomeuta*), K.V. Akad. Handl. 1852: 102. (NRM: Syntypes); not given (1♂ in NRM).
Type locality: South Africa – Natal, Caffraria.
Distribution: Ethiopian – Burundi, Cameroon, Gold Coast, Kenya, Mozambique, South Africa (Natal).

> *perficitellis*; Janse, 1917 (*Hyponomeuta*), Checklist S. Afr. Lepid. Het.: 199. [misspelling of *perficitellus*].
>
> *perficitellus* Walker, 1863 (*Hyponomeuta*), List Lepid. Ins. BM 28: 531. (BMNH: Lectotype ♀ [designated here]); described from 3 specimens (BMNH).
> Type locality: South Africa – Natal.
> Note: *Hyponomeuta perficitellus* was described from three syntypes, marked as 'a, b and c' by Walker (1863). Of these, only one specimen is still extant in the BMNH and often erroneously referred to as the holotype. For species clarification, this specimen is designated as the lectotype of *perficitellus* here. The pin data labels read: "Lecto- | type" [circular label with indigo margin], "Type" [circular label with red margin], "Pt Natal | 55.96.," "*Hyponomeuta* | *perficitellus* Wlkr. | Cat. Lep. BM. 28. 531 (1863) || TYPE ♀" [hand-written on a rectangular label with thick, black margins], "LECTOTYPE | *Yponomeuta* | *perficitellus* W. | teste J. Sohn, 2014."

subplumbellus Walsingham, 1881a (*Hyponomeuta*), Trans. Ent. Soc. Lond. 1881(2): 248. (BMNH: Holotype (not found)); described from 1 specimen.
Type locality: South Africa – Natal, Victoria County, [probably] Spring Vale.
Distribution: Ethiopian – Democratic Republic of the Congo, Namibia, South Africa (Natal), Zimbabwe.

tokyonella Matsumura, 1931 (*Hyponomeuta*), 6000 Ill. Ins. Jap.: 1098. (EIHU: Syntypes); not given (1♀ in EIHU).
Type locality: Japan – Honshu, Tokyo.
Distribution: Oriental – China (Jiangxi, Shanghai); Palearctic – China (Anhui, Beijing, Hebei, Heilongjiang, Henan, Jiangsu, Liaoning, Ningxia, Tianjin), Japan (Honshu, Kyushu), South Korea.
Host: Celastraceae – *Euonymus hamiltonianus* ssp. *sieboldianus* (Blume) H. Hara (Moriuti, 1977a).

triangularis Möschler, 1890 (*Hyponomeuta*), Abh. Senck. Nat. Ges. 16: 339. (Type not found); described from 2 specimens.
Type locality: Puerto Rico.
Distribution: Neotropical – British Virgin Islands (Anegada, Guana, Tortola, Virgin Gorda), Cuba, Mexico (Veracruz), Puerto Rico.
Host: Celastraceae – *Cassine* [or *Elaeodendron*] *xylocarpa* (Vent.).

> *eusoma* Walsingham, 1914 (*Hyponomeuta*), Biol. Centr. Amer. 4:325. (BMNH: Holotype ♀); described from 2 specimens (BMNH; USNM).
> Type locality: Mexico – Veracruz, Jalapa.
> Distribution: Neotropical – Mexico (Vera Cruz).

uttaranchalensis Pathania & Rose, 2003 (*Yponomeuta*), J. Nat. Conserv. 15(2): 532. (DZPU: Holotype ♂); described from 1 specimen.
Type locality: India – Dehra Dun, FRI Dehra Dun, Uttaranchal.
Distribution: Oriental – India (Dehra Dun).

yanagawanus Matsumura, 1931 (*Hyponomeuta*), 6000 Ill. Ins. Jap.: 1098. (EIHU: Lectotype ♂ [designated by Moriuti, 1977a: 198]); not given (5 specimens in EIHU).
Type locality: Japan – Fukuoka Prefecture, Yanagawa.
Distribution: Palearctic – Japan (Honshu, Kyushu, Shikoku), South Korea.
Hosts: Celastraceae – *Euonymus alatus* (Thunb.) Siebold, *E. japonicus* Thunb., *Euonymus* sp. (Herrebout et al. 1982; Gershenson & Ulenberg, 1998).

> *mochlocrossa* Meyrick, 1935c (*Hyponomeuta*), Exot. Micr. 4(19): 602. (BMNH: Lectotype ♂ [designated by Clarke, 1965: 331]); described from 3 specimens (2♂ in BMNH).
> Type locality: Japan – Tokyo.

zebra Sohn & Wu, 2010 (*Yponomeuta*), In: Sohn et al., J. Nat. Hist. 44(45/46): 2807 (IZAS: Holotype ♀); described from 1 specimen.

Type locality: China – Tibet, Zhangmukoan [=Zhu Feng Nature Reserve].
Distribution: Palearctic – China (Tibet), Nepal.

ZELLERIA Stainton, 1849c, Syst. Cat. Brit. Tin. & Pteroph. 8: 22.
Type species: *Zelleria hepariella* Stainton, 1849c, by subsequent designation by Meyrick, 1915b, Trans. N. Z. Inst. 47: 227.

EUDRYMOPA Lower, 1908, Trans. R. Soc. S. Austr. 32: 118.
Type species: *Eudrymopa cyanoleuca* Lower, 1908, by monotypy.

abisella Chrétien, 1910 (*Argyresthia*), Naturaliste 32: 261. (MNHN: Syntypes); stated as "several specimens" together with var. *alterella*.
Type localities: France – Hautes-Alpes and Isère.
Distribution: Palearctic – France.

alterella Chrétien, 1910 (*Argyresthia abisella* var.) Naturaliste 32: 261. (MNHN: Syntypes); stated as "several specimens."
Type locality: France – Hautes-Alpes and Isère.

aphrospora Meyrick, 1892 [1893] (*Zelleria*), Proc. Linn. Soc. N.S. Wales for the year 1892. Second Series. 7: 584. (BMNH: Lectotype ♂ [designated here]); described from 5 specimens (BMNH).
Type locality: Australia – South Australia, Port Lincoln.
Distribution: Australasian – Australia (South Australia).
Note: In 1966, I. F. B. Common selected a lectotype for *aphrospora* Meyrick, but the designation has never been published. We here make this designation valid. The pin data labels read: "Lecto- | type" [circular label with indigo margin.] "Port Lincoln | S. Australia | 5/11/82 [hand-written]," "*aphrospora* Meyr. [hand-written on one line]," "Meyrick Coll. | B.M. 1938-290" | "*Zelleria* | *aphrospora* | 1/5 Meyr. | E. Meyrick det. | in Meyrick Coll.," "LECTOTYPE ♂ | *Zelleria* | *aphrospora* Meyr. | selected by 1892. | I. F. B. Common, 1966" | "teste J. Sohn, 2014."

araeodes Meyrick, 1892 [1893] (*Zelleria*), Proc. Linn. Soc. N.S. Wales for the year 1892. Second Series. 7: 582. (BMNH: Lectotype ♂ [designated here]); described from 5 specimens (BMNH).
Type locality: Australia – West Australia, Geraldton.
Distribution: Australasian – Australia (New South Wales, West Australia).
Note: In 1966, I. F. B. Common selected a lectotype for *araeodes* Meyrick, but the designation has never been published. We here make this designation

valid. The pin data labels read: "Lecto- | type" [circular label with indigo margin.] | "Geraldton | W. Australia | 18/10/86 [hand-written]," "*araeodes* Meyr. [hand-written on one line]," "Meyrick Coll. | B.M. 1938-290," "*Zelleria* | *araeodes* "| 1/5 Meyr. | E. Meyrick det. | in Meyrick Coll.," "B.M. ♂ | Genitalia Slide | IFBC | No. 13731," "LECTOTYPE ♂ | *Zelleria* | *araeodes* Meyr. | selected by 1892. | I. F. B. Common, 1966" | teste J. Sohn, 2014."

arizonica Braun, 1940 (*Zelleria*), Trans. Am. Ent. Soc. 66: 279. (ANSP: Holotype ♂); described from 4♂, 7♀ (ANSP; USNM).
Type locality: USA – Arizona, Flagstaff.
Distribution: Nearctic – USA (Arizona).

callidoxa Meyrick, 1892 [1893] (*Zelleria*), Proc. Linn. Soc. N.S. Wales for the year 1892. Second Series. 7: 584. (BMNH: Lectotype ♂ [designated here]); described from 7 specimens (BMNH).
Type locality: Australia – South Australia, Port Lincoln.
Distribution: Australasian – Australia (Queensland, South Australia).
Note: In 1966, I. F. B. Common selected a lectotype for *Zelleria callidoxa* Meyrick, but the designation has never been published. We here make this designation valid. The pin data labels read: "LECTO- | TYPE" [circular label with indigo margin], "Port Lincoln | S. Australia | 6/11/82 [hand-written]," "*callidoxa* Meyr. [hand-written on one line]," "Meyrick Coll. | B.M. 1938-290," "*Zelleria* | *callidoxa* | 1/7 Meyr. | E. Meyrick det. | in Meyrick Coll.," "LECTOTYPE | *Zelleria callidoxa* | Meyrick | teste J. Sohn, 2014."

celastrusella Kearfott, 1903 (*Zelleria*), Jour. N. Y. Ent. Soc. 11: 150. (USNM: Lectotype ♂ [designated by Sohn & Lewis, 2012: 6]); described from about 200 specimens (AMNH; BMNH; MCZ; USNM).
Type locality: USA – New Jersey, Essex Co., Caldwell.
Distribution: Nearctic – Canada (Ontario), USA (Illinois, Maryland, Michigan, New Jersey, New York, Ohio, Texas).
HOST: Celastraceae – *Celastrus scandens* L. (Kearfott, 1903).

> *celastrella* Meyrick, 1914c (*Xyrosaris*), Lep. Cat. 19: 14. [unjustified emendation of *celastrusella*].

cirrhoscia Meyrick, 1931c (*Zelleria*), Ann. Mus. Nac. Hist. Nat. Buenos Aires 36: 398. (BMNH: Holotype ♂); described from 1 specimen.
Type locality: Argentina – Rio Negro.
Distribution: Neotropical – Argentina (Rio Negro).

citrina Meyrick, 1892 [1893] (*Zelleria*), Proc. Linn. Soc. N.S. Wales for the year 1892. Second Series. 7: 586. (BMNH: Lectotype ♂ [designated here]); described from 9 specimens (BMNH).
Type locality: Australia – New South Wales, Sydney.
Distribution: Australasian – Australia (New South Wales).
Note: In 1966, I. F. B. Common selected a lectotype of *citrina* Meyrick, but the designation has never been published. We here make this designation valid. Pin data labels read: "Lecto- | type" [circular label with indigo margin.] | "Sydney | N. S. Wales | 6/9/84 [hand-written]," "*citrina* Meyr. [hand-written on one line]," "Meyrick Coll. | B.M. 1938-290," "*Zelleria* | *citrina* | 1/5 Meyr. | E. Meyrick det. | in Meyrick Coll.," "LECTOTYPE | *Zelleria citrina* | Meyrick | teste J. Sohn, 2014."

coniostrepta Caradja & Meyrick, 1938 (*Zelleria*), In: Caradja & Meyrick, Dt. Ent. Z. Iris 52(1): 17. (BMNH: Holotype ♂); described from 1 specimen.
Type locality: China – Yunnan, Likiang [=Lijiang], Yülingshanmass [possibly Yulongxueshan].
Distribution: Oriental – China (Yunnan).

cremnospila Lower, 1900 (*Zelleria*), Proc. Linn. Soc. N.S. Wales 25: 421. (SAMA: Syntypes); described from 3 specimens (SAMA).
Type locality: Australia – South Australia, Port Victor.
Distribution: Australasian – Australia (South Australia).

cryptica Meyrick, 1913a (*Zelleria*), Exot. Micr. 1(5): 136. (DEMV: Holotype ♂); described from 1 specimen.
Type locality: Australia – Victoria, Gisborne.
Distribution: Australasian – Australia (Victoria).
Host: Myrtaceae – *Eucalyptus* sp. (Gershenson & Ulenberg, 1998).

cyanoleuca Lower, 1908 (*Eudrymopa*), Trans. R. Soc. S. Austr. 32: 118. (DEMV: Holotype ♂); described from 1 specimen.
Type locality: Australia – Victoria, Gisborne.
Distribution: Australasian – Australia (Victoria).
Host: Myrtaceae – Pupae found under bark of *Eucalyptus* (smooth gum) trunks (Turner, 1923).

> *perimeces* Turner, 1923 (*Zelleria*), Trans. R. Soc. S. Austr. 47: 167. (DEMV: Syntypes); described from 5 specimens (DEMV).
> Type locality: Australia – Victoria, near Gisborne, Mt. Macedon.

cynetica Meyrick, 1892 [1893] (*Zelleria*), Proc. Linn. Soc. N.S. Wales for the year 1892. Second Series. 7: 582. (BMNH: Lectotype ♂ [designated here]); described from 10 specimens (BMNH).
Type Localities: Australia – Tasmania, Deloraine.
Distribution: Australasian – Australia (New South Wales, Queensland, Tasmania, Victoria).
Note: In 1966, I. F. B. Common selected a lectotype for *cynetica* Meyrick, but the designation has never been published. We here make this designation valid. The pin data labels read: "Lecto- | type" [circular label with indigo margin.] | "Deloraine | Tasmania | 25/12/84 [hand-written]" | "*cynetica* Meyr. [hand-written on one line]," "Meyrick Coll." | B.M. 1938–290," "*Zelleria* | *cynetica* | 2/10 Meyr. | E. Meyrick det. | in Meyrick Coll.," || "LECTOTYPE | *Zelleria cynetica* | Meyrick | teste J. Sohn, 2014."

euthysema Turner, 1923 (*Zelleria*), Trans. R. Soc. S. Austr. 47: 167. (ANIC: Syntypes); described from 2 specimens (ANIC).
Type Localities: Australia – Queensland, Crow's Nest near Toowoomba; Stanthorpe.
Distribution: Australasian – Australia (New South Wales, Queensland, Tasmania).
Host: Ericaceae – *Melichrus urceolatus* R. Br. (Common, 1990).

haimbachi Busck, 1915 (*Zelleria*), Proc. Ent. Soc. Wash. 17(2): 91. (USNM: Syntypes); not given (1♂ in USNM).
Type locality: USA – New Jersey, Gloucester Co., Wenonah.
Distribution: Nearctic – Canada (Alberta, British Columbia, Manitoba, New Brunswick, Nova Scotia, Ontario, Québec, Saskatchewan), USA (Arizona, California, Colorado, Indiana, Kentucky, Maine, Maryland, Michigan, Minnesota, Nebraska, New Jersey, New York, North Carolina, Oregon, South Carolina, Washington, West Virginia, Wyoming).
Hosts: Pinaceae – *Pinus banksiana* Lambert, *P. contorta* Douglas ex Loudon, *P. jeffreyi* A. Murray, *P. ponderosa* Douglas & C. Lawson, *P. resinosa* Aiton, *P. sylvestris* L., *Pseudotsuga menziesii* (Mirb.) Franco, *Ps. macrocarpa* (Vasey) Mayr (Powell & Opler, 2009; Prentice, 1965; Furniss & Carolin, 1977).

hemixipha Lower, 1900 (*Zelleria*), Proc. Linn. Soc. N.S. Wales 25: 421. (SAMA: Syntypes); described from 2 specimens (SAMA).
Type locality: Australia – South Australia, Highbury.
Distribution: Australasian – Australia (South Australia).

hepariella Stainton, 1849c (*Zelleria*), Syst. Cat. Brit. Tin. & Pteroph. 8: 22. (Type not found); not given.

Type locality: England.
Distribution: Palearctic – Algeria, Austria, Belgium, Corsica, Denmark, England, France, Germany, Greece, Ireland, Italy, Japan (Honshu), Latvia, Netherlands, Norway, Poland, Romania, Russia (Primorsky), Sardinia, Sicily, Slovakia, Spain, Sweden, Switzerland, Turkey, [former] Yugoslavia.
Hosts: Oleaceae – *Fraxinus excelsior* L., *Fraxinus* sp., *Ligustrum tschonskii* Decne, *Phillyrea angustifolia* L. (Friese, 1960; Gershenson & Ulenberg, 1998); records on *Lonicera caerulea* L. (Caprifoliaceae) and *Artemisia vulgaris* L. (Compositae) are possibly erroneous (see Friese, 1960: 89; Ulenberg, 2009: 191).
Note: *Zelleria hepariella* was a Mann manuscript name made nomenclaturally available by Stainton.

fusca Stainton, 1876 (*Zelleria*), Ent. M. Mag. 13:1. (Type not found); described from 11 specimens.
Type locality: England – Scotland, Paisley.

insignipennella Stainton, 1849c (*Zelleria*), Syst. Cat. Brit. Tin. & Pteroph. 8: 22. ("Mr. Bedell and Mr. Weir" collections [private] [as cited by Stainton]; types not found); not given.
Type locality: England.

joannisella Maneval, 1934 (*Zelleria*), Amat. Papill. 7: 55. (Maneval collection [private] [not found]: Holotype ♂); described from 1 specimen.
Type locality: France – Haute-Loire, Tence Canton, Chenereilles.

phillyrella Millière, 1867 (*Zelleria*), Ann. Soc. Linn. Lyon 2(18): 286. (Possibly lost); may have been destroyed in World War II, (G. Luquet via pers. comm. to J. Minet); not given.
Type locality: France – [Provence], Henriette villa [in Antibes].

taxella Herrich-Schäffer, [1855] (*Swammerdamia*), Syst. Bearb. Schmett. Eur. 5: 289 [see Herrich-Schäffer, 1853–1855]. (Type lost); not given.
Type locality: Switzerland – Zurich.

taxi Frey, 1856 (*Gracilaria*), Tin. Pter. Schweiz: 234. [unjustified emendation of *taxella*].

isopyrrha Meyrick, 1921b (*Zelleria*), Exot. Micr. 2(14): 437. (BMNH: Holotype ♂); described from 1 specimen.

Type locality: Australia – West Australia, Yallingua.
Distribution: Australasian – Australia (West Australia).

japonicella Moriuti, 1977a (*Zelleria*), Fauna Jap. Ypon. s. l.: 221. (UOP: Holotype ♂); described from 3♂, 4♀ (UOP).
Type locality: Japan – Hokkaido, Yamabe, Kamikawa.
Distribution: Palearctic – China (Heilongjiang), Japan (Hokkaido, Honshu).
Hosts: Oleaceae – *Ligustrum tschonoskii* var. *glabrescens* Koidzumi, *Syringa persica* L. (Moriuti, 1977a; Sohn et al. 2010).

kesslerioides Gibeaux, 1985b (*Zelleria*), Nouv. Revue Ent. (N.S.) 2(3): 277. (MNHN: Holotype ♂); described from 1 specimen.
Type locality: Madagascar – Centre, Betroka.
Distribution: Ethiopian – Madagascar.

leucoschista Meyrick, 1931c (*Zelleria*), Ann. Mus. Nac. Hist. Nat. Buenos Aires 36: 399. (BMNH: Holotype ♂); described from 1 specimen.
Type locality: Argentina – Rio Negro, Bariloche.
Distribution: Neotropical – Argentina (Rio Negro).

leucostrota Meyrick, 1928e (*Zelleria*), Trans. Ent. Soc. Lond. 76: 504. (BMNH: Lectotype ♀ [labeled as ♂; designated by Clarke, 1965: 342]); described from 4 specimens (BMNH).
Type locality: French Polynesia – Marquesas Islands, Nuka Hiva, Tahuata.
Distribution: Australasian – French Polynesia (Marquesas Islands).

leucostra; Viette, 1949 (*Zelleria*), Pacif. Sci. 3(4): 317. [misspelling of *leucostrota*].

loranthivora Meyrick, 1930c (*Zelleria*), Exot. Micr. 3(19): 591. (BMNH: Lectotype ♂ [designated by Clarke, 1965: 411]); described from 9 specimens (BMNH).
Type locality: India – Pusa.
Distribution: Oriental – India (Bihar, Pusa).
Host: Loranthaceae – *Loranthus* sp. (Meyrick, 1930c).

maculata Philpott, 1930 (*Zelleria*), Rec. Auckland Inst. Mus. 1(1): 13. (AMNZ: Syntypes); not given, stated as "several" specimens (1♀ in AMNZ).
Type locality: New Zealand – Mt. Maungatua, Otago.
Distribution: Australasian – New Zealand.
Host: Loranthaceae – *Peraxilla* sp. (Gershenson & Ulenberg, 1998).

malacodes Turner, 1939 (*Zelleria*), Pap. Proc. R. Soc. Tasm. 1938: 98. (ANIC: Holotype ♀); described from 1 specimen.
Type locality: Australia – Tasmania, Cradle Mt.
Distribution: Australasian – Australia (Tasmania).

memorella Meyrick, 1892 [1893] (*Zelleria*), Proc. Linn. Soc. N.S. Wales for the year 1892. Second Series. 7: 583. (BMNH: Lectotype ♂ [designated here]); described from 10 specimens (BMNH).
Type locality: Australia – Tasmania, Hobart.
Distribution: Australasian – Australia (New South Wales, West Australia, Victoria, Tasmania).
Note: One specimen in the type series of *memorella* Meyrick has a lectotype label. This lectotype designation seems to have been made by I. F. B. Common, but it has never been published. We here make this designation valid. The pin data labels read: "Lecto- | type" [circular label with indigo margin.], "Hobart | Tasmania | 2/12/82 [hand-written]," "*memorella* Meyr." [hand-written on one line], "Meyrick Coll. | B.M. 1938-290," "*Zelleria* | *memorella* | 3/10 Meyr. | E. Meyrick det. | in Meyrick Coll.," "LECTOTYPE | *Zelleria memorella* | Meyrick | teste J. Sohn, 2014."

metriopa Meyrick, 1933 (*Zelleria*), Exot. Micr. 4(12): 373. (BMNH: Holotype ♀); described from 1 specimen.
Type locality: Indonesia – Java, Seneng.
Distribution: Oriental – Indonesia (Java).
Host: Rhamnaceae – *Ventilago madaraspatana* Gaertn. (Gershenson & Ulenberg, 1998).

mystarcha Meyrick, 1892 [1893] (*Zelleria*), Proc. Linn. Soc. N.S. Wales for the year 1892. Second Series. 7: 586. (BMNH: Lectotype ♀ [designated here]); described from 9 specimens (BMNH).
Type locality: Australia – Tasmania, Campbelltown.
Distribution: Australasian – Australia (Tasmania).
Note: One specimen in the type series of *mystarcha* Meyrick has a lectotype label. This lectotype designation seems to have been made by I. F. B. Common, but it has never been published. We here make this designation valid. The pin data labels read: "Lecto- | type" [circular label with indigo margin.], "Campbelltown | Tasmania | 29/12/84 [hand-written]," "*mystarcha* Meyr. [hand-written on one line]," "Meyrick Coll. | B.M. 1938-290" | "*Zelleria* | *mystarcha* | 7/1 Meyr. | E. Meyrick det. | in Meyrick Coll.," "LECTOTYPE | *Zelleria mystarcha* | Meyrick | teste J. Sohn, 2014."

namibiensis Mey, 2011 (*Zelleria*), Esperiana Mem. 6: 181. (NMNW: Holotype ♂); described from 4♂, 2♀ (NMNW; ZMB).
Type locality: Namibia – Kaokoveld, Joubertpass.
Distribution: Ethiopian – Namibia.

ochroplagiata Braun, 1918 (*Xyrosaris*), Can. Ent. 50: 232. (ANSP: Syntypes: [sex undeterminable, per J. Weintraub, pers. comm. to JCS]); not given (1 specimen in ANSP).
Type locality: USA – Louisiana, Winnfield.
Distribution: Nearctic – USA (Louisiana).

oleastrella Millière, 1864 (*Tinea*), Ann. Soc. Linn. Lyon 2(12): 42. (Possibly lost; may have been destroyed in World War II, per G. Luquet via pers. comm. to J. Minet); not given.
Type locality: France – Hyères.
Distribution: Palearctic – Corsica, Crete, Denmark, England, France, Greece, Italy, Madeira Archipelago, Malta, Morocco, Portugal, Sardinia, Sicily, Spain, Sweden, Tunisia, Turkey, [former] Yugoslavia.
Hosts: Oleaceae – *Olea europaea* L., O. *oleaster* Hoffmanns & Link (Martelli, 1915; Friese, 1960; Balachowsky, 1966); records of *Elaeagnus angustifolia* L. (Elaeagnaceae) by Kaçar & Ulusoy (2011) are incorrect due to a translation error of 'oleaster', which means wild olives.

> *scambota* Meyrick, 1928a (*Xyrosaris*), Bull. Hill Mus. 2(3): 237. (BMNH: Syntypes); described from 2 specimens (BMNH).
> Type localities: Morocco – Tinmel and Arround.

orthopleura Turner, 1923 (*Zelleria*), Trans. R. Soc. S. Austr. 47: 167. (ANIC: Syntypes); described from 4 specimens (ANIC).
Type Localities: Australia – Queensland: Brisbane, Charleville, and Coolangatta.
Distribution: Australasian – Australia (Queensland).

panceuthes Turner, 1923 (*Zelleria*), Trans. R. Soc. S. Austr. 47: 167. (ANIC: Holotype ♂); described from 1 specimen.
Type locality: Australia – Queensland, National Park.
Distribution: Australasian – Australia (Queensland).

pistopis Meyrick, 1931c (*Zelleria*), Ann. Mus. Nac. Hist. Nat. Buenos Aires 36: 398. (BMNH: Lectotype ♀ [designated by Clarke, 1965: 412]); described from 6 specimens (BMNH).

Type locality: Argentina – Rio Negro, Lake Nahuel Huapi.
Distribution: Neotropical – Argentina (Rio Negro).
Note: Clarke (1965: 412) selected one female from the type series and designated it as the "Type," meaning Lectotype, although he did not use the word lectotype.

plumbeella Staudinger, 1870 [1871] (*Zelleria*), Berlin. Ent. Ztschr. 14(3/4): 320. (ZMB: Lectotype ♀ [designated by Friese, 1960: 91]); not given (3♀ in ZMB).
Type locality: Spain – Andalusia, Málaga.
Distribution: Palearctic – Spain.
Note: Staudinger did not specify the type status of the syntypes of *Zelleria plumbeella*. Friese (1960) reported that the type series is composed of 3 specimens, one "hololectotype" and two "paratypes," without any intention of designating a lectotype. We regard Friese's statement as virtually equal to the publication of the lectotypy.

porphyraula Meyrick, 1927b (*Zelleria*), Trans. Proc. N. Z. Inst. 58(3): 314. (BMNH: Holotype ♂); described from 1 specimen.
Type locality: New Zealand – Wellington.
Distribution: Australasian – New Zealand.
Host: Oleaceae – *Nestegis* spp. (Gershenson & Ulenberg, 1998).

proterospila Meyrick, 1892 [1893] (*Zelleria*), Proc. Linn. Soc. N.S. Wales for the year 1892. Second Series. 7: 584. (BMNH: Lectotype ♂ [designated here]); described from 11 specimens (BMNH).
Type locality: Australia – West Australia, Geraldton.
Distribution: Australasian – Australia (West Australia).
Note: One specimen in the type series of *proterospila* Meyrick has a lectotype label. This lectotype designation seems to have been made by I. F. B. Common, but it has never been published. We here make this designation valid. The pin data labels read: "Lecto- | type" [circular label with indigo margin], "Geraldton | W. Australia | 5/11/86 [hand-written]," "*proterospila* Meyr. [hand-written on one line]," "Meyrick Coll. | B.M. 1938-290," "*Zelleria* | *proterospila* | 1/11 Meyr. | E. Meyrick det. | in Meyrick Coll.,""LECTOTYPE | *Zelleria* | *proterospila* M. | teste J. Sohn, 2014."

pyri Clarke, 1942 (*Zelleria*), Proc. USNM 92(3149): 275. (USNM: Holotype ♂); described from 5♂, 8♀ (USNM).
Type locality: USA – Washington, Pierce Co., Puyallup.
Distribution: Nearctic – Canada (British Columbia), USA (Washington).
Host: Rosaceae – *Pyrus* sp. (Clarke, 1942).

pyroleuca Meyrick, 1892 [1893] (*Zelleria*), Proc. Linn. Soc. N.S. Wales for the year 1892 . Second Series. 7: 585. (BMNH: Holotype ♂); described from 1 specimen.
Type locality: Australia – New South Wales, Bathurst.
Distribution: Australasian – Australia (New South Wales).

retiniella Forbes, 1923 (*Zelleria*), Lepid. N. Y. & Neigh. States 1: 348. (USNM: Syntypes); not given (2♂, 1♀ in USNM).
Type locality: USA – New Jersey, Ocean Co., Lakehurst.
Distribution: Nearctic – USA (Florida, New Jersey).
Note: Forbes cited Kearfott as author, validating a Kearfott manuscript name, making the name nomenclaturally available. One syntype in the type series of *retiniella* Forbes bears a holotype label, reported by Sohn & Lewis (2012). However, Forbes (1923) did not designate a holotype, making the 3 specimens in the USNM syntypes.

rorida Philpott, 1918 (*Zelleria*), Trans. Proc. N. Z. Inst. 50: 130. (MNZ: Syntypes); described from 1♂, 1♀ (MNZ).
Type localities: New Zealand: Southland; Bluff and Invercargill.
Distribution: Australasian – New Zealand (Southland).
Hosts: Loranthaceae – *Lorantha* spp. (Ulenberg, 2009).
Note: Although Philpott (1918) reported both a male and a female in the type series, the original description was based only on the male. We believe Philpott (1918) considered the male specimen as the holotype but failed to designate it. This specimen has been labeled as holotype in the MNZ collection, but is technically a syntype.

semitincta Meyrick, 1930b (*Zelleria*), Exot. Micr. 3(18): 562. (BMNH: Lectotype ♂ [designated by Clarke, 1965: 412]); described from 2 specimens (BMNH).
Type locality: USA – Texas, Jeff Davis Co., Fort Davis.
Distribution: Nearctic – USA (Texas).

sigillata Meyrick, 1892 [1893] (*Zelleria*), Proc. Linn. Soc. N.S. Wales for the year 1892. Second Series. 7: 587. (BMNH: Lectotype ♂ [designated here]); described from 6 specimens (BMNH).
Type locality: Australia – New South Wales, Sydney.
Distribution: Australasian – Australia (New South Wales).
Note: One specimen in the type series of *sigillata* Meyrick has a lectotype label. This lectotype designation seems to have been made by I. F. B. Common, but it has never been published. We here make this designation valid. The pin data

labels read: "Lecto- | type" [circular label with indigo margin], "Sydney | N. S. Wales | 4/1/78" [hand-written], "*sigillata* Meyr." [hand-written on one line], "Meyrick Coll. | B.M. 1938-290," "*Zelleria* | *sigillata* | 1/5 Meyr. | E. Meyrick det. | in Meyrick Coll.," "LECTOTYPE | *Zelleria sigillata* | Meyrick | teste J. Sohn, 2014."

silvicolella Moriuti, 1977a (*Zelleria*), Fauna Jap. Ypon. s. l.: 222. (UOP: Holotype ♀); described from 3♀ (UOP).
Type locality: Japan – Honshu, Iwate Pref., Gando.
Distribution: Palearctic – Japan (Honshu), Russia (Primorsky).
Host: Oleaceae – *Syringa reticulata* (Blume) H. Hara (Moriuti, 1977a).

sphenota Meyrick, 1889 (*Hofmannia*), Trans. N. Z. Inst. 21: 162. (BMNH: Holotype ♂); described from 1 specimen.
Type locality: New Zealand – Christchurch, Mid Canterbury.
Distribution: Australasian – New Zealand.
Hosts: Loranthaceae – *Ileostylus micranthus* Tiegh., *Loranthus* sp. (Sullivan et al. 1995; Ulenberg, 2009).

strophaea Meyrick, 1914f (*Zelleria*), Jour. Bomb. Nat. Hist. Soc. 23(1): 124. (BMNH: Lectotype ♂ [designated by Clarke, 1965: 412]); described from 11 specimens (BMNH).
Type locality: Sri Lanka – Central Prov., Maskeliya.
Distribution: Oriental – China (Yunnan), Sri Lanka.

stylograpta Meyrick, 1907d (*Zelleria*), Proc. Linn. Soc. N.S. Wales 32(1): 71. (BMNH: Holotype ♀); described from 1 specimen.
Type locality: Australia – Victoria, Mt. Macedon.
Distribution: Australasian – Australia (Victoria).

wolffi Klimesch, 1983 (*Zelleria*), Nota lepid. 6(2–3): 121. (NHMW: Holotype ♂); described from 20♂, 9♀ (Klimesch collection [private], NHMW, ZMUC).
Type locality: Canary Islands – Tenerife, Barranco de Ruiz.
Distribution: Palearctic – Canary Islands, Madeira archipelago.
Hosts: Celastraceae – *Catha dryandri* Lowe, *Maytenus umbellata* (R. Br.) Mabb. (Gershenson & Ulenberg, 1998; Aguiar & Karsholt, 2006).

Genera *incertae sedis*

This section includes the genera whose placement in the yponomeutoid families treated in this catalogue is uncertain.

ANGOONOPTERYX Moriuti, 1983. Bull. Univ. Osaka Pref. (B) 35: 11.
Type species: *Angoonopteryx siamensis* Moriuti, 1983, by monotypy.

siamensis Moriuti, 1983 (*Angoonopteryx*), Bull. Univ. Osaka Pref. (B) 35: 14. (UOP: Holotype ♂); described from 6♂, 6♀ (UOP).
Type locality: Thailand – Nakorn Nayok, Kao Yai.
Distribution: Oriental – Thailand, Vietnam.
Note: Moriuti (1983) reported the possible affinity of *Angoonopteryx* with *Argyresthia*.

BETHARGA Walker, 1866a, List Lepid. Ins. BM 34, Supp. 4: 1154.
Type species: *Betharga lycoides* Walker, 1866a, by monotypy.

lycoides Walker, 1866a (*Betharga*), List Lepid. Ins. BM 34, Supp. 4: 1154. (OUM: Holotype ♀); described from 1 specimen.
Type locality: New Guinea.
Distribution: Australasian – New Guinea.
Note: We think the yponomeutoid association of *Betharga* is unlikely considering the figure of *B. lycioides* examined from the BMNH, but retain it as Yponomeutidae incertae sedis until a specimen is available for examination.

DIAPHRAGMISTIS Meyrick, 1914f, Jour. Bomb. Nat. Hist. Soc. 23(1): 126.
Type species: *Diaphragmistis macroglena* Meyrick, 1914f, by monotypy.
Note: Meyrick (1914f) stated that *Diaphragmistis* has ocelli on the head. However, when JCS re-examined the type specimen, he found no ocelli present.

macroglena Meyrick, 1914f (*Diaphragmistis*), Jour. Bomb. Nat. Hist. Soc. 23(1): 126. (BMNH: Holotype ♂); described from 1 specimen.
Type locality: India – Khasis.
Distribution: Oriental – India (Khasis).

trapezia Wang, 2013 (*Diaphragmistis*), In: Jin, Q., Fan, X. & Wang, S. Ent. News 123(3): 238. (NUIC: Holotype ♂); not given.
Type locality: China – Hainan Province, Nanchahe, Bawangling [Natural Reserve].
Distribution: Palearctic – China (Hainan).

 | DOI 10.1163/9789004264267_004

ventiprocessa Wang, 2013 (*Diaphragmistis*), Ent. News 123(3): 238. (NUIC: Holotype ♂); not given.
Type locality: China – Yunnan Province, Kunming.
Distribution: Palearctic – China (Yunnan).

DISTAGMOS Herrich-Schäffer, [1853], Syst. Bearb. Schmett. Eur. 5: 8 (key) [see Herrich-Schäffer, 1853–1855].
Type species: *Distagmos ledereri* Herrich-Schäffer, [1853], by monotypy.

ARTENACIA Chrétien, 1905 (*Artenacia*), Naturaliste 2nd Series 19: 30.
Type species: *Artenacia jaurella* Chrétien, 1905, by monotypy.

ledereri Herrich-Schäffer, [1853] (*Distagmos*), Syst. Bearb. Schmett. Eur. 5: 90 [see Herrich-Schäffer, 1853–1855]. (ZMB: Holotype ♀); described from one specimen.
Type locality: Spain – Málaga Prov., Ronda.
Distribution: Palearctic – France, Portugal, Spain.
Host: Rhamnaceae – *Rhamnus saxatilis* Jacq. (Chrétien, 1905).
Notes: Herrich-Schäffer ([1853]) mentioned "the type specimen was from *Quercus ilex*." This statement very likely was based on an adult association, since no rearing record is cited from the label data of the type specimen.
Friese (1960) moved *Distagmos* from Yponomeutidae to Plutellidae. Later, Karsholt & Razowski (1996) tentatively placed the genus under Praydidae without explanation.

jaurella Chrétien, 1905 (*Artenacia*), Naturaliste 2nd Series 19: 30. (MNHN: Syntypes); not given [both sexes mentioned in the description].
Type locality: France – Hérault, Saint-Pons-de-Thomières.

EUARNE Möschler, 1890, Abh. Senck. Nat. Ges. 16: 340.
Type species: *Euarne obligatella* Möschler, 1890, by original designation.

obligatella Möschler, 1890 (*Euarne*), Abh. Senck. Nat. Ges. 16: 340. (Type not found); described from 1 specimen.
Type locality: Puerto Rico.
Distribution: Neotropical – Puerto Rico.

LISSOCHROA Turner, 1923, Trans. R. Soc. S. Austr. 47: 170.
Type species: *Lissochroa argostola* Turner, 1923, by monotypy.
Note: Turner (1923) stated that *Lissochroa* is intermediate between *Chionogenes* Meyrick, 1913a and *Sphenograptis* Meyrick, 1913a. *Chionogenes* is excluded from

Yponomeutidae due to the presence of ocelli on the head. This characteristic disputes its association with Yponomeutidae as defined by Kyrki (1990) and redefined by Sohn et al. (2013), and *Sphenograptis* currently is assigned to the Roeslerstammiidae. Since there are no statements regarding ocelli for *Lissochroa*, we tentatively retain it in the Yponomeutidae until specimens become available for examination.

argostola Turner, 1923 (*Lissochroa*), Trans. R. Soc. S. Austr. 47: 171. (ANIC: Syntypes); described from 2 specimens (ANIC).
Type localities: Australia – New South Wales, Lismore; Queensland, Brisbane.
Distribution: Australasian – Australia (New South Wales, Queensland).

LITANEUTIS Meyrick, 1913a, Exot. Micr. 1(5): 143.
Type species: *Litaneutis sacrifica* Meyrick, 1913a, by monotypy.
Note: There is no statement about the presence of ocelli on the head in the original description of *Litaneutis*. We could not find any characters in the description that would justify its exclusion or inclusion with the families treated here and thus left it incertae sedis. Gershenson & Ulenberg (1998) stated that the genus cannot be placed within Yponomeutinae due to the presence of long, upcurved labial palpi and the scale tuft on the hind tibia, two characters that are not very useful to define taxonomic position in Yponomeutoidea.

sacrifica Meyrick, 1913a (*Litaneutis*), Exot. Micr. 1(5): 143. (BMNH: Lectotype ♂ [designated by Clarke, 1965: 343]); described from 6 specimens, only one of which is still available in the BMNH collection.
Type locality: India – Assam, Khasi Hills.
Distribution: Australasian – Australia (Queensland), Samoa (introduced?); Oriental – India (Assam).

LYCOPHANTIS Meyrick, 1914f, Jour. Bomb. Nat. Hist. Soc. 23(1): 123.
Type species: *Lycophantis chalcoleuca* Meyrick, 1914f, by monotypy.
Note: Meyrick (1914f) stated that *Lycophantis* possesses ocelli on the head. This statement must be a mistake. We examined several species of *Lycophantis* and found no trace of ocelli. Moriuti (1977a) also did not indicate any ocelli in his figure or descriptions. Meyrick (1914f) thought that *Lycophantis* was allied to *Argyresthia* but this argument was simply based on superficial similarity. There is no character state that would directly connect *Lycophantis* with *Argyresthia*.

bradleyi Moriuti, 1963b (*Lycophantis*), Kontyû 31(4): 262. (UOP: Holotype ♂); described from 1 specimen.
Type locality: Japan – Amami-Osima Island, Santarotoge Pass.
Distribution: Oriental – Japan (Amami archipelago).

chalcoleuca Meyrick, 1914f (*Lycophantis*), Jour. Bomb. Nat. Hist. Soc. 23(1): 123. (BMNH: Lectotype ♂ [designated by Clarke, 1965: 344]); described from 12 specimens (BMNH).
Type locality: India – Assam, Khasi Hills.
Distribution: Oriental – India (Assam), North Vietnam, Taiwan.
Host: Iteaceae – *Itea chinensis subserrata* Maxim. (Moriuti, 1963b).

> *chalcoleura*; Sohn & Park, 2010 (*Lycophantis*), Oriental Insects 44: 232 (in figure legend). [misspelling of *chalcoleuca*].

elongata Moriuti, 1963b (*Lycophantis*), Kontyû 31(4): 262. (USNM: Holotype ♀); described from 2♀ (USNM).
Type locality: Taiwan – Marrepa.
Distribution: Oriental – Taiwan.

leviaurata Sohn & Park, 2010 (*Lycophantis*), Oriental Insects 44: 227. (MCGF: Holotype ♀); described from 3♀ (MCGF).
Type locality: North Vietnam – Vinh Phu, Tam Dao National Park.
Distribution: Oriental – North Vietnam.

moriutii Sohn & Park, 2010 (*Lycophantis*), Oriental Insects 44: 229. (MCGF: Holotype ♂); described from 3♂, 3♀ (MCGF).
Type locality: North Vietnam – Vinh Phu, Tam Dao National Park.
Distribution: Oriental – North Vietnam.

vietnamica Sohn & Park, 2010 (*Lycophantis*), Oriental Insects 44: 230. (MCGF: Holotype ♀); described from 1 specimen.
Type locality: North Vietnam – Vinh Phu, Tam Dao National Park.
Distribution: Oriental – North Vietnam.

ORTHOSARIS Meyrick, 1914f, Jour. Bomb. Nat. Hist. Soc. 23(1): 126.
Type species: *Orthosaris strictulata* Meyrick, 1914f, by monotypy.

strictulata Meyrick, 1914f (*Orthosaris*), Jour. Bomb. Nat. Hist. Soc. 23(1): 126. (BMNH: Holotype ♂); described from 1 specimen.
Type locality: Sri Lanka – Central Prov., Matale.
Distribution: Oriental – Sri Lanka.

PALLEURA Turner, 1926b, Trans. R. Soc. S. Austr. 50: 146.
Type species: *Palleura nitida* Turner, 1926b, by monotypy.
Note: We have not had a chance to examine specimens of *Palleura*. The original descriptions of the genus and the type species contain no characters that could be used to test its yponomeutoid association and the taxonomic position within the superfamily. Because of this, we retain the genus as incertae sedis.

nitida Turner, 1926b (*Palleura*), Trans. R. Soc. S. Austr. 50: 146. (ANIC: Holotype ♀); described from 1 specimen.
Type locality: Australia – Queensland, Bunya Mts.
Distribution: Australasian – Australia (Queensland).

PLEXIPPICA Meyrick, 1912, Ann. S. Afr. Mus. 10: 67.
Type species: *Plexippica verberata* Meyrick, 1912, by monotypy.
Note: *Plexippica* was included in the Yponomeutidae by Fletcher, 1929 and was erroneously transferred to the Gelechiidae by Clarke (1955). According to Sattler (1973: 239), Clarke confused *Brachmia verberata* Meyrick, 1912, (Gelechiidae) with *Plexippica verberata* Meyrick, 1912 (Yponomeutidae). The original description of *Plexippica* lacked information useful to determine its family association, so we retain it as incertae sedis.

verberata Meyrick, 1912 (*Plexippica*), Ann. S. Afr. Mus. 10: 67. (SAMC: Holotype ♂); described from 1 specimen.
Type locality: South Africa – Bushmanland, Jackals Water.
Distribution: Ethiopian – South Africa (Bushmanland).

PSEUDORINYMPHA Clarke, 1971, Smith. Contr. Zool. 56: 170.
Type species: *Pseudorinympha laeta* Clarke, 1971, by original designation.
Note: Clarke (1971) compared *Pseudorinympha* to *Orinympha*, a synonym of *Atemelia*.

laeta Clarke, 1971 (*Pseudorinympha*), Smith. Contr. Zool. 56: 170. (USNM: Holotype ♀); described from 2♀ (USNM).
Type locality: French Polynesia – Rapa Island, Teumukopuke.
Distribution: Australasian – French Polynesia (Rapa Island).

SPANIOPHYLLA Turner, 1917, Proc. R. Soc. Qd. 29: 89.
Type species: *Spaniophylla epiclithra* Turner, 1917, by monotypy.
Note: Turner (1917) stated that *Spaniophylla* is placed "nearest *Prays* Hb." The original description contains no useful information to test this statement, so we retain it as incertae sedis.

epiclithra Turner, 1917 (*Spaniophylla*), Proc. R. Soc. Qd. 29: 89. (ANIC: Syntypes); described from 4 specimens (ANIC).
Type locality: Queensland – Kuranda, near Cairns.
Distribution: Australasian – Australia (Queensland).

SYMPETALISTIS Meyrick, 1935b, Exot. Micr. 4(18): 556.
Type species: *Sympetalistis petrograpta* Meyrick, 1935b by monotypy.
Note: The original description of *Sympetalistis* contains no useful information to define its family position. We place the genus here until further evidence can be obtained.

petrograpta Meyrick, 1935b (*Sympetalistis*), Exot. Micr. 4(18): 557. (BMNH: Holotype ♂ [not found]); described from 1 specimen.
Type locality: Taiwan – Takima.
Distribution: Oriental – Taiwan.

TERTHROPTERA Clarke, 1971, Smith. Contr. Zool. 56: 172.
Type species: *Terthroptera eremosesia* Clarke, 1971, by monotypy.
Note: Clarke (1971) thought that *Terthroptera* is overall close to *Pseudorinympha*, a genus that he thought allied to *Orinympha*.

astochia Clarke, 1986 (*Terthroptera*), Smith. Contr. Zool. 416: 342. (USNM: Holotype ♂); described from 1 specimen.
Type locality: French Polynesia – Marquesas Islands, Hiva Oa Island, Mt. Feani.
Distribution: Australasian – French Polynesia (Marquesas Islands).

eremosesia Clarke, 1971 (*Terthroptera*), Smith. Contr. Zool. 56: 172. (USNM: Holotype ♂); described from 1 specimen.
Type locality: French Polynesia – Rapa Island, Maugaoa.
Distribution: Australasian – French Polynesia (Rapa Island).

ZYGOGRAPHA Meyrick, 1917, Ann. S. Afr. Mus. 17: 12.
Type species: *Zygographa asaphochalca* Meyrick, 1917, by monotypy.
Note: The original description of *Zygographa* contains no useful information to determine its family position within Yponomeutoidea or other superfamilies. We have not examined the type species and thus retain this genus as incertae sedis.

asaphochalca Meyrick, 1917 (*Zygographa*), Ann. S. Afr. Mus. 17: 12. (SAMC: Holotype ♂); described from 1 specimen.
Type locality: South Africa – Cape Colony, Hottentots Holland Mts., Caledon.
Distribution: Ethiopian – South Africa (Cape Colony).

Species *incertae sedis*

oenothriella Diorio, 1865 (*Yponomeuta*), Atti Accad. Pont. Roma 18: 124. (Type not found); not given, but described from both sexes.
Type locality: Italy – Rome.
Distribution: Palearctic – Italy.
Note: Since Herrich-Schäffer (1866: 142) introduced Diorio's original records, the name *oenothriella* has never been used again. Its true identity remains uncertain. Diorio (1865) reported that it is a pest on grapes, which no known larvae of Yponomeutidae feed on.

pruni Stainton, 1858 (Swammerdamia), Ent. Weekly Intell. 4(93): 118. (Type not found); described from one specimen.
Type locality: England, Exeter [=Devon].
Distribution: Palearctic – England.
Host: Rosaceae – *Prunus* sp. [plum] (Stainton, 1859).
Note: This species appeared in two publications by Stainton (1858, and 1859). Since then, it has never been cited again. Stainton (1870) himself did not cite this species when reviewing the species of *Swammerdamia*. Stainton (1858) mentioned that this species closely resembles *S. griseocapitella* (=*caesiella* Hübner, [1796]). It's true identity remains uncertain.

 | DOI 10.1163/9789004264267_005

Nomina nuda

Argyresthia glaberrimella Zeller [from label of *A. illuminatella*]. See Bengtsson & Johansson (2011b) for invalidation.

Argyresthia visaliella Chambers, 1875b, Can. Ent. 7(8): 145. See Covell (1999) for invalidation.

Yponomeuta hebeiensis Liu, 1984, unpublished name cited in Hua, 2005, without description. List Chin. Ins. 3: 11. Declared nomen nudum Sohn et al. (2010: 2815).

Yponomeuta nigrofasciatus Yang, 1979; in Hua, 2005, List Chin. Ins. 3: 11. Declared nomen nudum Sohn et al. (2010: 2815).

Yponomeuta shansiella Caradja [from label]. See Gershenson & Ulenberg (1998: 125) for invalidation.

Yponomeuta ulingensis Yang, 1979; in Hua, 2005, List Chin. Ins. 3: 11. Declared nomen nudum Sohn et al. (2010: 2815).

 | DOI 10.1163/9789004264267_006

Genera and Their Component Species Removed from the Families Treated Here

Note: Descriptions of most component species listed here are not cited in the references of this catalogue. Only those which are in publications for taxa included in this catalogue are given. Species in these genera may be detailed in later editions of the catalogue series; we list them here to illustrate their current placement within the classification. Defining the affinity of removed genera as defined by Kyrki (1990) and redefined by Sohn et al. (2013) with other yponomeutoid or non-yponomeutoid families will need further scrutiny.

ABACISTIS Meyrick, 1913b, Ann. Transv. Mus. 3: 318.
Type species: *Abacistis hexanoma* Meyrick, 1913b, by monotypy.
Included species: *hexanoma* Meyrick, 1913b (*Abacistis*); *teligera* Meyrick, 1914b (*Abacistis*).
Current status: This genus is excluded from Yponomeutoidea here. JCS examined one non-type specimen of *Abacistis teligera* determined by Meyrick and found that it has a scaled proboscis and exposed genitalia without pleural lobes, two characters disputing its association with Yponomeutoidea. In addition, Meyrick (1913b) mentioned that this genus is very close to *Aetherastis* and *Comocritis*, which are no longer yponomeutoids. We make no alternative family assignment for the genus until further evidence becomes available.

ACMOSARA Meyrick, 1887a, Proc. Linn. Soc. N.S. Wales. Second Series 1: 1043.
Type species: *Acmosara polyxena* Meyrick, 1887a, by monotypy.
Included species: monotypic.
Current status: "*Heliocosma*" group (unsettled non-yponomeutoid lineage) (Horak & Common, 1996).

ACRATAULA Meyrick, 1921a, Ann. Transv. Mus. 8: 118.
Type species: *Acrataula catapachna* Meyrick, 1921a, by monotypy.
Included species: monotypic.
Current status: The original description of *Acrataula* stated that the genus or type species has ocelli on the head. This character disputes its association with Yponomeutidae, as defined by Kyrki (1990) and redefined by Sohn et al. (2012). Therefore, we exclude *Acrataula* from the yponomeutoid families included here.

 | DOI 10.1163/9789004264267_007

AEMYLURGIS Meyrick, 1936a, Arb. Morph. Taxon. Ent. Berl. 3(2): 107.
Type species: *Aemylurgis xanthoclina* Meyrick, 1936a, by monotypy.
Included species: monotypic.
Current status: The original description of *Aemylurgis* stated that the genus or type species has ocelli on the head. This character disputes its association with Yponomeutidae, as defined by Kyrki (1990) and redefined by Sohn et al. (2012). Therefore, we exclude *Aemylurgis* from the yponomeutoid families included here.

AGRIOCEROS Meyrick, 1928c, Exot. Micr. 3(14): 418.
Type species: *Agrioceros platycypha* Meyrick, 1928c, by monotypy.
Synonym: *Chrysethmia* Diakonoff, 1966, Tijdschr. v. Ent. 109: 83 (Type species: *Chrysethmia hypomelas* Diakonoff, 1966, by original designation).
Included species: *hypomelas* Diakonoff, 1966 (*Chrysethmia*); *magnificella* Sauber, 1902 (*Psecadia*); *neogena* Diakonoff, 1966 (*Chrysethmia*); *platycypha* Meyrick, 1928c (*Agrioceros*); *subnota* Diakonoff, 1966 (*Agrioceros*); *zelaea* Meyrick, 1906 (*Ethmia*).
Current status: Elachistidae, Ethmiinae (Nye & Fletcher (eds.), 1991).

AICTIS Turner, 1926b, Trans. R. Soc. S. Austr. 50: 145.
Type species: *Aictis erythrozona* Turner, 1926b, by monotypy.
Included species: monotypic.
Current status: Lacturidae (Heppner, 1995).

AMALTHINA Meyrick, 1914b, Ann. Transv. Mus. 4(4): 200.
Type species: *Amalthina lacteata* Meyrick, 1914b, by monotypy.
Included species: monotypic.
Current status: The original description of *Amalthina* stated that the genus or type species has ocelli on the head. This character disputes its association with Yponomeutidae, as defined by Kyrki (1990) and redefined by Sohn et al. (2012). Therefore, we exclude *Amalthina* from the yponomeutoid families included here. Meyrick (1914b) allied it with *Ethmia*.

AMIANTASTIS Meyrick, 1932c, Exot. Micr. 4(11): 347.
Type species: *Amiantastis manicola* Meyrick, 1932c, by original designation.
Synonym: *Autocnaptis* Meyrick, 1935b, Exot. Micr. 4(18): 576 (Type species: *Autocnaptis sciospora* Meyrick, 1935c).
Included species: *brachycasis* Meyrick, 1932c (*Amiantastis*); *manicola* Meyrick, 1932c (*Amiantastis*); *sciospora* Meyrick, 1935c (*Autocnaptis*).

Current status: Pyschidae, in Becker (1984) and Davis & Robinson (1998), possibly in the subfamily Typhoniinae.

ANAPHANTIS Meyrick, 1907d, Proc. Linn. Soc. N.S. Wales 32(1): 90.
Type species: *Anaphantis isochrysa* Meyrick, 1907d, by monotypy.
Included species: *aurifraga* Diakonoff, 1948 (*Anaphantis*); *isochrysa* Meyrick, 1907d (*Anaphantis*); *protona* Meyrick, 1910b (*Anaphantis*); *zonotorna* Meyrick, 1925a (*Melodryas*).
Current status: Kyrki In: Nye and Fletcher (1991), disputed the association of *Anaphantis* with Yponomeutoidea but did not place it elsewhere.

ANCHIMACHETA Walsingham, 1914, Biol. Centr. Amer. 4: 323.
Type species: *Anchimacheta capnodes* Walsingham, 1914, by original designation.
Included species: *iodes* Walsingham, 1914 (*Anchimacheta*) = *capnodes* Walsingham, 1914 (*Anchimacheta*) = *tolmetes* Walsingham, 1914 (*Anchimacheta*).
Current status: Urodidae (Kyrki, 1988).

ANOISTA Turner, 1939, Pap. Proc. R. Soc. Tasm. 1938: 99.
Type species: *Anoista insolita* Turner, 1939, by monotypy.
Included species: monotypic.
Current status: The original description of *Anoista* stated that the genus or type species has ocelli on the head. This character disputes its association with Yponomeutidae, as defined by Kyrki (1990) and redefined by Sohn et al. (2012). Therefore, we exclude *Anoista* from the yponomeutoid families included here.

ANTICRATES Meyrick, 1905, Jour. Bomb. Nat. Hist. Soc. 16(4): 612.
Type species: *Anticrates chrysantha* Meyrick, 1905, by monotypy.
Synonym: *Pyrozela* Meyrick, 1906, Jour. Bomb. Nat. Hist. Soc. 17(2): 414 (Type species: *Pyrozela xanthomima* Meyrick, 1906).
Included species: *agrippina* Meyrick, 1930c (*Anticrates*); *anticlina* Meyrick, 1907a (*Pyrozela*); *argyroplintha* Meyrick, 1938c (*Anticrates*) = *chrysaema* Diakonoff, 1955 (*Anticrates argyroplintha* ssp.); *autobrocha* Meyrick, 1922a (*Anticrates*); *chrysantha* Meyrick, 1905 (*Anticrates*); *crocophaea* Meyrick, 1921a (*Anticrates*); *decaplaca* Meyrick, 1914f (*Anticrates*); *denticulata* Diakonoff, 1967 (*Anticrates*); *despotica* Meyrick, 1926c (*Anticrates*); *difflua* Meyrick, 1927c (*Anticrates*) = *hebridensis* Bradley, 1962 (*Anticrates difflua* ssp.) = *rennellensis* Bradley, 1957 (*Anticrates difflua* ssp.); *digitosa* Diakonoff, 1967 (*Anticrates*); *electropis* Meyrick, 1921a (*Anticrates*); *erythromima* Meyrick,

1906 (*Pyrozela*); *eulima* Meyrick, 1913a (*Anticrates*); *haematantha* Meyrick, 1924 (*Anticrates*); *hygraema* Meyrick, 1927a (*Anticrates*); *lucifera* Meyrick, 1914f (*Anticrates*); *mesopercna* Meyrick, 1926c (*Anticrates*); *miltochorda* Meyrick, 1914f (*Anticrates*); *paratarsa* Meyrick, 1913a (*Anticrates*); *paraxantha* Meyrick, 1907d (*Anticrates*); *phaedima* Turner, 1913 (*Anticrates*); *rhodometra* Diakonoff, 1967 (*Anticrates*); *rutilella* Pagenstecher, 1900 (*Enaemia*); *sanguinicornis* Walsingham, 1900 (*Mieza*?); *sulfurata* Meyrick, 1907d (*Anticrates*); *thermastris* Meyrick, 1914f (*Anticrates*); *tridelta* Meyrick, 1907a (*Pyrozela*); *venatrix* Meyrick, 1914f (*Anticrates*); *xanthomima* Meyrick, 1906 (*Pyrozela*); *zapyra* Meyrick, 1907d (*Anticrates*).
Current status: Lacturidae (Heppner, 1995).

ASTAROPOLA Meyrick, 1936a, Arb. Morph. Taxon. Ent. Berl. 3(2): 107.
Type species: *Astaropola magicosema* Meyrick, 1936a, by monotypy.
Included species: monotypic.
Current status: A synonym of *Archaeoses* Turner, 1932 (Cossidae) (Edwards, 1996).

ASTOXENA Meyrick, 1930c, Exot. Micr. 3(19): 593.
Type species: *Astoxena clotho* Meyrick, 1930c, by monotypy.
Included species: monotypic.
Current status: Oecophoridae, Oecophorinae (Becker, 1984).

BALANOPTICA Meyrick, 1930c, Exot. Micr. 3(19): 592.
Type species: *Cyme orbicularis* Felder & Rogenhofer, 1875, by monotypy.
Included species: monotypic.
Current status: Meyrick (1930c) compared this genus with *Melodryas,* which is currently placed in Tineidae. The original description of *Balanoptica* stated the genus or type species has ocelli on the head. This character disputes its association with Yponomeutidae, as defined by Kyrki (1990) and redefined by Sohn et al. (2012). Therefore, we exclude *Balanoptica* from the yponomeutoid families included here.

CALAMOTIS Meyrick, 1918a, Exot. Micr. 2(6): 188.
Type species: *Calamotis prophracta* Meyrick, 1918a, by monotypy.
Included species: monotypic.
Current status: The original description of *Calamotis* stated the genus or type species has ocelli on the head. This character disputes its association with Yponomeutidae, as defined by Kyrki (1990) and redefined by Sohn et al. (2012). Therefore, we exclude *Calamotis* from the yponomeutoid families included here.

CAMINOPHANTIS Meyrick, 1933, Exot. Micr. 4(12): 373.
Type species: *Caminophantis mystolitha* Meyrick, 1933, by monotypy.
Included species: monotypic.
Current status: The original description of *Caminophantis* stated the genus or type species has ocelli on the head. This character disputes its association with Yponomeutidae, as defined by Kyrki (1990) and redefined by Sohn et al. (2012). Therefore, we exclude *Caminophantis* from the yponomeutoid families included here.

CALLITHRINCA Meyrick, 1913a, Exot. Micr. 1(5): 140.
Type species: *Psecadia evocatella* Walker, 1863, by original designation.
Included species: *angoonae* Moriuti, 1982b (*Callithrinca*); *evocatella* Walker, 1863 (*Psecadia*); *niphopyrrha* Meyrick, 1927a (*Callithrinca*); *sphendonista* Meyrick, 1927c (*Callithrinca*).
Current status: Lacturidae (Heppner, 1995).

CAMPBELLANA Salmon & Bradley, 1956, Rec. Dominion Mus. 3(1): 68.
Type species: *Campbellana attenuata* Salmon & Bradley, 1956, by original designation.
Included species: monotypic.
Current status: Carposinidae (Dugdale, 1989).

CARMENTINA Meyrick, 1930c, Exot. Micr. 3(19): 598.
Type species: *Carmentina iridesma* Meyrick, 1930c, by monotypy.
Synonym: *Metapodistis* Meyrick, 1933, Exot. Micr. 4(12): 372 (Type species: *Metapodistis chrysosema* Meyrick, 1933, by monotypy).
Included species: *chrysosema* Meyrick, 1933 (*Metapodistis*); *iridesma* Meyrick, 1930c (*Carmentina*); *molybdotoma* Diakonoff & Arita, 1979 (*Metapodistis*); *perculta* Diakonoff, 1979 (*Metapodistis*); *polychrysa* Meyrick, 1934 (*Metapodistis*); *taiwanensis* Arita & Heppner, 1992 (*Carmentina*).
Current status: Glyphipterigidae, Glyphipteriginae (Heppner, 1982).

CHARICRITA Meyrick, 1913a, Exot. Micr. 1(5): 143.
Type species: *Charicrita citrozona* Meyrick, 1913a, by monotypy.
Synonym: *Charicrata*; Turner, 1923, Trans. Proc. R. Soc. S. Aust.47: 168 [misspelling of *Charicrita*].
Included species: *citrozona* Meyrick, 1913a (*Charicrita*); *othonina* Turner, 1927a (*Charicrita*); *sericoleuca* Turner, 1923 (*Charicrata*).
Current status: The original description of *Charicrita* stated that this genus or type species has ocelli on the head. This character disputes its association with

Yponomeutidae, as defined by Kyrki (1990) and redefined by Sohn et al. (2012). Therefore, we exclude *Charicrita* from the yponomeutoid families included here.

CHIONOGENES Meyrick, 1913a, Exot. Micr. 1(5): 144.
Type species: *Anticrates isanema* Meyrick, 1907d, by original designation.
Included species: *drosochlora* Meyrick, 1907d (*Anticrates*); *isanema* Meyrick, 1907d (*Anticrates*); *trimetra* Meyrick, 1913a (*Chionogenes*).
Current status: The original description of *Chionogenes* stated that this genus or type species has ocelli on the head. This character disputes its association with Yponomeutidae, as defined by Kyrki (1990) and redefined by Sohn et al. (2012). Therefore, we exclude *Chionogenes* from the yponomeutoid families included here.

CITRINARCHIS Meyrick, 1938b, Dt. Ent. Z. Iris 52(6): 88.
Type species: *Citrinarchis oxyphanta* Meyrick, 1938b, by monotypy.
Included species: monotypic.
Current status: An unknown, non-yponomeutoid genus (Kyrki In: Nye & Fletcher, 1991).

COMOCRITIS Meyrick, 1894, Trans. Ent. Soc. Lond.: 24.
Type species: *Comocritis olympia* Meyrick, 1894, by monotypy.
Synonym: *Aetherastis* Meyrick, 1909a, Jour. Bomb. Nat. Hist. Soc. 19: 422 (Type species: *Aetherastis uranias* Meyrick, 1909a, by monotypy).
Included species: *albicapilla* Moriuti, 1974a (*Comocritis*); *circulata* Meyrick, 1918a (*Aetherastis*); *constellata* Meyrick, 1914f (*Aetherastis*); *cyanobactra* Meyrick, 1922a (*Comocritis*); *enneora* Meyrick, 1914f (*Comocritis*); *heliconia* Meyrick, 1933 (*Comocritis*); *nephelista* Meyrick, 1914f (*Comocritis*); *olympia* Meyrick, 1894 (*Comocritis*); *pieria* Meyrick, 1906 (*Comocritis*); *pindarica* Meyrick, 1924 (*Comocritis*); *praecolor* Meyrick, 1914f (*Comocritis*); *thespias* Meyrick, 1909a (*Comocritis*); *uranias* Meyrick, 1909a (*Aetherastis*).
Current status: Xyloryctidae (Common, 1994).

CONCHIOPHORA Chrétien, 1915, Ann. Soc. Ent. Fr. 84: 349.
Type species: *Conchiophora spinosella* Chrétien, 1915, by monotypy.
Included species: monotypic.
Current status: Gelechioidea (Kyrki In: Nye & Fletcher (eds.), 1991). Chrétien (1915) originally compared *Conchiophora* with two other gelechioid genera: *Tetanocentria* Rebel, 1902 (Agonoxenidae) and *Ischnophanes* Meyrick, 1891 (Coleophoridae).

COPTOPROCTIS Zeller, 1852 [1854], K.V. Akad. Handl. 1852: 105.
Type species: *Coptoproctis languida* Zeller, 1852 [1854], by monotypy.
Included species: monotypic.
Current status: Zygaenidae (Kyrki In: Nye & Fletcher (eds.), 1991; Vári, et al. 2002).

CYMONYMPHA Meyrick, 1927c, Insects of Samoa 3 (2): 105.
Type species: *Cymonympha xantholeuca* Meyrick, 1927c, by monotypy.
Included species: monotypic.
Current status: The original description stated genus *Cymonympha* or the type species has ocelli on the head. This character disputes its association with Yponomeutidae, as defined by Kyrki (1990) and redefined by Sohn et al. (2012). Therefore, we exclude *Cymonympha* from the yponomeutoid families included here.

DASCIA Meyrick, 1892 [1893], Proc. Linn. Soc. N.S. Wales for the year 1892. Second Series. 7: 579.
Type species: *Dascia sagittifera* Meyrick, 1892 [1893], by monotypy.
Included species: *sagittifera* Meyrick, 1892 [1893] (*Dascia*) = *dasycerca* Turner, 1946 (*Eucosma*).
Current status: Lyonetiidae (Nielsen, 1996).

DITRIGONOPHORA Walsingham, 1897a, Proc. Zool. Soc. Lond. 1897: 117.
Type species: *Ditrigonophora marmoreipennis* Walsingham, 1897a, by original designation.
Included species: monotypic.
Current status: A synonym of *Dryadaula* Meyrick, 1892 [1893] (Tineidae) (Robinson & Nielsen, 1993).

DOCOSTOMA Diakonoff, 1955, Verh. K. ned. Akad. Wet. (2) 50(3): 46.
Type species: *Docostoma insignis* Diakonoff, 1955, by monotypy.
Included species: monotypic.
Current status: Coleophoridae, Blastobasinae (Nye & Fletcher (eds.), 1991).

EFTICHIA Clarke, 1986, Smith. Contr. Zool. 416: 339.
Type species: *Eftichia chrysoleuca* Clarke, 1986, by monotypy.
Included species: monotypic.
Current status: Clarke (1986) placed this genus into Yponomeutidae *sensu lato*, which includes a separate family, the Plutellidae. According to the original description, *Eftichia* has ocelli on the head. This character disputes its

association with Yponomeutidae, as defined by Kyrki (1990) and redefined by Sohn et al. (2012). Therefore, we exclude *Eftichia* from the yponomeutoid families included here.

ELLABELLA Busck, 1925, Proc. Ent. Soc. Wash. 27(3): 46.
Type species: *Ellabella editha* Busck, 1925, by original designation.
Synonyms: *Probolacma* Meyrick, 1927a, Exot. Micr. 3(12): 362 (Type species: *Probolacma melanoclista* Meyrick, 1927a, by monotypy); *Spilogenes* Meyrick, 1938a, In: Caradja & Meyrick, Dt. Ent. Z. Iris 52(1): 19 (Type species: *Spilogenes chalazombra* Meyrick, 1938a, by monotypy).
Included species: *bayensis* Heppner, 1984b (*Ellabella*); *chalazombra* Meyrick, 1938a (*Spilogenes*); *editha* Busck, 1925 (*Ellabella*); *johnstoni* Heppner, 1984b (*Ellabella*); *melanoclista* Meyrick, 1927a (*Probolacma*).
Current status: Copromorphidae (Heppner, 1984b).

EMBRYONOPSIS Eaton, 1875, Ent. M. Mag. 12: 61.
Type species: *Embryonopsis halticella* Eaton, 1875, by monotypy.
Included species: *halticella* Eaton, 1875 (*Embryonopsis*) = *heardensis* Brown, 1964 (*Pringleophaga*).
Current status: Plutellidae (Common, 1970; Dugdale et al. 1998).

ENTRICHIRIA Meyrick, 1921d, Zool. Meded. Leiden 6: 187.
Type species: *Entrichiria amphiphracta* Meyrick, 1921d, by monotypy.
Included species: monotypic.
Current status: The original description of *Entrichiria* stated that the genus or type species has ocelli on the head. This character disputes its association with Yponomeutidae, as defined by Kyrki (1990) and redefined by Sohn et al. (2012). Therefore, we exclude *Entrichiria* from the yponomeutoid families included here. Meyrick (1921d) mentioned that this genus is allied to *Prays*. However, Praydidae do not possess ocelli on the head.

EPACTOSARIS Meyrick, 1934, Exot. Micr. 4(15): 458.
Type species: *Anarsia longipalpella* Rebel, 1907 by monotypy.
Included species: monotypic.
Current status: The original description of *Epactosaris* stated that the genus or type species has ocelli on the head. This character disputes its association with Yponomeutidae, as defined by Kyrki (1990) and redefined by Sohn et al. (2012). Therefore, we exclude *Epactosaris* from the yponomeutoid families included here.

EPICHTHONODES Meyrick, 1938c, Trans. Roy. Ent. Soc. Lond. 87: 525.
Type species: *Epichthonodes caustopola* Meyrick, 1938c, by original designation.
Included species: monotypic.
Current status: An unknown non-yponomeutoid genus (Kyrki In: Nye & Fletcher (eds.), 1991).

EPOPSIA Turner, 1903, Proc. Linn. Soc. N.S. Wales. 28: 89.
Type species: *Epopsia metreta* Turner, 1903, by monotypy.
Included species: monotypic.
Current status: Meyrick (1914c: 32) questionably cited *Epopsia* as a synonym of *Anticrates* (1905) although *Epopsia* (1903) is a senior name. This tentative synonymy was followed by Nielsen & Edwards (1996). Considering the nomenclatural complication in the synonymy, we prefer treating *Epopsia* and *Anticrates* as two separate genera, following Heppner (1995).

EXANTHICA Meyrick, 1913b, Ann. Transv. Mus. 3: 320.
Type species: *Semioscopis trigonella* Felder & Rogenhofer, 1875, by original designation.
Synonym: *Exanthia*; Gershenson & Ulenberg, 1998, Kon. Ned. Akad. Weten. Verh. Afdel. Nat., Tweede Reeks 99: 41 [misspelling of *Exanthica*].
Included species: *atelacma* Meyrick, 1926a (*Exanthica*); *trigonella* Felder & Rogenhofer, 1875 (*Semiscopis*).
Current status: Limacodidae (Vári et al. 2002).

EXAULISTIS Meyrick, 1911c, Ann. Transv. Mus. 3(1): 77.
Type species: *Exaulistis trichogramma* Meyrick, 1911c, by monotypy.
Included species: monotypic.
Current status: Meyrick (1911c) allied this genus with *Mieza*, a junior synonym of *Lactura*. The original description of *Exaulistis* stated that the genus or type species has ocelli on the head. This character disputes its association with Yponomeutidae, as defined by Kyrki (1990) and redefined by Sohn et al. (2012). Therefore, we exclude *Exaulistis* from the yponomeutoid families included here.

GALACTICA Walsingham, 1911, Ent. M. Mag. 47: 14.
Type species: *Galactica caradjae* Walsingham, 1911, by original designation.
Included species: *caradjae* Walsingham, 1911 (*Galactica*); *bootella* Turati, 1926 (*Calantica*); *inornata* Walsingham, 1900 (*Mieza*); *variinotella* Chrétien, 1915 (*Psecadia*); *walsinghami* Caradja, 1920 (*Galactica*).
Current status: Galacticidae (Minet, 1986).

GYMNOGRAMMA Zeller, 1852 [1854], K.V. Akad. Handl. 1852: 104.
Type species: *Psecadia* (*Gymnogramma*) *rufiventris* Zeller, 1852, by monotypy.
Synonym: *Eremothyris* Walsingham, 1897b, Trans. Ent. Soc. Lond. 1897: 47 (Type species: *Eremothyris hollandi* Walsingham, 1897b, by original designation).
Included species: *atmocycla* Meyrick, 1918a (*Gymnogramma*); *candidella* Viette, 1962 (*Eremothyris*); *cyanea* Meyrick, 1912 (*Gymnogramma*); *eoxantha* Meyrick, 1921a (*Gymnogramma*); *flavivitella* Walsingham, 1881a (*Eustixis*); *griveaudi* Gibeaux, 1982 (*Eremothyris*); *hollandi* Walsingham, 1897b (*Eremothyris*); *hutchinsoni* Walsingham, 1891b (*Gymnogramma*); *iambiodella* Viette, 1958 (*Eremothyris*); *luctuosa* Gibeaux, 1982 (*Eremothyris*); *plagiula* Meyrick, 1923c (*Gymnogramma*); *privata* Meyrick, 1924 (*Gymnogramma*); *psyllodecta* Meyrick, 1924 (*Gymnogramma*); *pyrozancla* Meyrick, 1912 (*Gymnogramma*); *racemosa* Meyrick, 1918b (*Gymnogramma*); *ratovosoni* Gibeaux, 1982 (*Eremothyris*); *rhodoneura* Meyrick, 1909c (*Gymnogramma*); *rufiventris* Zeller, 1852 (*Psecadia* (*Gymnogramma*)) = *acclinatella* Walker, 1864c (*Lithosia*?); *sphaerobola* Meyrick, 1924 (*Gymnogramma*); *tabulatrix* Meyrick, 1930c (*Gymnogramma*); *toulgoeti* Gibeaux, 1982 (*Eremothyris*); *viettei* Gibeaux, 1982 (*Eremothyris*).
Current status: Lacturidae (Heppner, 1995).

HESPERARCHA Meyrick, 1918b, Ann. Transv. Mus. 6: 38.
Type species: *Hesperarcha pericentra* Meyrick, 1918b, by monotypy.
Included species: monotypic.
Current status: Zygaenidae (Kyrki In: Nye & Fletcher (eds.), 1991).

HIERODRYAS Meyrick, 1931a, Exot. Micr. 4(3): 88.
Type species: *Hierodryas eriochiras* Meyrick, 1931a, by monotypy.
Included species: monotypic.
Current status: The original description of *Hierodryas* stated the genus or type species has ocelli on the head. This character disputes its association with Yponomeutidae, as defined by Kyrki (1990) and redefined by Sohn et al. (2012). Therefore, we exclude *Hierodryas* from the yponomeutoid families included here.

HOMADAULA Lower, 1899, Proc. Linn. Soc. N.S. Wales 24 : 115.
Type species: *Homadaula lasiochroa* Lower, 1899, by monotypy.
Synonyms: *Paraprays* Rebel, 1910b, Dt. Ent. Z. Iris 24: 13 (Type species : *Paraprays punctigera* Rebel, 1910b, by monotypy); *Stichotactis* Meyrick, 1930b, Exot. Micr. 3(18): 562 (Type species: *Stichotactis calamitosa* Meyrick, 1930b, by monotypy).
Included species: *albida* Mey, 2004 (*Homadaula*); *anisocentra* Meyrick, 1922b (*Homadaula*) = *usuguronis* Matsumura, 1931 (*Hyponomeuta*) = *albizziae* Clarke,

1943 (*Homadaula*); *calamitosa* Meyrick, 1930b (*Stichotactis*); *coscinopa* Lower, 1990 (*Homadaula*); *dispertita* Meyrick, 1922a (*Homadaula*); *lasiochroa* Lower, 1899 (*Homadaula*); *martitima* Mey, 2007 (*Homadaula*); *montana* Mey, 2007 (*Homadaula*); *myriospila* Meyrick, 1907d (*Homadaula*); *poliodes* Meyrick, 1907d (*Homadaula*); *punctigera* Rebel, 1910b (*Paraprays*); *ravula* Mey, 2004 (*Homadaula*); *submontana* Mey, 2007 (*Homadaula*); *watamomaritima* Mey, 2007 (*Homadaula*).
Current status: Galacticidae (Common, 1990).

ILYCHYTIS Meyrick, 1937, Exot. Micr. 5(5): 143.
Type species: *Ilychytis anaemopa* Meyrick, 1937, by monotypy.
Included species: monotypic.
Current status: The original description of *Ilychytis* stated the genus or type species has ocelli on the head. This character disputes its association with Yponomeutidae, as defined by Kyrki (1990) and redefined by Sohn et al. (2012). Therefore, we exclude *Ilychytis* from the yponomeutoid families included here.

IRIANIA Diakonoff, 1955, Verh. K. ned. Akad. Wet. (2) 50(3): 39.
Type species: *Iriania mystica* Diakonoff, 1955, by original designation.
Included species: *anisoptera* Diakonoff, 1955 (*Iriania*); *auriflua* Diakonoff, 1955 (*Iriania*); *lutescens* Diakonoff, 1955 (*Iriania*); *minor* Diakonoff, 1955 (*Iriania*); *mystica* Diakonoff, 1955 (*Iriania*); *ochlodes* Diakonoff, 1955 (*Iriania*); *tricosma* Diakonoff, 1955 (*Iriania*).
Current status: This genus was tentatively associated with Gelechioidea (Kyrki In: Nye & Fletcher, 1991). The head illustrated by Diakonoff (1955) shows a scaled proboscis, which supports this association.

IRIDOSTOMA Meyrick, 1909a, Jour. Bomb. Nat. Hist. Soc. 19: 425.
Type species: *Iridostoma ichthyopa* Meyrick, 1909a, by monotypy.
Included species: *diana* Bradley, 1959 (*Iridostoma ichthyopa* ssp.); *ichthyopa* Meyrick, 1909a (*Iridostoma*).
Current status: *Iridostoma* was included in the Glyphipterygidae by Fletcher (1929) and transferred to the Yponomeutidae by Heppner (In: Heppner & Duckworth, 1981) without any explanation. The original description of *Iridostoma* indicated the genus or type species has ocelli on the head. This character disputes its association with Yponomeutidae, as defined by Kyrki (1990) and redefined by Sohn et al. (2012). Therefore, we exclude *Iridostoma* from the yponomeutoid families included here.

ISOTORNIS Meyrick, 1935a, in Caradja & Meyrick, Mat. Microlep.-Fauna Chin. Prov. Kiangsu, Chekiang, Hunan: 89.
Type species: *Isotornis nephelobathra* Meyrick, 1935a, by monotypy.
Included species: monotypic.
Current status: The original description of *Isotornis* indicated the genus or type species has ocelli on the head. This character disputes its association with Yponomeutidae, as defined by Kyrki (1990) and redefined by Sohn et al. (2012). Therefore, we exclude *Isotornis* from the yponomeutoid families included here.

ITHUTOMUS Butler, 1883, Trans. Ent. Soc. Lond. 1883: 84.
Type species: *Ithutomus formosus* Butler, 1883, by monotypy.
Synonym: *Ithytomus* Meyrick, 1914c, Lep. Cat. 19: 20 [An unjustified emendation of *Ithutomus*].
Included species: *formosus* Butler, 1883 (*Ithutomus*); *valdivianus* Beéche & Parra, 1998 (*Ithutomus*).
Current status: Putative Plutellidae (Davis, 1991).

LACTURA Walker, 1854, List Lepid. Ins. 2: 485.
Type species: *Lactura dives* Walker, 1854, by monotypy.
Synonyms: *Buxeta* Walker, 1866b, List Lepid. Ins. BM 35: 1982 [An unnecessary replacement name for *Sarbena*]; *Cyptasia* Walker, 1866b, List Lepid. Ins. BM 35: 1836 (Type species: *Cyptasia egregiella* Walker, 1866b, by monotypy); *Enaemia* Zeller, 1872, Verh. Zool.-Bot. Ges. Wien 22: 562 (Type species: *Enaemia psammitis* Zeller, 1872, by subsequent designation by Walsingham, 1914, Biol. Centr. Amer. (Zool.) Lepid.-Heter. 4: 325); *Epidictica* Turner, 1903, Proc. Linn. Soc. N.S. Wales 28: 81 (Type species: *Epidictica calliphylla* Turner, 1903, by original designation); *Eriopyrrha* Meyrick, 1913a, Exot. Micr. 1(5): 141 (Type species: *Mieza colabristis* Meyrick, 1907d, by original designation); *Eustixis* Hübner, 1825 [1831], Zuträge Samml. Exot. Schmett. 3: 24 (Type species: *Eustixis pupula* Hübner, 1825 [1831], by subsequent designation by Grote, 1874, Bull. Buff. Soc. N.S. 2 : 152) [An unused senior synonym, suppressed by Nye & Fletcher (1991)]; *Dianasa* Walker, 1854, List Lepid. Ins. BM 2: 488 (Type species: *Dianasa suffusa* Walker, 1854, by monotypy); *Hedycharis* Turner, 1903, Proc. Linn. Soc. N.S. Wales 28: 90 (Type species: *Hedycharis phoenobapta* Turner, 1903, by monotypy); *Mieza* Walker, 1854, List Lepid. Ins. BM 2: 527 (Type species: *Mieza igninix* Walker, 1854, by subsequent designation by Walsingham, 1914, Biol. Centr. Amer. (Zool.) Lepid.-Heter. 4: 325).
Physicampa Swainson, 1851, The Sydney Morning Herald 30 August 1851: 2 [Type species: *Physicampa sapotearum* Swainson, 1851, by monotypy]; *Pseudocaprima*

Walsingham, 1900, In: Swinhoe et al., Cat. East. Aust. Lepid. Het. Oxford. Univ. Mus. 2: 563 (Type species: *Pseudocaprima callopisma* Walsingham, 1900, by original designation); *Pseudotalara* Druce, [1885], Biol. Centr. Amer. (Zool.) Lepid. Heter. 1: 126 (Type species: *Pseudotalara chrysippa* Druce, 1885, by monotypy); *Sarbena* Walker, [1865], List Lepid. Ins. BM 31: 256 (Type species: *Sarbena conflagrans* Walker, [1865], by monotypy) [A junior homonym of *Sarbena* Walker, 1865 (Nolidae)]; *Themiscyra* Walker, 1864c [1865], List Lepid. Ins. BM 31 Supplement: 258 (Type species: *Themiscyra laetifera* Walker, 1864 [1865], by monotypy).

Included species: *acrantha* Diakonoff, 1955 (*Lactura*); *aglaodora* Turner, 1941 (*Lactura*); *albofimbriata* Walsingham, 1900 (*Pseudocaprima*); *anaemoptila* Diakonoff, 1955 (*Lactura*); *anthina* Durrant, 1916 (*Lactura*); *apoplectica* Meyrick, 1930c (*Lactura*); *atrolinea* Barnes, 1913 (*Mieza*); *aureocuprea* Walker, 1866b (*Lithosia*); *aurosa* Diakonoff, 1955 (*Lactura*); *autocosma* Meyrick, 1924 (*Lactura*); *basistriga* Barnes, 1913 (*Mieza*); *bisecta* Diakonoff, 1955 (*Lactura*); *britomartis* Meyrick, 1924 (*Lactura*); *calliphylla* Turner, 1903 (*Epidictica*) = *tenuilinea* Turner, 1903 (*Epidictica*); *callipyra* Meyrick, 1928b (*Lactura*); *callopisma* Walsingham, 1900 (*Pseudocaprima*); *caminaea* Meyrick, 1887a (*Enaemia*); *captatrix* Meyrick, 1924 (*Lactura*); *citrina* Busck, 1913 (*Mieza*); *coacervata* Meyrick, 1938c (*Lactura*); *coleoxantha* Meyrick, 1924 (*Lactura*); *conflagrans* Walker, 1864c (*Sarbena*); *coronopis* Meyrick, 1928b (*Lactura*); *dichroa* Durrant, 1915 (*Lactura*); *dives* Walker, 1854 (*Lactura*); *eclipticopa* Meyrick, 1928b (*Lactura*); *egregiella* Walker, 1866b (*Cyptasia*) = *erythocera* Felder & Rogenhöfer, 1875 (*Mieza*) = *erythrocera* [sic]; Turner, 1903 (*Mieza*?); *empedartha* Meyrick, 1924 (*Lactura*); *erythractis* Meyrick, 1887a (*Enaemia*); *erythrodendron* Diakonoff, 1955 (*Lactura*); *erythrodesma* Meyrick, 1930c (*Lactura*); *eurycrates* Meyrick, 1924 (*Lactura*); *eurylyca* Meyrick, 1928b (*Lactura*); *eurypercna* Meyrick, 1930c (*Lactura*); *euthoracica* Schaus, 1912 (*Lactura*); *floricoma* Meyrick, 1924 (*Lactura*); *haplochroa* Turner, 1932 (*Lactura*); *heliantha* Meyrick, 1918a (*Lactura*); *infervescens* Meyrick, 1930c (*Lactura*); *insecutrix* Meyrick, 1924 (*Lactura*); *irrorata* Busck, 1913 (*Mieza*); *laetifera* Walker, [1865] (*Themiscyra*) = *pyrochrysa* Lower, 1894 (*Enaemia*); *laeta* Hübner, [1826] (*Eustixis*) = *pupula* Hübner, [1826] (*Eustixis*) [preoccupied] = *igninix* Walker, 1854 (*Mieza*) = *crassivenella* Zeller, 1872 (*Enaemia*); *lateralis* Dyar, 1912 (*Lactura*); *leucophthalma* Meyrick, 1907d (*Mieza*) = *clitodes* Turner, 1932 (*Lactura*); *mactata* Felder & Rogenhofer, [1875] (*Mieza*) = *varicosa* Butler, 1876 (*Themiscyra*); *magocosma* Meyrick, 1930c (*Anticrates*); *marsyas* Meyrick, 1938c (*Lactura*); *ophioglossa* Meyrick, 1924 (*Lactura*); *ophiucha* Meyrick, 1924 (*Lactura*) = *xanthodes* Diakonoff, 1955 (*Lactura ophiucha* ssp.); *oroglypta* Meyrick, 1930c (*Lactura*); *panopsia* Turner, 1926b (*Lactura*); *parallela* Meyrick, 1889 (*Enaemia*) = *eupoecila* Turner, 1903 (*Lactura*); *persicopa* Meyrick, 1928b (*Lactura*); *phlogopa* Meyrick, 1886

(*Eustixis*); *phoenobapta* Turner, 1903 (*Hedycharis*); *picta* Felder & Rogenhofer, 1875 (*Mieza*) = *cristata* Butler, 1886 (*Cyptasia*) = *callianthes* Lower, 1894 (*Enaemia*) = *mixoleuca* Turner, 1900 (*Enaemia*); *pilcheri* Lucas, 1891 (*Calligenia*); *platyorma* Meyrick, 1924 (*Lactura*); *plectica* Diakonoff, 1955 (*Lactura*); *psammitis* Zeller, 1872 (*Enaemia*); *pseudophile* Dyar, 1912 (*Lactura*); *pteropoecila* Turner, 1913 (*Lactura*); *pyrilampis* Meyrick, 1887a (*Enaemia*); *pyronympha* Meyrick, 1924 (*Lactura*); *pyrothorax* Meyrick, 1928b (*Atteva*) **n.comb.**; *quadrifrenis* Meyrick, 1936a (*Lactura*); *rhodocentra* Meyrick, 1913a (*Mieza*); *rhodographa* Meyrick, 1928b (*Lactura*); *rhodomochla* Turner, 1942 (*Lactura*); *rubriflora* Meyrick, 1930c (*Lactura*); *rubritexta* Meyrick, 1913a (*Mieza*); *rutila* Bartlett-Calvert, 1890 (*Hyponomeuta*); *sanguiflua* Meyrick, 1924 (*Lactura*); *sapotearum* Swainson, 1851 (*Physicampa*) = *suffusa* Walker, 1854 (*Dianasa*) = *haematopus* Felder & Rogenhöfer, 1875 (*Hypoprepia*) = *obscura* Butler, 1877 (*Dianasa*); *schausia* Busck, 1920 (*Lactura*); *schenoxantha* Schaus, 1912 (*Lactura*); *spatula* Busck, 1913 (*Mieza*); *subfervens* Walker, 1854 (*Mieza*); *thiospila* Turner, 1903 (*Epidictica*); *vulnerosa* Diakonoff, 1955 (*Lactura*).
Current status: Lacturidae (Heppner, 1995).
Note: *Lactura pyrothorax* (Meyrick, 1928b) was originally described under *Atteva*. JCS examined the type series of this species in the BMNH, where one male and one female specimens were found. Mr. Kevin Tuck of the BMNH provisionally designated the male as lectotype and transferred this species to Lacturidae. His argument that this species is very closely related to *Lactura eurylyca* Meyrick, 1928b (personal comment to JCS). Agreeing with Mr. Tuck's provisional treatment, we formalize his lectotype designation and hereby transfer the species to *Lactura*. The lectotype labels read: "New Hannover/ Feb.-March, 1923/ (A. S. Meek.)," "pyrothorax Meyr. [hand written]," "Meyrick Coll./ B.M. 1988-290," "Atteva/ pyrothorax/ Meyr./ E. Meyrick det./ in Meyrick Coll.," "LECTOTYPE | *Atteva pyrothorax* | Meyrick | teste J. Sohn, 2014."

LOXOZYGA Meyrick, 1938a, in Caradja & Meyrick. Dt. Ent. Z. Iris 52(1): 19.
Type species: *Loxozyga eurynephes* Meyrick, 1938, by monotypy.
Included species: monotypic.
Current status: Oecophoridae (Clarke, 1955).

LYSIPHRAGMA Meyrick, 1887b, Trans. N. Z. Inst. 20: 104.
Type species: *Lysiphragma mixochlora* Meyrick, 1887b, by subsequent designation by Meyrick, 1915b, Trans. N. Z. Inst. 47: 239.
Included species: *adreptella* Walker, 1864a (*Gelechia*); *argentaria* Salmon, 1948 (*Lysiphragma*); *epixyla* Meyrick, 1887b (*Lysiphragma*); *howesii* Quail, 1901 (*Lysiphragma*); *mixochlora* Meyrick, 1887b (*Lysiphragma*).
Current status: Tineidae (Robinson, 2009).

LYTROPHILA Meyrick, 1913b, Ann. Transv. Mus. 3: 319.
Type speices: *Lytrophila panarga* Meyrick, 1913b, by original designation.
Included species: *humida* Meyrick, 1913b (*Lytrophila*); *panarga* Meyrick, 1913b (*Lytrophila*).
Current status: Psychidae (Gozmany & Vári, 1973).

MACROCIRCA Meyrick, 1931d, Mitt. Münch. Ent. Ges. 21: 38.
Type species: *Macrocirca strabo* Meyrick, 1931d, by monotypy.
Included species: monotypic.
Current status: Elachistidae, Ethmiinae (Nye & Fletcher (eds.), 1991).

MAPSIDIUS Walsingham, 1907c, Fauna Hawaiiensis 1(5): 650.
Type species: *Mapsidius auspicata* Walsingham, 1907c, by original designation.
Included species: *auspicata* Walsingham, 1907c (*Mapsidius*); *charpentierii* Swezey, 1932 (*Mapsidius*); *chenopodii* Swezey, 1947 (*Mapsidius*); *iridescens* Walsingham, 1907c (*Mapsidius*); *quadridentata* Walsingham, 1907c (*Mapsidius*).
Current status: Scythrididae (Zimmerman, 1978).

MAUGINIA Turati, 1924, Atti Soc. Ital. Sci. Nat. 63: 158.
Type species: *Mauginia kruegeri* Turati, 1924, by monotypy.
Current status: Tortricidae – A junior synonym of *Oxypteron* Staudinger, 1870 [1871] (Tuck In: Nye & Fletcher (eds.), 1991).

MELODRYAS Meyrick, 1910b, Trans. Ent. Soc. Lond. 1910: 472.
Type species: *Melodryas doris* Meyrick, 1910b, by monotypy.
Included species: monotypic.
Current status: Tineidae (Robinson, 2009).

METHARMOSTIS Meyrick, 1921b, Exot. Micr. 2(14): 439.
Type species: *Metharmostis asaphaula* Meyrick, 1921b, by monotypy.
Included species: *Metharmostis asaphaula* Meyrick, 1921b, *M. multilineata* Adamski, 2013.
Current status: Cosmopterigidae (Adamski et al. 2013).

MNEMOSES Durrant, 1922, Trans. Ent. Soc. Lond. 1921: 494.
Type species: *Mnemoses farquharsoni* Durrant, 1922, by original designation.
Included species: monotypic.
Current status: The exposed male genitalia of the type specimen of *Mnemoses farquharsoni* are not concealed by pleural lobes. The lack of pleural lobes disputes its association with Yponomeutoidea. Thus we exclude *Mnemoses* from the Yponomeutoidea.

MOROTRIPTA Meyrick, 1917, Ann. S. Afr. Mus. 17: 11.
Type species: *Morotripta fatagata* Meyrick, 1917, by monotypy.
Included species: monotypic.
Current status: According to the original description, the type species of genus *Morotripta* has ocelli on the head. This character disputes its association with Yponomeutidae, as defined by Kyrki (1990) and redefined by Sohn et al. (2012). Therefore, we exclude *Morotripta* from the yponomeutoid families included here.

MYCHONOA Meyrick, 1892 [1893], Proc. Linn. Soc. N.S. Wales for the year 1892. Second Series. 7: 559.
Type species: *Mychonoa mesozona* Meyrick, 1892 [1893], by monotypy.
Included species: monotypic.
Current status: Plutellidae (Edwards, 1996).

NEMATOBOLA Meyrick, 1892 [1893], Proc. Linn. Soc. N.S. Wales for the year 1892. Second Series. 7: 593.
Type species: *Nematobola candescens* Meyrick, 1892 [1893], by subsequent designation by Fletcher, 1929: Mem. Dep. Agric. India (Ent.) 11: 146.
Included species: monotypic.
Current status: Roeslerstammiidae (Nielsen, 1996).

NOSYMNA Walker, 1864a, List Lepid. Ins. BM 29: 831.
Type species: *Nosymna repletella* Walker, 1864a, by monotypy.
Synonym: *Androgyne* Walsingham, 1900, In: Swinhoe et al., Cat. East. Aust. Lepid. Het. Oxford. Univ. Mus. 2: 565 (Type species: *Androgyne punctata* Walsingham, 1900).
Included species: *lapillata* Meyrick, 1914f (*Nosymna*); *macrorrhyncha* Meyrick, 1930c (*Nosymna*); *obnubila* Durrant, 1916 (*Nosymna*); *ochrochorda* Meyrick, 1924 (*Nosymna*); *punctata* Walsingham, 1900 (*Androgyne*); *repletella* Walker, 1864a (*Nosymna*); *stipella* Snellen, 1903 (*Nosymna*).
Current status: Lacturidae (Heppner, 1995).

NYMPHONIA Meyrick, 1913a, Exot. Micr. 1(5): 136.
Type species: *Nymphonia zaleuca* Meyrick, 1913a, by monotypy.
Included species: monotypic.
Current status: The type species has ocelli on the head in the original description of *Nymphonia*. This character disputes its association with Yponomeutidae, as defined by Kyrki (1990) and redefined by Sohn et al. (2012). Therefore, we exclude *Nymphonia* from the yponomeutoid families included here.

ORIDRYAS Meyrick, 1938a, in Caradja & Meyrick, Dt. Ent. Z. Iris 52(1): 18.
Type species: *Oridryas isalopex* Meyrick, 1938a, by monotypy.
Included species: *angarensis* Caradja, 1939 (*Oridryas*); *isalopex* Meyrick, 1938a (*Oridryas*) = *mienshanensis* Caradja, 1939 (*Oridryas isalopex* var.).
Current status: The systematic position of *Oridryas* remains uncertain. The original description stated that the type species has ocelli on the head. This character disputes its association with Yponomeutidae, as defined by Kyrki (1990) and redefined by Sohn et al. (2012). Therefore, we exclude *Oridryas* from the yponomeutoid families included here.

ORSOCOMA Meyrick, 1921b, Exot. Micr. 2(14): 438.
Type species: *Orsocoma macrogona* Meyrick, 1921b, by monotypy.
Included species: monotypic.
Current status: The original description of *Orsocoma* stated the type species has ocelli on the head. This character disputes its association with Yponomeutidae, as defined by Kyrki (1990) and redefined by Sohn et al. (2012). Therefore, we exclude *Orsocoma* from the yponomeutoid families included here.

PALAETHETA Meyrick, 1909b, Ann. Transv. Mus. 2: 23.
Type species: *Palaetheta ischnozona* Meyrick, 1909b, by monotypy.
Included species: *innocua* Meyrick, 1912 (*Palaetheta*); *ischnozona* Meyrick, 1909b (*Palaetheta*) = *ischnozena* Janse, 1917.
Current status: Gelechioidea (Kyrki In: Nye & Fletcher, 1991).

PARAZELOTA Meyrick, 1913b, Ann. Transv. Mus. 3: 319.
Type species: *Parazelota dryotoma* Meyrick, 1913b, by monotypy.
Included species: monotypic.
Current status: The original description of the type species included ocelli on the head. This character disputes its association with Yponomeutidae, as defined by Kyrki (1990) and redefined by Sohn et al. (2012). Therefore, we exclude *Parazelota* from the yponomeutoid families included here.

PAREXAULA Meyrick, 1909c, Ann. S. Afr. Mus. 5: 356.
Type species: *Parexaula isomima* Meyrick, 1909c, by monotypy.
Included species: monotypic.
Current status: Gelechioidea (Kyrki In: Nye & Fletcher (eds.), 1991).

PAURIDIONEURA Turner, 1927b, Pap. Proc. R. Soc. Tasm. 1927: 29. An objective replacement name for *Pauroneura* Turner, 1927a.

Synonym: *Pauroneura* Turner, 1927a, Proc. R. Soc. Tas. 1926: 158 (Type species: *Pauroneura acrospila* Turner, 1927a, by monotypy) [A junior homonym of *Pauroneura* Turner, 1919, Gelechiidae].
Included species: monotypic.
Current status: A junior synonym of *Tritymba* Lower, 1894, which was assigned to Plutellidae (Edwards, 1996) and to Bucculatricidae or an unknown family nearby (Mutanen et al. 2010).

PERISCEPTIS Meyrick, 1931e, J. Linn. Soc. Lond. 37: 283.
Type species: *Perisceptis horiarcha* Meyrick, 1931e, by monotypy.
Included species: monotypic.
Current status: Psychidae (Becker, 1984).

PHASMATOGRAPHA Meyrick, 1928c, Exot. Micr. 3(14): 419.
Type species: *Phasmatographa neurotypa* Meyrick, 1928c, by monotypy.
Included species: monotypic.
Current status: Zygaenidae (Kyrki In: Nye & Fletcher (eds.), 1991).

PODIASA Busck, 1900b, Proc. USNM 23: 240.
Type species: *Podiasa chiococcella* Busck, 1900b, by monotypy.
Included species: monotypic.
Current status: Excluded from Yponomeutoidea here because the male genitalia of *Podiasa chiococcella* are not concealed by pleural lobes.
Note: The date of publication of the genus and species is 15 October 1900. There is a note on the bottom of the contents page of the 1901 volume which gives the date of publication of the descriptions.

PORPHYROCRATES Diakonoff, 1955, Verh. K. ned. Akad. Wet. (2) 50(3): 37.
Type species: *Porphyrocrates aurostricta* Diakonoff, 1955, by monotypy.
Included species: Monotypic.
Current Status: Diakonoff (1955) compared this genus with *Lamyristis* Meyrick, 1911b, which is now associated with Psychidae. The genitalia illustrated by him shows gelechioid-like characteristics.

PRONOMEUTA Meyrick, 1905, Jour. Bomb. Nat. Hist. Soc. 16: 608.
Type species: *Pronomeuta sarcopis* Meyrick, 1905, by monotypy.
Included species: *lemniscata* Meyrick, 1922c (*Pronomeuta*); *sarcopis* Meyrick, 1905 (*Pronomeuta*).
Current Status: This genus is currently associated with the Zygaenidae (Kyrki In: Nye & Fletcher (eds.), 1991).

PROTONOMA Meyrick, 1938a, In: Caradja & Meyrick, Dt. Ent. Z. Iris 52(1): 17.
Type species: *Protonoma glomeratrix* Meyrick, 1938a, by monotypy.
Included species: monotypic.
Current status: The original description of *Protonoma* stated that the type species has ocelli on the head. This character disputes its association with Yponomeutidae, as defined by Kyrki (1990) and redefined by Sohn et al. (2012). Therefore, we exclude *Protonoma* from the yponomeutoid families included here.

SAPTHA Walker, 1864b, List Lepid. Ins. BM 30: 1015.
Type species: *Saptha divitiosa* Walker, 1864b, by monotypy.
Synonyms: *Badera* Walker, 1866b, List Lepid. Ins. BM 35: 1819 (Type species: *Badera pretiosa* Walker, 1866b), by subsequent designation by Meyrick, 1914a, In: Wytsman, Genera Insect.: 164: 18; *Chordates* Snellen, 1877, Tijdschr. v. Ent. 20: 49 (Type species: *Simaethis pronubana* Snellen, 1877, by subsequent designation by Fletcher, 1929, Mem. Dep. Agric. India (Ent.) 11: 47); *Choredates* [sic]: Pagenstecher, 1884, Jahr. Nass. Ver. Natur. 37: 289; *Saphtha* [sic]: Walsingham & Durrant, 1900 In: Swinhoe et al., Cat. East. Aust. Lepid. Het. Oxford. Univ. Mus. 2: 567.
Included species: *aeolodoxa* Meyrick, 1928c (*Tortyra*); *angustistriata* Issiki, 1930 (*Tortyra*); *beryllitis* Meyrick, 1910b (*Tortyra*); *chrysoprasitis* Meyrick, 1936b (*Tortyra*); *cypridia* Meyrick, 1910b (*Tortyra*); *divitiosa* Walker, 1864b (*Saptha*) = *prodigella* Walker, 1866b (*Badera*) = *nobilis* Felder & Rogenhofer, 1875 (*Badera*) = *striana* Snellen, 1885 (*Choregia*); *elegans* Walsingham, 1900 (*Saphtha* [sic]); *exanthista* Meyrick, 1910b (*Tortyra*) = *prodigella* Meyrick, 1907d (*Tortyra*); *iridopa* Meyrick, 1907d (*Tortyra*); *libanota* Meyrick, 1910b (*Tortyra*); *macrospila* Diakonoff, [1968] (*Tortyra*); *paradelpha* Meyrick, 1907d (*Tortyra*); *prasochalca* Meyrick, 1907d (*Tortyra*); *pretiosa* Walker, 1866b (*Badera*) = *basalis* Snellen, 1885 (*Choregia*) = *formosana* Matsumura, 1931 (*Tortyra*); *pronubana* Snellen, 1877 (*Simaethis*); *smaragditis* Meyrick, 1905 (*Saptha*) = *centrifuga* Meyrick, 1912 (*Tortyra*); *sybaritis* Meyrick, 1912 (*Tortyra*); *tabulatia* Meyrick, 1912 (*Tortyra*).
Current status: Choreutidae (Heppner & Duckworth, 1981).

SPHENOGRAPTIS Meyrick, 1913a, Exot. Micr. 1(5): 145.
Type species: *Sphenograptis celetica* Meyrick, 1913a, by monotypy.
Included species: monotypic.
Current status: Roeslerstammiidae (Moriuti, 1978, Bull. Univ. Osaka Pref. (B) 30: 16). Note: Moriuti placed the genus in Amphitheridae, now considered a subfamily of Roeslerstammiidae.

SPILADARCHA Meyrick, 1913a, Exot. Micr. 1(5): 139.
Type species: *Spiladarcha derelicta* Meyrick, 1913a, by monotypy.
Included species: *adamskii* Sohn, 2012 (*Spiladarcha*); *derelicta* Meyrick, 1913a (*Spiladarcha*).
Current status: Urodidae (Kyrki, 1988).

SPORADARCHIS Meyrick, 1935c, Exot. Micr. 4(19): 601.
Type species: *Sporadarchis galactombra* Meyrick, 1935c, by monotypy.
Included species: monotypic.
Current status: Friese (1960) placed it under "Genera of doubtful position" and mentioned that diagnostic characters agree with *Conchiophora*, possibly representing a synonym. *Conchiophora* is excluded from Yponomeutoidea (Kyrki In: Nye & Fletcher (eds.), 1991).

STASIPHRON Meyrick, 1931b, Exot. Micr. 4(6): 171.
Type species: *Stasiphron cryptomorpha* Meyrick, 1931b, by monotypy.
Included species: monotypic.
Current status: Gelechioidea (Kyrki In: Nye & Fletcher (eds.), 1991).

STEGANOSTICHA Meyrick, 1921a, Ann. Transv. Mus. 8: 118.
Type species: *Steganosticha remigera* Meyrick, 1921a, by monotypy.
Included species: monotypic.
Current status: The original description of *Steganosticha* stated type species has ocelli on the head. This character disputes its association with Yponomeutidae, as defined by Kyrki (1990) and redefined by Sohn et al. (2012). Therefore, we exclude *Steganosticha* from the yponomeutoid families included here. Meyrick (1921) linked this genus with *Calamotis,* also excluded here from the Yponomeutoidea families treated.

STRYPHNAULA Meyrick, 1938a, In: Caradja & Meyrick, Dt. Ent. Z. Iris 52(1): 20.
Type species: *Stryphnaula capnanthes* Meyrick, 1938a, by monotypy.
Included species: monotypic.
Current status: In the original description, Meyrick (1938a) stated the type species has ocelli on the head. This character disputes its association with Yponomeutidae, as defined by Kyrki (1990) and redefined by Sohn et al. (2012). Therefore, we exclude *Stryphnaula* from the yponomeutoid families included here.

SYNCATHARTIS Meyrick, 1921d, Zool. Meded. Leiden 6: 188.
Type species: *Syncathartis argestis* Meyrick, 1921d, by monotypy.

Included species: monotypic.
Current status: Meyrick (1921d) thought, probably based on superficial similarity, that *Syncathartis* is allied to *Prays* Hübner. However, according to the original description, the type species has ocelli on the head. This character disputes its association with Yponomeutidae, as defined by Kyrki (1990) and redefined by Sohn et al. (2012). Therefore, we exclude *Syncathartis* from the yponomeutoid families included here.

SYNCERASTIS Meyrick, 1931c, Ann. Mus. Nac. Hist. Nat. Buenos Aires 36: 410.
Type species: *Syncerastis ptisanopa* Meyrick, 1931c, by monotypy.
Included species: monotypic.
Current status: This genus was originally established in the Tineidae but later transferred to Yponomeutidae by Heppner (1984a). According to the original description, the type species has ocelli on the head. This character disputes its association with Yponomeutidae, as defined by Kyrki (1990) and redefined by Sohn et al. (2012). Therefore, we exclude *Syncerastis* from the yponomeutoid families included here.

SYNCROTAULELLA Fletcher, 1940, Ent. Rec. J. Var. 52(10): 109. A replacement name for *Syncrotaula* Meyrick, 1937.
Synonyms: *Syncrotaula* Meyrick, 1937, Exot. Micr. 5(5): 142 (Type species: *Syncrotaula strepsicentra* Meyrick, 1937, by monotypy) [A junior homonym of *Syncrotaula* Meyrick, 1933, Pyralidae].
Included species: monotypic.
Current status: Gelechioidea (Kyrki, 1990).

TANAOCTENA Turner, 1913, Proc. Linn. Soc. N.S. Wales 38(1): 204.
Type species: *Tanaoctena ooptila* Turner, 1913, by monotypy.
Synonyms: *Cylicophora* Turner, 1927a, Pap. Roy. Soc. Tasmania 1926: 156 (Type species: *Cylicophora collina* Turner, 1927a, by monotypy); *Nesotropha* Turner, 1926a, Pap. Roy. Soc. Tasmania 1925: 110 (Type species: *Nesotropha pygmaeodes* Turner, 1926a, by monotypy); *Tanaoctenota* Meyrick, 1918a, Exot. Micr. 2(6): 188 [Unnecessary replacement name].
Included species: *dubia* Philpott, 1931 (*Tanaoctena*); *indubitata* Clarke, 1971 (*Tanaoctena*); *ooptila* Turner, 1913 (*Tanaoctena*); *pygmaeodes* Turner, 1926a (*Nesotropha*) = *collina* Turner, 1927a (*Cylicophora*).
Note: Galacticidae (Nielsen et al. 1996).

TARPHYSCELIS Meyrick, 1913a, Exot. Micr. 1(5): 144.
Type species: *Tarphyscelis palaeota* Meyrick, 1913a, by monotypy.

Included species: *cirrhozona* Meyrick, 1921d (*Tarphyscelis*); *palaeota* Meyrick, 1913a (*Tarphyscelis*).
Current status: According to the original description, the type species has ocelli on the head. This character disputes its association with Yponomeutidae, as defined by Kyrki (1990) and redefined by Sohn et al. (2012). Therefore, we exclude *Tarphyscelis* from the yponomeutoid families included here.

TELOSPHRANTIS Meyrick, 1932d, Trans. Ent. Soc. Lond. 80(1): 116.
Type species: *Telosphrantis aethiopica* Meyrick, 1932d, by monotypy.
Included species: monotypic.
Current status: Choreutidae (Heppner, 1985).

THYRIDECTIS Meyrick, 1887a, Proc. Linn. Soc. N.S. Wales. Second Series 1: 1045
Type species: *Thyridectis psephonoma* Meyrick, 1887a, by monotypy.
Included species: *psephonoma* Meyrick, 1887a (*Thyridectis*) = *hesychima* Turner, 1923 (*Atteva*).
Current status: Lacturidae (Heppner, 1995).

THYRSOTARSA Meyrick, 1921d, Zool. Meded. Leiden 6: 188.
Type species: *Thyrsotarsa platybyrsa* Meyrick, 1921d, by monotypy.
Included species: monotypic.
Current status: According to the original description, the type species has ocelli on the head. This character disputes its association with Yponomeutidae, as defined by Kyrki (1990) and redefined by Sohn et al. (2012). Therefore, we exclude *Thyrsotarsa* from the yponomeutoid families included here.

TINEARUPA Salmon & Bradley, 1956, Rec. Dominion Mus. 3: 66.
Type species: *Tinearupa sorenseni* Salmon & Bradley, 1956, by original designation.
Included species: *oleariae* Dugdale, 1971 (*Izatha*); *sorenseni* Salmon & Bradley, 1956 (*Tinearupa*) = *aucklandica* Dugdale, 1971 (*Tinearupa sorenseni* ssp.).
Current status: **Oecophoridae (Hoare, 2005).**

TOECORHYCHIA Butler, 1883, Trans. Ent. Soc. Lond. 1883: 74.
Type species: *Toecorhychia cinerea* Butler, 1883, by monotypy.
Included species: monotypic.
Current status: Unknown non-yponomeutoid genus (Kyrki In: Nye & Fletcher, 1991).

TOIANA Walker, 1866b, List Lepid. Ins. BM 35: 1732.
Type species: *Toiana venosella* Walker, 1866b, by monotypy.

Included species: monotypic.
Current status: Lacturidae (Heppner, 1955).

TOXOPEIA Diakonoff, 1955, Verh. K. ned. Akad. Wet. (2) 50(3): 48.
Type species: *Toxopeia demodes* Diakonoff, 1955, by monotypy.
Included species: monotypic.
Current status: Glyphipterigidae (Nye & Fletcher (eds.), 1991).

TROCHASTICA Meyrick, 1913b, Ann. Transv. Mus. 3: 321.
Type species: *Trochastica albifrenis* Meyrick, 1913b, by monotypy.
Included: monotypic.
Current status: According to the original description, the type species has ocelli on the head. This character disputes its association with Yponomeutidae, as defined by Kyrki (1990) and redefined by Sohn et al. (2012). Therefore, we exclude *Trochastica* from the yponomeutoid families included here.

TRYCHNOMERA Turner, 1913, Proc. Linn. Soc. N.S. Wales 38(1): 199.
Type species: *Trychnomera anthemis* Turner, 1913, by monotypy.
Included species: monotypic.
Current status: Lacturidae (Heppner, 1995).

TURNERIELLA Seven & Ozdikmen, 2005, N.Z.J. Zool. 32: 235. A replacement name for *Amblyzancla* Turner, 1939.
Synonym: *Amblyzancla* Turner, 1939, Pap. Proc. R. Soc. Tasm. 1938: 98 (Type species: *Amblyzancla araeoptila* Turner, 1939, by monotypy) [A junior homonym of *Amblyzancla* Turner, 1936, Noctuidae].
Included species: monotypic.
Current status: According to the original description, the type species has ocelli on the head. This character disputes its association with Yponomeutidae, as defined by Kyrki (1990) and redefined by Sohn et al. (2012). Therefore, we exclude *Turneriella* from the yponomeutoid families included here.

TYPHOGENES Meyrick, 1919, Exot. Micr. 2(8): 256.
Type species: *Epaleura psapharota* Meyrick, 1917, by monotypy.
Included species: monotypic.
Current status: The genus was moved from Tineidae to Yponomeutidae by Robinson & Tuck (1996). According to the original description, the type species of this genus has ocelli on the head. This character disputes its association with Yponomeutidae, as defined by Kyrki (1990) and redefined by Sohn et al. (2012). Therefore, we exclude *Typhogenes* from the yponomeutoid families treated here.

URODUS Herrich-Schäffer, [1854], Samml. aussereurop. Schmett. 1: wrapper, pl. 42, figs. 219, 220 [see Herrich-Schäffer, 1850–1858].
Type species: *Urodus monura* Herrich-Schäffer, [1854], by subsequent designation (Kirby, 1892, Syn. Cat. Lepid. Het. 1: 112).
Synonyms: *Aperla* Walker, 1856, List Lepid. Ins. BM 7: 1598 (Type species: *Aperla tineiformis* Walker, 1856, by monotypy); *Paratiquadra* Walsingham, 1897a, Proc. Zool. Soc. Lond. 1897: 116 (Type species: *Paratiquadra forficulella* Walsingham, 1897a, by monotypy); *Pexicnemidia* Möschler, 1890, Abh. Senck. Nat. Ges. 16: 337 (Type species: *Pexicnemidia mirella* Möschler, 1890, by monotypy); *Pygmocrates* Meyrick, 1932b, Exot. Micr. 4(9): 283 (Type species: *Pygmocrates lissopeda* Meyrick, 1932b, by monotypy); *Trichostibas* Zeller, 1863, Stett. Ent. Zeit. 24: 150 (Type species: *Trichostibas fumosa* Zeller, 1863, by monotypy); *Yrodus;* Kyrki, 1984, Ent. Scand. 15: 82 [a misspeling of *Urodus*].
Included species: *amphilocha* Meyrick, 1924 (*Urodus*); *aphanoptis* Meyrick, 1930a (*Urodus*); *aphrogama* Meyrick, 1936a (*Urodus*); *auchmera* Walsingham, 1914 (*Urodus*); *brachyanches* Meyrick, 1931a (*Urodus*); *calligera* Zeller, 1877 (*Trichostibas*); *carabopa* Meyrick, 1925a (*Urodus*); *chiquita* Busck, 1910 (*Trichostibas*); *chrysoconis* Meyrick, 1932b (*Urodus*); *costaricae* Busck, 1910 (*Trichostibas*); *cumulans* Walsingham, 1914 (*Urodus*); *cyanombra* Meyrick, 1913c (*Trichostibas*); *cyclopica* Meyrick, 1930a (*Urodus*); *decens* Meyrick, 1925a (*Urodus*); *distincta* Strand, 1911 (*Trichostibas*); *favigera* Meyrick, 1913c (*Trichostibas*); *fonteboae* Strand, 1911 (*Trichostibas*); *forficulella* Walsingham, 1897a (*Paratiquadra*); *fulminalis* Meyrick, 1931a (*Urodus*); *fumosa* Zeller, 1863 (*Trichostibas*); *hephaestiella* Zeller, 1877 (*Trichostibas*); *hexacentris* Meyrick, 1931a (*Urodus*); *hypsicrates* Meyrick, 1925a (*Urodus*); *imitans* Felder & Rogenhofer, 1875 (*Trichostibas*); *imitata* Druce, 1884 (*Urodus*); *iophlebia* Zeller, 1877 (*Trichostibas*); *isoxesta* Meyrick, 1932b (*Urodus*); *isthmiella* Busck, 1910 (*Trichostibas*); *lissopeda* Meyrick, 1932b (*Pygmocrates*); *lithophaea* Meyrick, 1913a (*Trichostibas*); *marantica* Walsingham, 1914 (*Urodus*); *merida* Strand, 1911 (*Trichostibas*); *mirella* Möschler, 1890 (*Pexicnemidia*); *modesta* Druce, 1884 (*Urodus*); *monura* Herrich-Schäffer, [1854] (*Urodus*); *niphatma* Meyrick, 1925a (*Urodus*); *opticosema* Meyrick, 1930a (*Urodus*); *ovata* Zeller, 1877 (*Trichostibas*); *pallidicostella* Walsingham, 1897a (*Trichostibas*); *pamporphyra* Meyrick, 1936a (*Urodus*); *parvula* Edwards, 1881 (*Penthetria*); *perischias* Meyrick, 1925a (*Urodus*); *porphyrina* Meyrick, 1932b (*Urodus*); *praetextata* Meyrick, 1913a (*Trichostibas*); *procridias* Meyrick, 1936a (*Urodus*); *pulvinata* Meyrick, 1924 (*Urodus*); *santipaulensis* Strand, 1911 (*Trichostibas*); *scythrochalca* Meyrick, 1932b (*Urodus*); *sordidata* Zeller, 1877 (*Trichostibas*); *spumescens* Meyrick, 1925a (*Urodus*); *staphylina* Meyrick, 1932b (*Urodus*); *subcaerulea* Dognin, 1910 (*Urodus*); *sympiestis* Meyrick, 1925a (*Urodus*); *tineiformis* Walker, 1856 (*Aperla*); *transverseguttata*

Zeller, 1877 (*Trichostibas*); *triancycla* Meyrick, 1931a (*Urodus*); *venatella* Busck, 1910 (*Trichostibas*); *xiphura* Meyrick, 1931a (*Urodus*).
Current status: Urodidae (Kyrki, 1988).

WOCKIA Heinemann, 1870, Schmett. Deutsch. 2(1)1: 102.
Type species: *Wockia funebrella* Heinemann, 1870, by monotypy.
Synonyms: *Patula* Bruand, 1849 [1850], Mém. Soc. Emul. Doubs 3(3): 50 (Type species: *Patula asperipunctella* Bruand, 1849 [1850], by monotypy) [preoccupied]; *Wockea*: Reutti, 1898, Verh. Natur. Vereins Karlsruhe 12: 291 [unjustified emendation of *Wockia*]; *Wockeia*: Spuler, 1910, Schmett. Eur. 2: 443 [unjustified emendation of *Wockia*].
Included species: *asperipunctella* Bruand, 1849 [1850] (*Patula*) = *funebrella* Heinemann, 1870 (*Wockia*); *balikpapanella* Kyrki, 1986 (*Wockia*); *chewbacca* Adamski, 2009 (Wockia); *koreana* Sohn, 2008 (*Wockia*); *mexicana* Adamski, 2009 (*Wockia*); *pelotas* Heppner, 2008 (*Wockia*); *variata* Sohn & Park, 2009 (*Wockia*).
Current status: Urodidae (Kyrki, 1988).

ZARCINIA Chrétien, 1915, Ann. Soc. Ent. Fr. 84 : 310.
Type species: *Zarcinia nigrosignatella* Chrétien, 1915, by monotypy.
Included species: *melanozestas* Meyrick, 1935b (*Zarcinia*); *nigrosignatella* Chrétien, 1915 (*Zarcinia*); *sacra* Meyrick, 1925d (*Zarcinia*).
Current status: Galacticidae (Kyrki, 1990).

Species Removed from Yponomeutidae

adspersella Kollar, 1832 (*Haemilis*), Beitr. Landesk. Osterr. Enns. 2: 92.
Current status: Depressariidae (Depressariinae) – *Agonopterix adspersella* (Kollar, 1832) (Schütze, 1931: 146).

apicipunctella Chambers, 1875a (*Hyponomeuta*), Can. Ent. 7(1): 8.
Current status: Elachistidae (Ethmiinae) – *Ethmia apicipunctella* (Chambers, 1875a) (Dyar, 1900a: 38).

assamensis Butler, 1879b (*Hyponomeuta*), Trans. Ent. Soc. Lond. 1879: 6.
Current status: Elachistidae (Ethmiinae) – *Ethmia assamensis* (Butler, 1879b) (Meyrick, 1914c).

aurisquamosa Butler, 1881 (*Argyresthia*), Ann. Mag. Nat. Hist. 5th Series 7: 403.
Current status: Tineidae – *Hieroxestis aurisquamosa* (Butler, 1881) (Meyrick, 1922a).

castella Walker, 1863 (*Prays*), List Lepid. Ins. BM 28: 541.
Current status: Tineidae – *Metapherna castella* (Walker, 1863) (Robinson & Nielsen, 1993)

circumdatellus Walker, 1863 (*Hyponomeuta*), List Lepid. Ins. BM 28: 532.
Current status: Elachistidae (Ethmiinae) – *Ethmia circumdatella* (Walker, 1863) (Meyrick, 1914c).

confusella Walker, 1863 (*Hyponomeuta*), List Lepid. Ins. BM 28: 531.
Current status: Elachistidae (Ethmiinae) – *Ethmia confusella* (Walker, 1863) (Bugnion, 1891).

conjunctella Walker, 1863 (*Incurvaria*), List Lepid. Ins. BM 28: 491.
Current status: Psychidae – *Piestoceros conjunctella* (Walker, 1863) (Nielsen et al. 1996).

contrariella Zeller, 1877 (*Atemelia*), Hor. Soc. Ent. Ross. 13: 244.
Current status: Tineidae (Tineinae) – *Niditinea contrariella* (Zeller, 1877) (Robinson, 2009).

crocostacta Meyrick, 1912 (*Scythropia*), Ann. S. Afr. Mus. 10: 68.

 | DOI 10.1163/9789004264267_008

Current status:. Psychidae – *Picrospora crocostacta* (Meyrick, 1912) (Vári et al. 2002).

cuprea Meyrick, 1901 (*Yponomeuta*), Trans. Ent. Soc. Lond. 1901: 575.
Current status: Plutellidae – *Orthenches cuprea* (Meyrick, 1901) (Meyrick, 1921c).

cyaneimarmorella Millière, 1854 (*Argyresthia*), Ann. Soc. Ent. Fr. 1854: 64.
Current status: Tineidae – *Stenoptinea cyaneimarmorella* Millière, 1854 (Robinson, 2002, Robinson, 2009).

elegantella Lucas, 1955 (*Prays*), Bull. Soc. Sci. Nat. Maroc 35: 257.
Current status: Scythrididae – *Scythris elegantella* (Lucas, 1950) (Gibeaux, 1985a).

euryphracta Meyrick, 1892 [1893] (*Eriocottis*), Proc. Linn. Soc. N.S. Wales for the year 1892. Second Series. 7: 514.
Current status: Plutellidae – *Diathryptica euryphracta* (Meyrick, 1892 [1893]) (Robinson & Nielsen, 1993).

eximiella Rebel, 1899 (*Calantica*), Verh. Zool.-Bot. Ges. Wien 49: 175.
Current status: Ypsolophidae – *Phrealcia eximiella* (Rebel, 1899) (Kyrki, 1990: 37).

gemmella Linnaeus, 1758 (*Phalaena Tinea*), Syst. Nat. 10(1): 539.
Current status: Gelechiidae – *Stenolechia gemmella* (Linnaeus, 1758) (Robinson & Nielsen, 1983).

halidora Meyrick, 1908b (*Oxygrapha*), Jour. Bomb. Nat. Hist. Soc. 18: 624.
Current status: Tortricidae – *Lopharcha halidora* (Meyrick, 1908b) (Diakonoff, 1974).

hilarella Snellen, 1878 (*Prays?*), Tijdschr. v. Ent. 21: 133.
Current status: Noctuidae – A junior synonym of *Cydosia nobilitella* Cramer, 1779 (Poole, 1989).

lapidellus Walsingham, 1880 (*Hyponomeuta*), Proc. Zool. Soc. Lond. 1880: 86.
Current status: Elachistidae (Ethmiinae) – *Ethmia lapidella* (Walsingham, 1880) (Sattler, 1967).

lineatonotella Moore, 1867 (*Hyponomeuta*), Proc. Zool. Soc. Lond. 1867: 669.

Current status: Elachistidae (Ethmiinae) – *Ethmia lineatonotella* (Moore, 1867) (Meyrick, 1914c).

longimaculella Chambers, 1872 (*Hyponomeuta*), Can. Ent. 4: 43.
Current status: Elachistidae (Ethmiinae) – *Ethmia longimaculella* (Chambers, 1872) (Dyar, 1902a).

maculatella Turati, 1930 (*Swammerdamia*), Atti Soc. Ital. Milano 69: 78.
Current status: Glyphipterigidae (Acrolepiinae) – A junior synonym of *Digitivalva eglanteriella* (Mann, 1855) (Gaedike, 1979).

nebulosus Turner, 1942 (*Schistocyttara*), Proc. R. Soc. Qd. 53: 96.
Current status: Copromorphidae – *Schistocyttara nebulosus* Turner, 1942 (Common, 1996).

oxydelta Meyrick, 1913a (*Calantica*), Exot. Micr. 1(5): 147.
Current status: Plutellidae – *Anthonympha oxydelta* (Meyrick, 1913a) (Moriuti, 1971b).

paucella Walker, 1863 (*Hyponomeuta*), List Lepid. Ins. BM 28: 530.
Current status: Elachistidae (Ethmiinae) – *Ethmia paucella* (Walker, 1863) (Bugnion, 1891).

plagiferella Walker, 1863 (*Prays*?), List Lepid. Ins. BM 28: 540.
Current status: Gelechiidae – A junior synonym of *Dactylethrella candida* (Stainton, 1859).
(Meyrick, 1925b: 164).

quinquepunctella Chambers, 1875a (*Hyponomeuta*), Can. Ent. 7: 7.
Current status: Prodoxidae – *Prodoxus quinquepunctella* (Chambers, 1875a) (Dyar, 1903).

rustica Haworth, 1828 (*Recurvaria*), Lep. Brit. 4: 548.
Current status: Tineidae (Tineinae) – A junior synonym of *Monopis laevigella* (Denis & Schiffermüller, 1775) (Robinson, 2002).

stilbella Doubleday, 1843 (*Argyrosetia*), in Dieffenbach, Trav. N.Z. 2: 289.
Current status: Tineidae – *Erechthias stilbella* (Doubleday, 1843) (Meyrick, 1880).

suzukiella Matsumura, 1931 (*Argyresthia*), 6000 Ill. Ins. Jap.: 1096.
Current status: Glyphipterigidae (Acrolepiinae) – *Acrolepiopsis suzukiella* (Matsumura, 1931) (Gaedike, 1970).

texanella Chambers, 1880 (*Hyponomeuta*), J. Cincinn. Soc. 2: 180.
Current status: Elachistidae (Ethmiinae) – A junior synonym of *Ethmia zelleriella* Chambers, 1878 (Dyar, 1902a).

transversella Walker, 1864b (*Argyresthia*), List Lepid. Ins. BM 30: 849.
Current status: Glyphipterigidae (Glyphipteriginae) – *Glyphipterix transversella* (Walker, 1864b) (Meyrick, 1914c).

viduata Walker, 1869 (*Hyponomeuta*?), Chars. Undesc.: 85.
Current status: Xyloryctidae – unplaced species (Common, 1996).

zebrina Butler, 1881 (*Argyresthia*), Ann. Mag. Nat. Hist. 5th Series 7: 403.
Current status: Tineidae – *Erechthias zebrina* (Butler, 1881) (Meyrick, 1915a).

zelleriella Chambers, 1878 (*Hyponomeuta*), Bull. U.S. Geol. Surv. 4: 80.
Current status: *Ethmia zelleriella* (Chambers, 1878) (Dyar, 1902a).

References

Note: Some references and citations are cited with two years, one enclosed in brackets. The year without brackets (usually an older year) indicates the year that was given within the original work; the year enclosed in brackets (usually a more recent year) is used when the year of publication comes from bibliographic sources that differ from the original publication itself. For example, Hübner, 1816 [1825] was dated by the author in 1816 but was not printed and distributed until 1825. References and citations cited only by a year enclosed in brackets represents a dated, disarticulated and truncated portion of a larger work published over a period of several years. For example, Hübner, [1813] is a dated, disarticulated, truncated, portion published in 1813, as part of the larger work by him, entitled "Sammlung europäischer Schmetterlinge," spanning the period 1796–1838.

Adamski, D., Makinson, J.R., Brown, B.T., Wright, S.A., Pratt, P.D. & Brown, J.W. 2013. Description and evaluation of Metharmostis multilineata and Idiophantis soreuta for biocontrol of Downy Rose Myrtle, Rhodomyrtus tomentosa. Journal of the Lepidopterists' Society 67(2): 111–127.

Agassiz, D. 1981. Swammerdamia passerella (Zetterstedt) (Lepidoptera: Yponomeutidae), a valid species resident in Britain. Entomologist's Gazette 32(4): 217–224.

——. 1996. Yponomeutidae, glyphipterigidae, and roeslerstammiidae. In: Emmet, A.M. (ed.), The Moths and Butterflies of Great Britain and Ireland, Vol. 3. Yponomeutidae–Elachistidae. pp. 39–114. Harley Books, Stenstrup, Denmark.

——. 2007. Prays peregrina sp. n. a presumed adventive species in Greater London. Nota lepidopterologica 30(2): 407–410.

Agassiz, D. & Friese, G. 1996. Yponomeutidae. In: Karsholt, O. & Razowski, J. (eds.), The Lepidoptera of Europe. pp. 55–58. Apollo Books, Stenstrup, Denmark.

Agassiz, D.J.L. & Tuck, K.R. 1999. The cypress tip moth Argyresthia cupressella Walsingham, 1890 (Lepidoptera: Yponomeutidae) new to Britain. Entomologist's Gazette 50(1): 11–16.

Agenjo, R. 1960. Dos plagas de los álamos (Populus) originadas por "arañuelos" (Yponomeuta Latr.). Boletin del Servicio de Plagas Forestales 3(5): 97–114, pls. 2–4.

——. 1963. Presencia en España de la Ocnerostoma copiosella (Frey, 1856), plaga de las acículas del Pinus silvestris L. Boletin del Servicio de Plagas Forestales 6(11): 1–4.

Aguiar, A.M.F. & Karsholt, O. 2006. Systematic catalogue of the entomofauna of the Madeira Archipelago and Selvagens Islands. Lepidoptera, Vol. 1. Boletim do Museu Municipal do Funchal (Historia Natural), Supplement 9: 5–139.

Ahmad, R. 1976. Studies on Prays acmonias and its natural enemies in Pakistan. Entomophaga 21(3): 265–268.

Alford, D.V. 2007. Pests of Fruit Crops: A Color Handbook. Manson Publishing Ltd., London, England. 461 pp.

Alonso, C., Vuorisalo, T., Wilsey, B. & Honkanen, T. 2000. Yponomeuta evonymellus outbreaks in southern Finland: spatial synchrony but different local magnitudes. Annales Zoologici Fennici 37: 178–188.

Arduino, P. & Bullini, L. 1994 [1995]. A new species of small Ermine moth (Yponomeuta padellus complex) from Japan, formerly assigned to Y. malinellus. Bollettino dell Instituto di Entomologia "Guido Grandi" della Universita degli studi di Bologna 49: 131–154. Note: The title page is dated 1994 in the 1995 volume.

Bailey, L.H. & Bailey, E.Z. 1976. Hortus Third A Concise Dictionary of Plants Cultivated in the United States and Canada, Second Edition. Macmillan Publishing Company, New York, usa. 1290 pp.

Balachowsky, A.S. 1966. Entomologie appliquée à l'agriculture, Tome II, Lépidoptères. Mason et Cie, Paris, France. 1057 pp.

Bankes, E.R. 1896. On a new species of the genus Argyresthia, Hb., from England. The Entomologist's Monthly Magazine, Second Series 2(32): 25–26.

——. 1898. Notes on the unique "Argyresthia" decimella, Stn. The Entomologist's Monthly Magazine, Second Series 9(34): 263–264.

Baraniak, E. 1988. Species of the genus Kessleria Now. (Yponomeutidae) in Poland. Polskie Pismo Entomologiczne 58: 537–545.

——. 1993. Distribution of Argyresthiidae (Lepidoptera) in Poland. Polskie Pismo Entomologiczne 62: 53–82.

Baraniak, E., Kulfan, J. & Patočka, J. 2003. Argyresthia tatrica sp. n. – a new Lepidoptera species feeding on Larix decidua in the Tatra Mts. Mitteilungen aus dem Museum für Naturkunde in Berlin, Deutsche Entomologische Zeitschrift 50(2): 231–236.

Baraniak, E. & Junnilainen, J. 2011. Contribution to the knowledge of Yponomeutoidea. III. Argyresthia (Blastotere) uralensis sp. n., a new species from Russia. Polish Journal of Entomology 80: 499–503.

Baraniak, E. & Walczak, U. 2003. Contribution to the knowledge of Yponomeutoidea (Lepidoptera). I. Argyresthia trifasciata (Staudinger, 1871) (Lepidoptera: Argyresthiidae) – a moth new to the fauna of Poland. Wiadomości Entomologiczne 21(4): 223–227.

Beavan, S.D. & Heckford, R.J. 2012. Atemelia torquatella (Lienig & Zeller, 1846) (Lepidoptera: Yponomeutidae) reared from Myrica gale L. on the Isle of Mull, Mid Ebudes, Scotland. Entomologist's Gazette 63: 48.

Beccaloni, G., Scoble, M., Kitching, I., Simonsen, T., Robinson, G., Pitkin, B., Hine, A. & Lyal, C. 2005. The Global Lepidoptera Names Index. http://www.nhm.ac.uk/research-curation/research/projects/lepindex/ [accessed: Feb. 2014].

Bechstein, J.M. & Scharfenberg, G.L. 1805. Vollständige Naturgeschichte der Schädlichen Forstinsecten nebst einem Nachtrag der schonenswerthen Inseckten, welche die schädlichen vertilgen helfen. Ein Handbuch fur Forstmanner, Cameralisten un Dekonomen. 3. C.F.E. Richter, Leipzig. pp. 605–1046.

Becker, V.O. 1984. Taxonomic notes on the Neotropical Microlepidoptera. Revista Brasileira de Entomologia 28(2): 129–201.

——. 2009. A review of the New World Atteva Walker moths. Revista Brasileira de Entomologia 53(3): 349–355.

——. 2013. The Neotropical Yponomeuta Latreille, 1796 moths. shilap Revista de Lepidopterologia 41(163): 305–310.

Beeson, C.F.C. 1941. The Ecology and Control of Forest Insects of India and Adjoining Countries. Vasant Press, Dehradun, India. 1007 pp.

Beirne, B.P. 1942. The biology and natural control of the Larch-shoot Moth, Argyresthia laevigatella, H.-S. The Economic Proceedings of the Royal Dublin Society 3(10–12): 130–149.

Bengtsson, B.Å. & Johansson, R. 2011a. Nationalnyckeln till Sveriges flora och fauna. Fjärilar: Bronsmalar-rullvingemalar. Lepidoptera: Roeslerstammiidae-Lyonetiidae. Art Databanken, slu, Uppsala, Sweden. 494 pp.

——. 2011b. Review of the unicolorous species of the subgenus Blastotere with descriptions of Argyresthia svenssoni sp. n. and A. kulfani sp. n. Entomologisk Tidskrift 132(4): 257–274.

Berg, C. 1880. Observaciones acerca de la Familia Hyponomeutidae. Anales de la Sociedad Científica Argentina 10: 99–109.

Bernard, M. 1788. Des insectes qui vivent sur l'olivier. Mémoires pour servir à l'histoire naturelle de la Provence 2: 265–318.

Biezanko, C.M., Ruffinelli, A. & Carbonell, C.S. 1957. Lepidoptera del Uruguay. Revista de la Facultad de Agronomía, Montevideo 46: 1–152.

Bigot, L., Boumezzough, A. & Fels, A.E. 1988. Contribution à l'étude des insectes liés au genévrier thurifère (Juniperus thurifera L.) dans le Haut Atlas marocain: biologie de deux microlépidoptères ravageurs dominants des galbules et des graines: Argyresthia reticulata (Yponomeutidae) et Pammene juniperana (Tortricidae), espèces nouvelles pour le Maroc. Bulletin de la Société entomologique de France 93(7/8): 233–238.

Billberg, G.J. 1820. Enumeratio insectorum in Museo Gustav Johan Billberg. Gadelianis, Stockholm, Sweden. 138 pp.

Bjerkander, C. 1784. Insect – Calender, för år 1784. Kongliga Vetenskaps Academiens Nya Handlingar 5: 319–325.

Bolton, H.C. 1897. A Catalogue of Scientific and Technical Periodicals. 1665–1895 together with Chronological Tables and a Library Check-List. Smithsonian Institution, District of Columbia, usa. 1247 pp.

Börner, C. 1939. Die Grundlagen meines Lepidopterensystems. Verhandlungen der VII Internationaler Kongress für Entomologie, Berlin 2: 1372–1424.

Bradley, J.D. 1962. Microlepidoptera from the New Hybrides. Records and descripitons of Microlepidoptera collected on the island of Aneityum by Miss Evelyn Cheesman, O.B.E. Bulletin of the British Museum (Natural History), Entomology 12(5): 249–271, pls. 14–27.

——. 1965. Microlepidoptera. Ruwenzori Expedition 1952 2(12): 81–148.

——. 1966a. Type Specimens of Microlepidoptera in the University Museum, Oxford, described by Haworth. Entomologist's Gazette 17(3): 129–140.

——. 1966b. Some changes in the nomenclature of British Lepidoptera. Entomologist's Gazette 17: 213–235.

——. 1982. New species of Microlepidoptera from Norfolk Island. Journal of Natural History 16: 367–379.

Braun, A.F. 1910. New species of Tineina from California. Entomological News 21: 171–179.

——. 1912. Notes on some North American Tineina. The Canadian Entomologist 44(5): 159–161.

——. 1914. Notes on North American Tineina, with descriptions of new species. Entomological News 25: 113–117.

——. 1918. New species of microlepidoptera. The Canadian Entomologist 50: 229–236.

——. 1921. Notes on microlepidoptera with descriptions of new species. Entomological News 32: 8–18.

——. 1925. Microlepidoptera of northern Utah. Transactions of the American Entomological Society 51(3): 183–226.

——. 1940. Notes and new species in the Yponomeutoid group. Transactions of the American Entomological Society 66(1079): 273–282.

Briganti, V. 1822. Appendix. Atti del Reale Istituto d'Incoraggiamento alle Scienze Naturali di Napoli 3: 127–137.

Britton, D.R. & Couper, J. 2009. A larval food plant for Atteva albiguttata (Zeller). Australian Entomologist 36(1): 23–28.

Brock, J.P. 1971. A contribution towards an understanding of morphology and phylogeny of the ditrysian Lepidoptera. Journal of Natural History 5: 29–102.

Brower, A.E. 1953. Three new species of microlepidoptera. Annals of the Entomological Society of America 46(1): 95–98.

——. 1984. A List of the Lepidoptera of Maine, Part 2. The Microlepidoptera, Section 2, Cosmopterigidae through Hepialidae. Department of Entomology, Orono, Maine, usa. 70 pp.

Browne, F.G. 1968. Pests and Diseases of Forest Plantation Trees. Clarendon Press, Oxford, England. 1330 pp.

Bruand, T. 1849 [1850]. Catalogue systématique et synonymique des Lépidoptères du Département du Doubs. Tinéides. Mémoires de la Société d'émulation du Doubs 3(3): 23–68.

Bugnion, E. 1891. Note sur la résistance de la Teigne du fusain aux basses températures de l'hiver. Mitteilungen der Schweizerischen Entomologischen Gesellschaft 8(8): 319–321.

Burmann, K. 1956. Swammerdamia zimmermanni Now. (Lep., Yponomeutidae). Zeitschrift der Wiener Entomologischen Gesellschaft 67: 187–190.

——. 1973. Faunistik und Biologie der Kessleria-Arten (Lepidoptera: Yponomeutidae) Tirols. Berichte des naturwissenschaftlichen-medizinischen Verein Innsbruck 60: 151–158.

——. 1980. Eine neue Kessleria-Art aus dem Monte Baldogebiet in Italien (Lepidoptera, Yponomeutidae). Nachrichtenblatt der Bayerischen Entomologen 29(6): 105–107.

Busck, A. 1900a. New American Tineina. Journal of the New York Entomological Society 8: 234–247.

——. 1900b. New Species of Moths of the Superfamily Tineina from Florida. Proceedings of the United States National Museum 23: 225–254.

——. 1903. Notes on Brackenridge Clemens types of Tineina. Proceedings of the Entomological Society of Washington 5(3): 181–220.

——. 1904. Tineid moths from British Columbia, with descriptions of new species. Proceedings of the United States National Museum 27(1375): 745–778.

——. 1907. Revision of the American moths of the genus Argyresthia. Proceedings of the United States National Museum 32: 5–24.

——. 1908. Descriptions of North American Tineina. Proceedings of the Entomological Society of Washington 9(1–4): 85–86.

——. 1912. New Microlepidoptera from Mexico. Proceedings of the Entomological Society of Washington 14: 83–89.

——. 1914. Descriptions of new microlepidoptera of forest trees. Proceedings of the Entomological Society of Washington 16(4): 143–150.

——. 1915. Descriptions of new North American microlepidoptera. Proceedings of the Entomological Society of Washington 17(2): 79–94.

——. 1916. Descriptions of new North American microlepidoptera. Proceedings of the Entomological Society of Washington 18(3): 147–154.

——. 1921. Microlepidoptera from British Columbia. The Canadian Entomologist 53: 275–279.

——. 1925. A new North American genus of Microlepidoptera. Proceedings of the Entomological Society of Washington 27(3): 46–48.

Butler, A.G. 1877. Lepidoptera, Orthoptera, and Hemiptera. Proceedings of the Scientific Meetings of the Zoological Society of London 1877: 86–87.

——. 1879a. Illustrations of Typical Specimens of Lepidoptera Heterocera in the Collection of the British Museum, Part 3. London, England. 82 pp.

——. 1879b. On a collection of Lepidoptera from Cachar, N. E. India. Transactions of the Entomological Society of London 1879(1): 1–7.

——. 1881. On a collection of nocturnal Lepidoptera from the Hawaiian Islands. Annals and Magazine of Natural History 5th series 7: 392–408.

——. 1882. Descriptions of new Species of Lepidoptera, chiefly from Duke-of-York Island and New Britain. Annals and Magazine of Natural History 5(10): 226–238.

——. 1883. III. Heterocerous Lepidoptera collected in Chile by Thomas Edmonds, Esq. Transactions of the Entomological Society of London for the year 1883: 49–90.

——. 1887. Descriptions of two new species of Hyponomeutidae from the Solomon Islands. Annals and Magazine of Natural History, vol. 20 – 5th series: 414–415.

Byun, B.-K. & Bae, Y.-S. 2003. Systematic Review of the Genus Yponomeuta Latreille in Korea. Insecta Koreana 20(2): 227–237.

Byun, B.-K., Park, K.-T., Bae, Y.-S. & Lee, B.-W. 2009. A Checklist of the Microlepidoptera in Korea. Korea National Arboretum, samsunggad-com, Seoul, South Korea. 413 pp.

Caradja, A. 1920. Beitrag zur Kenntnis der geographischen Verbreitung der Mikrolepidopteren des palaearktischen Faunengebietes nebst Beschreibung neuer Formen. Deutsche Entomologische Zeitschrift "Iris" 34(1/2): 75–179.

Caradja, A. & Meyrick, E. 1935. Materialien zu Einer Microlepidopteren Fauna der Chinesischen Provinzen Kiangsu, Chekiang und Hunan. R. Friedlander & Sohn, Berlin, Germany. 96 pp.

——. 1938. Materialien zu einer Mikrolepidopterenfauna des Yülingshanmassivs (Provinz Yünnan). Deutsche Entomologische Zeitschrift "Iris" 52 (1): 1–29.

Carter, D.J. 1984. Pest Lepidoptera of Europe with Special Reference to the British Isles. W. Junk Publishers, Dordrecht, Boston & Lancaster, Dordrecht, The Netherlands. 431 pp.

Chambers, V.T. 1872. Micro-Lepidoptera. The Canadian Entomologist 4: 41–44, 65–69, 88–92.

——. 1874. Micro-Lepidoptera. The Canadian Entomologist 6(1): 8–11.

——. 1875a. Tineina from Texas. The Canadian Entomologist 7(1): 7–12.

——. 1875b. Tineina from Canada. The Canadian Entomologist 7(8): 124–128, 144–147.

——. 1875c. Tineina of Colorado. The Cincinnati Quarterly Journal of Science 2(4): 289–305.

——. 1877. The Tineina of Colorado. Bulletin of the United States Geological Survey 3(1): 121–142.

——. 1878. Index to the described Tineina of the United States and Canada. Bulletin of the United States Geological Survey 4(1): 125–167.

——. 1879. Tineina. The Canadian Entomologist 11: 143–146.

——. 1880. Descriptions of some new Tineina, with notes on a few old species. The Journal of the Cincinnati Society of Natural History 2(4): 179–194.

Cho, S., Zwick, A., Regier, J., Mitter, C., Cummings, M., Yao, J., Du, Z., Hong, Z., Kawahara, A., Weller, S., Davis, D., Baixeras, J., Brown, J. & Parr, C. 2011. Can deliberately incomplete gene sample augmentation improve a phylogeny estimate for the advanced moths and butterflies (Hexapoda: Lepidoptera)? Systematic Biology 60: 782–796.

Chrétien, P. 1905. Les chenilles du Rhamnus infectorius. Le Naturaliste 2nd Series 19: 29–31.

——. 1907 [1908]. Description de deux espèces nouvelles de Microlépidoptères d'Algérie. Bulletin de la Société entomologique de France 20: 339–340.

——. 1910. Lépidoptères nouveaux pour la faune française. Le Naturaliste 32: 260–262.

——. 1915. Contribution a la connaissance des Lépidoptères du Nord de l'Afrique. Annales de la Société entomologique de France 84(3): 289–374.

Christoph, H. 1882. Neue lepidopteren des amurgebietes. Bulletin de la Société Impériale des naturalistes de Moscou 57(1): 5–47.

——. 1888. Diagnosen zu einigen neuen Lepidopteren des palaearctischen Faunengebietes. Horae Societatis Entomologicae Rossicae 22: 308–314.

Clarke, J.F.G. 1942. Notes and new species of microlepidoptera from Washington State. Proceedings of the United States National Museum 92(3149): 267–282.

——. 1953. Meyrick Types of Microlepidoptera in the British Museum (Natural History) London. I. Jarrold and Sons Ltd., Norwich, England. 332 pp.

——. 1955. Catalog of the type specimens of Microlepidoptera in the British Museum (Natural History) described by Edward Meyrick, Vol. 1. Trustees of the British Museum (Natural History), London, England. 332 pp.

——. 1965. Catalog of the type specimens of Microlepidoptera in the British Museum (Natural History) described by Edward Meyrick, Vol. 5. Trustees of the British Museum (Natural History), London, England. 581 pp.

——. 1971. The Lepidoptera of Rapa Island. Smithsonian Contributions to Zoology No. 56. Smithsonian Institution Press, Wash., d.c., usa. 282 pp.

——. 1986. Pyralidae and Microlepidoptera of the Marquesas Archipelago. Smithsonian Contributions to Zoology, No. 416. Smithsonian Institution Press, Washington, d.c., usa. 485 pp.

Clemens, B. 1861a. Contributions to American Lepidopterology – No. 3. Proceedings of the Academy of Natural Sciences of Philadelphia 12: 4–15.

——. 1861b. Contributions to American Lepidopterology – No. 7. Proceedings of the Academy of Natural Sciences of Philadelphia 12: 522–547.

Clerck, C. 1759–[1764]. Icones insectorum rariorum. Sectio Prima. Icones Insectorum rariorum cum nominibus eorum trivialibus locisque e C. Linnaei systema naturae allegatis. Stockholm, Sweden. 21 pp., 55 pls.

Common, I.F.B. 1970 Lepidoptera. In: Mackerras, I.M. (ed.), The Insects of Australia. pp. 765–866. csiro, Canberra, Australia.

——. 1990. Moths of Australia. Melbourne University Press, Carlton, Australia. 535 pp.

——. 1994. Oecophorine Genera of Australia I. The Wingia Group. Monographs of Australian Lepidoptera 3. 390 pp. csiro, Canberra, Australia.

——. 1996. Oecophoridae, Copromorphidae. In: Nielsen, E.S., Edwards, E.D. & Rangsi, T.V. (eds.), Checklist of the Lepidoptera of Australia. Monographs on Australian Lepidoptera 4: 59–89, 151. csiro, Victoria, Australia.

Coquillett, D.W. 1891. The cypress twig borer (Argyresthia cupressella Wlsm.). Insect Life 3: 116–117.

Costa, O.-G. 1836. Fauna del Regno di Napoli ossia Enumerazione di Tutti gli Animali, Lepidotteri. Napoli, Dai Torchi del Tramater, Italy. 446 pp.

——. 1839. Degl' insetti ospitanti sull' Ulivo, e nelle olive. Ordine V.; Genere Oecophora. Corrispondenza Zoologica Destinata a Diffondere nel Regno delle due Sicilie tutto cio che si va Discuoprendo Entro e Fuori Europa (e vice versa) Risguardante la Zoologia in Generale 1: 118–121.

Covell, C.V. Jr. 1999. Argyresthia visaliella Chambers 1875 (Argyresthiidae), a nomen nudum. Journal of the Lepidopterists' Society 53(2): 76.

Csóka, G. 2001. Recent invasions of five species of leafmining Lepidoptera in Hungary. In: Liebhold, A.M., McManus, M.L., Otvos, I.S. & Fosbroke, S.L.C. (eds.), Proceedings: Integrated Management and Dynamics of Forest Defoliating Insects. pp. 31–36. Northeastern Research Station, Victoria, Australia.

Curtis, J. 1833. Art. XIX – Characters of some undescribed Genera and Species, indicated in the "Guide to an Arrangement of British Insects." The Entomological Magazine 1(2): 186–199.

Dallas, W.S. 1866. Insecta. Zoological Record 3: 268–576.

Davis, D.R. 1991. Neotropical Microlepidoptera XXIV. Description and biological observations of Ithutomus formosus Butler webbing leaves of Drimys winteri in Chile. Proceedings of the Entomological Society of Washington 93(3): 690–702.

Davis, D.R. & Robinson, G.S. 1998. The Tineoidea and Gracillarioidea. In: Kristensen, N.P. (ed.), Lepidoptera, Moths and Butterflies. Vol. 1: Evolution, Systematics, and Biogeography. Handbook of Zoology 4: 91–117. Walter de Gruyter, Berlin & New York.

De Geer, C. 1771 [1783]. Mémoires pour servir à l'histoire des insectes; précédés de discours sur les insectes en général, Tome second, Première partie. Papiliones. Pierre Hesselberg, Stockholm, Sweden. 250 pp.

——. 1783. Genera et species insectorum e generosissimi auctoris scriptis extraxit, digessit, Latine quoad partem reddidit, et terminologiam insectorum Linneanam addidit. S.L. Crusium, Lipsiae. 220 pp.

de Joannis, J. 1900. Description d'une nouvelle espèce de Microlépidoptère de France, Zelleria ribesiella. Bulletin de la Société entomologique de France 1900: 391–393.

Delgado, C. & Couturier, G. 2008. Atteva punctella Cramer (Lepidoptera: Yponomeutidae), plaga de Simarouba amara Aubl. (Simaroubaceae), en Peru. Boletin de la Sociedad Entomologica Aragonesa 43: 519–521.

Denis, M. & Schiffermüller, I. 1775. Ankündigung eines Systematischen Werkes von den Schmetterlingen der Wienergegend herausgegeben von einigen Lehrern. Bernardi, Vienna, Austria. 322 pp.

Desmarest, E. 1857. Papillons nocturnes. In: Chenu, J.C. (ed.), Encyclopédie d'histoire naturelle ou traité complet de cette science d'après les travaux des naturalistes les plus éminents de tous les pays et de toutes les époques, Dixième Tome. Harvard, Paris, France. 312 pp.

de Villers, C. 1789. Caroli Linnæi entomologia faunæ Suecicæ descriptonibus acuta; 2 Lugduni, Piestre et Delamolliere, Sweden. 656 pp.

Diakonoff, A. 1955. Microlepidoptera of New Guinea results of the third Archbold expedition (American – Netherlands Indean expedition 1938–1939) Part V. Verhandelingen der Koninklijke Nederlandse Akademie van Wetenschappen, afd. Naturkunde (2) 50(3): 1–210.

——. 1966. Records and descriptions of South Asiatic Microlepidoptera. Tijdschrift voor Entomologie 109: 49–88.

——. 1967. Microlepidoptera of the Philippine Islands. United States National Museum Bulletin 257. Smithsonian Institution Press, Washington, d.c., usa. 484 pp.

——. 1974. The South Asiatic Polyorthini with notes on species of Polyortha Dognin (Lepidoptera, Tortricidae). Zoologische Verhandelingen 131: 1–86.

Diorio, V. 1865. Di alcuni insetti i quali hanno danneggiato talune vigne della provincia Romana di Marittima e Campagna nel cessato anno. Atti dell'Accademia Pontificia de'nuovi Lincei (Roma) 18: 124–126.

Doğanlar, M. 1979. A new species of Swammerdamia Hbn. from British Columbia. The Canadian Entomologist 111: 967–969.

Donovan, E. 1793. The Natural History of British insects; explaining them in their several states, with the periods of their transformations, their food, œconomy, &c. together with the history of such minute insects as require investigation by the microscope, the whole illustrated by coloured figures, designed and executed from living specimens, Vol. 2. Printed for the author, London, England. 96 pp.

Doubleday, E. 1843. List of the Annulose Animals hitherto recorded as found in New Zealand, Lepidoptera. In: Dieffenbach, E. (ed.), Travels in New Zealand; with Contribution to the Geography, Geology, Botany, and Natural History of that Country, Vol. 2. John Murray, London, England. pp. 283–289.

Druce, H. [1885]. Biologia Centrali-Americana. Insecta. Lepidoptera-Heterocera 1. 1881–1900. London. 490 pp.

Duckworth, W.D. 1965. North American moths of the genus Swammerdamia. Proceedings of the United States National Museum 116(3507): 549–556.

——. 1967. A new species of Atteva from Central America. Proceedings of the Entomological Society of Washington 69(1): 69–72.

Dufrane, A. 1960. Microlepidopteres de la faune belge (Neuvieme note). Bulletin de l'Institut Royal des Sciences Naturelles de Belgique 36(29): 1–16.

Dugdale, J.S. 1971. Entomology of the Aucklands and other islands south of New Zealand: Lepidoptera, excluding non-crambine Pyralidae. Pacific Insects Monograph 27: 55–172.

——. 1989. Lepidoptera – annotated catalogue, and keys to family-group taxa. Fauna of New Zealand 14: 1–262.

Dugdale, J.S., Kristensen, N.P., Robinson, G.S. & Scoble, M.J. 1998. The Yponomeutoidea. In: Kristensen, N.P. (ed.), Lepidoptera, Moths and Butterflies. Vol. 1: Evolution, Systematics, and Biogeography. Handbook of Zoology 4: 119–130. Walter de Gruyter, Berlin & New York.

Duponchel, P.A.J. 1838 [1840]. Nocturnes VIII. In: Godart, J.B. & Duponchel, P.A. (eds.), Histoire naturelle des lépidoptères ou papillons de France, Tome 11. Méquignon-Marvis, Paris, France. 720 pp.

——. 1842 [1843]. Nocturnes. In: Godart, J.B. & Duponchel, P.A. (eds.), Histoire naturelle des lépidoptères ou papillons de France, Supplement 4. Méquignon-Marvis, Paris, France. 554 pp.

Durrant, J.H. 1900. Pterophoridae and Tineina. In: Swinhoe, C.C., Walsingham, R.H.L. & Durrant, J.H. (eds.), Catalogue of Eastern and Australian Lepidoptera Heterocera in the Collection of the Oxford University Museum, Part 2. pp. 541–585. Clarendon Press, Oxford.

——. 1916. Microlepidoptera (Pterophorina and Tineina) collected by the British Ornithologists' Union and Wollaston expeditions in the Snow Mountains, southern Dutch New Guinea. Report of the British Ornithologists' Union Expedition and the Wollaston Expedition in Dutch New Guinea, 1910–1913 2: 149–162.

——. 1921 [1922]. Description of a new Genus and Species of Tineina from Southern Nigeria. Transactions of the Entomological Society of London 1921(3–4): 494–495.

Dyar, H.G. 1900a. Notes on some North American Yponomeutidae. The Canadian Entomologist 32(1): 37–41.

——. 1900b. Notes on some North American Yponomeutidae. The Canadian Entomologist 32(3): 84–86.

——. 1902a. A review of the Genus Ethmia with descriptions of new species. Journal of the New York Entomological Society 10: 202–208.

——. 1902b. A new Yponomeuta. Journal of the New York Entomological Society 10: 208.

——. 1903. A list of North American Lepidoptera and key to literature of this order of insects. United States National Museum Bulletin 52. Smithsonian Institution Press, Washington, d.c., usa. 723 pp.

Eaton, A.E. 1875. Breves Dipterarum uniusque Lepidopterarum Insule Kerguelensi indigenarum diagnoses. Entomologist's Monthly Magazine 12: 58–68.

Edwards, E.D. 1996. Yponomeutidae, Argyresthiidae & Plutellidae. In: Nielsen, E.S., Edwards, E.D. & Rangsi, T.V. (eds.), Checklist of the Lepidoptera of Australia. Monographs on Australian Lepidoptera 4: 50–53. csiro, Victoria, Australia.

Efremov, V.F. 1969. An account of the morphological characters of some eastern species of the Yponomeutidae. Biological sbornik Blagoveshchensk Pedagogical Institute, Blagoveschensk [no volume or issue numbers]: 25–33.

——. 1976. [A new Ermine moth (Lepidoptera, Yponomeutidae) of Primorye]. Zhivotny mir Sovetskogo Dalnego Vostoka, Blagoveshchensk [Animals of the Far East] 1: 108–111.

Eidt, D.C. 1967. The immature stages of Yponomeuta carteri Walsingham. Bulletin of the Entomological Society of Nigeria 1: 13–17.

Eidt, D.C. & Sippell, W.L. 1961. The life history, parasites, and economic status of the Larch Shoot Moth, Argyresthia laricella Kft., and comparisons with A. laevigatella. The Canadian Entomologist 93(1): 7–24.

Emmet, A.M. 1982. Argyresthia trifasciata Staudinger, 1871 (Lep.: Yponomeutidae). Entomologist's Record and Journal of Variation 94: 180–182.

Esper, E.J.E. 1791. Berchreibung einiger der praechtigften Schmetterlinge von den Eleinften Urten. Der Naturforscher 25: 39–51.

Eversmann, E. 1844. Lepidopterorum species quas per viginti quinque annos in provinciis Volgan fluvium inter et montes Uralenses sitis observavit et descripsit Eduardus Eversmann. Fauna Lepidopterologica Volgo-Uralensis. Typis Universitatis, Casani. 633 pp.

Fabricius, J.C. 1775. Systema Entomologicae, sistens insectorum classes, ordines, genera, species, adiectis synonymis, locis, descriptionibus, observationibus. Flensburgi et Lipsiae [=Leipzig, Germany]. 832 pp.

——. 1776. Genera Insectorum eorumque characteres naturales secundum numerum, figuram, situm et proportionem omnium partium oris adiecta mantissa specierum nuper detectarum. Cologne. 310 pp.

——. 1781. Species insectorum exhibentes eorum differentias specificas, synonyma auctorum, loca natalia, metamorphosin adiectis observationibus, descriptionibus, Vol. 2. Carol Ernest Bohnii, Hamburgi & Kilonii. 494 pp.

——. 1787. Mantissa Insectorum sistens species nuper detectas adiectis synonymis, observationibus, descriptionibus, emendationibus. C.G. Proft, Hafniae [=Copenhagen, Denmark]. 382 pp.

——. 1794. Entomologia systematica emendata et aucta, secundum classes, ordines, genera, species, adjectis synonymis, locis, observationibus, descriptionibus. Supplement 3(2): 1–349.

——. 1798. Supplementum Entomologiae Systematicae. Proft & Storch, Hafniae [=Copenhagen, Denmark]. 572 pp.

Fan, X., Jin, Q. & Li, H. 2008. Seven new species and a checklist of the genus Thecobathra Meyrick from China. Zootaxa 1821: 13–24.

Felder, C. & Felder, R. 1862. Observationes de Lepidopteris nonnullis Chinae centralis et Japoniae. Wiener Entomologische Monatsschrift 6(2): 33–40.

Felder, C., Felder, R. & Rogenhofer, A.F. 1875. Reise der Österreichischen Fregatte Novara um die Erde in den Jahren 1857, 1858, 1859 unter den behilfen des Commodore B. von Wüllerstorf-Urbair. Zoologischer Theil (2) 2 [Lepidoptera]. Vienna. 5 pts., 140 pls.

Felder, C. & Rogenhofer, A.F. 1875. In: Felder, C., Felder, R. & Rogenhofer, A.F. 1875.

Ferguson, D.C., Hilburn, D.J. & Wright, B. 1991. The Lepidoptera of Bermuda: Their food plants, biogeography, and means of dispersal. Memoirs of the Entomological Society of Canada 158: 1–105.

Fischer von Röslerstamm, J.E. 1834–1843. Abbildungen zur Berichtigung und Ergänzung der Schmetterlingskunde, besonders der Microlepidopterologie als Supplement zu Treitschke's und Hübner's europaeischen Schmetterlingen, mit erläuterndem Text. Leipzig. 308 pp., 100 pls.

Fissetchko, R.N. 1970. New species of moths of the genus Argyresthia Hb. form [sic!] fruits of the savin in Talassky Alatau. Zoologischeskii Zhurnal 49(5): 792–795.

Fitch, A. 1856. Insects infesting fruit trees. Supplement. Transactions of the New York State Agricultural Society 16: 478–487.

Fletcher, T.B. 1914. Some South Indian insects and other animals of importance. The Superintendent, Government Press, Madras, India. 565 pp.

——. 1928. Catalogue of Indian Insects. Part 17 – Yponomeutidae. Goverment of India Central Publication Branch, Calcutta, India. 26 pp.

——. 1929. A list of the generic names used for Microlepidoptera. Memoirs of the Department of Agriculture in India (Entomology) 11: 1–244.

——. 1940. New Generic Names for Microlepidoptera. Entomologist's Record and Journal of Variation 52(10): 109.

Foggo, A. 1992. A possible role for the ash bud moth Prays fraxinella (Lep.: Yponomeutidae) in the dieback of hedgerow ash trees. Oxford Forestry Institute, Occasional Papers 41: 24–30.

Forbes, W.T.M. 1923. Lepidoptera of New York and Neighboring States Part 1. Cornell University Agricultural Experiment Station Memoir 68. Cornell University Press, Ithaca, New York, usa. 729 pp.

Forstén, A., Silfverberg, H. & Terhivuo, J. 1998. Accessions to the Zoological Museum of the Finnish Museum of Natural History, University of Helsinki/Helsingfors in 1997. Memoranda Societatis pro Fauna et Flora Fennica 74: 89–91.

Fracker, S.B. 1915. The classification of lepidopterous larvae. Illinois Biological Monographs 2(1): 1–161.

Freeman, T.N. 1960. Needle-mining Lepidoptera of pine in North America. The Canadian Entomologist 92 (Suppl. 16): 1–51.

——. 1967. Annotated keys to some Nearctic leaf-mining Lepidoptera on conifers. The Canadian Entomologist 99(4): 419–435.

——. 1972. The coniferous feeding species of Argyresthia in Canada. The Canadian Entomologist 104(5): 687–697.

Frey, H. 1856. Die Tineen und Pterophoren der Schweiz. Meyer und Zeller, Zürich. 430 pp.

——. 1870 [1871]. Ein Beitrag zur Kenntniss der Microlepidopteren. Mitteilungen der Schweizerischen Entomologischen Gesellschaft 3(5): 244–256.

——. 1880. Die Lepidopteren der Schweiz. Englemann, Leipzig, Germany. 454 pp.

——. 1882. Zweiter nachtrag zur Lepidopteren-fauna der Schweiz. Mitteilungen der Schweizerischen Entomologischen Gesellschaft 6(7): 349–375.

Freyer, C.F. 1839–1842. Neuere Beiträge Schmetterlingskunde mit Abbildungen nach der Natur, Vierter Band. Carl Kollman, Augsburg, Germany. 167 pp.

Friese, G. 1960. Revision der Paläarktischen Yponomeutidae unter besonderer Berücksichtigung der Genitalien (Lepidoptera). Beiträge zur Entomologie 10(1/2): 1–131.

——. 1962. Beitrag zur Kenntnis der Ostpaläarktischen Yponomeutidae. Beiträge zur Entomologie 12(3/4): 299–331.

——. 1963. Zoologische Ergebnisse der Mazedonienreisen Friedrich Kasys, IV. Teil Lepidoptera: Argyresthiidae. Österreichische Akademie der Wissenschaften Mathematisch-Naturwissenschaftliche Klasse Abhandlungen I 172(9/10): 403–412.

——. 1966. Ergebnisse der Albanien-Expedition 1961 des Deutschen Entomologischen Institutes. Beiträge zur Entomologie 16(3/4): 447–459.

——. 1969. Beiträge zur Insekten-Fauna der ddr: Lepidoptera – Argyresthiidae. Beiträge zur Entomologie 19(7/8): 693–752.

Friese, G. & Moriuti, S. 1968. Two new species of the Argyresthia from Japan. Transactions of the Lepidopterological Society of Japan (Tyō To Ga) 19(1): 13–15.

Furniss, R.L. & Carolin, V.M. 1977. Western Forest Insects. Miscellaneous Publication No. 1339. u.s. Department of Agriculture, Forest Service. Washington, d.c., usa. 654 pp.

Gaedike, R. 1970. Revision der paläarktischen Acrolepiidae. Entomologische Abhandlungen 38(1): 1–54.

——. 1979. Nomenklatorische Bemerkungen zu palaearktischen Acrolepiiden (Lepidoptera). Beiträge zur Entomologie 29(2): 413–414.

Gaj, A.J. 1954. Euhyponomeutoides albithoracellus gen. nov., spec. nov. Entomologische Berichten (Amsterdam) 15(1): 11–12.

Gal, T. & Szeoke, K. 2001. Appearance of Argyresthia trifasciata Staudinger, 1871 (Lepidoptera, Yponomeutidae) on Juniperus in Hungary. Novenyvedelem 36(6): 301–304.

Gardner, O.E., Markin, G.P. & Hodges, C.S. Jr. 1988. Survey for potential biological control agents for Myrica faya in the Azores and Madeira. Cooperative National Park Resources Studies Unit, University of Hawaii at Manoa, National Park Service, Technical Report 63: 1–18.

Gaumer, G.F. 1875. Observation on the habits of certain larvae. Transactions of the Kansas Academy of Science 4: 23–24.

——. 1878. Hyponomeuta wakarusa, new species. Observer of Nature 2(1): 3.

Gershenson, Z.S. 1969. A new species of the genus Yponomeuta Latr. from the Far East. Entomologicheskoye Obozreniye 48(1): 199–200.

——. 1972. New species of ermine moths of the genus Yponomeuta Latr. from Central Asia. Entomologicheskoye Obozreniye 51(3): 635–637.

——. 1974. Yponomeutidae & Argyresthiidae. Fauna Ukraini 15(6): 1–132.

——. 1977. Ermine moths of the genus Yponomeuta Latr. in the Far East. Entomologicheskoye Obozreniye 56(1): 150–155.

——. 1979. A new Yponomeutidae moth from Armenia. Trudy Vsesoyuznogo Entomologicheskogo Obshchestva 61: 103–104.

——. 1981. Yponomeutidae. In: Medvedev, G.S. (ed.), Keys to the insects of the European Part of the USSR, Vol. 4 Lepidoptera, Part 2. pp. 331–347. Nauka, Leningrad.

——. 1985. A new and little-known species of Yponomeuta Latreille from the soviet Far East. Entomologicheskoye Obozreniye 64(3): 580–583.

——. 1986. Annotated list of Small Ermine Moths (Lepidoptera, Yponomeutidae) of the USSR. Trudy Vesesoiuznogo Entomologischeskogo Obshchestva 67: 10–19.

——. 1988. A new species of the moth genus Argyresthia from the Far East. Vestnik Zoologii 1988(5): 84–86.

——. 2003. Two new species of yponomeutoid moths from Madagascar. Vestnik Zoologii 37(3): 93–95.

Gershenson, Z.S. & Kozhevnikova, V.A. 2004. Type specimens of Yponomeutoid moths in the collection of the Zoological Institute, Russian Academy of Sciences (St. Petersburg). Entomologicheskoye Obozreniye 83(3): 640–644.

——. 2005. Types of the Yponomeutoid moth species in the collection of Schmalhausen Institute of Zoology National Academy of Sciences, Ukraine. Entomologicheskoye Obozreniye 84(3): 576–578.

Gershenson, Z.S. & Ulenberg, S.A. 1998. Yponomeutinae of the World exclusive of the Americas. Koninklijke Nederlandse Akademic van Wetenschappen Verhandelingen Afdeling Natuurkunde, Tweede Reeks. Deel 99. Amsterdam, The Netherlands. 202 pp.

Gibeaux, C. 1982. Description d'une forme nouvelle de Kessleria egregiella Duponchel. Alexanor 12(8): 371–372.

——. 1983a. Révision des Argyresthiidae de France et de Belgique et description d'Argyresthia pruniella granonensis ssp. nova. Linneana Belgica 9(3): 146–162.

——. 1983b. Étude des microlépidoptères de Madagascar II: Description de neuf espèces nouvelles d'Argyresthia. Revue française d'Entomologie (Nouvelle série) 5(3): 120–128.

——. 1984. Étude des microlépidoptères de Madagascar III: Description de Swammerdamia villiersi, n. sp. Revue française d'Entomologie (Nouvelle série) 6(2): 83–84.

——. 1985a. Scythris elegantella (Lucas, 1955), comb. nov., =Scythris limbelloides Jäckh, 1977, syn. nov. Alexanor 14(1): 22–23.

——. 1985b. Étude des microlépidoptères de Madagascar IV: Description de trois espèces nouvelles d'Yponomeutides. Nouvelle Revue d'Entomologie, Nouvelle Série 2(3): 277–280.

——. 1992 [1993]. Description de deux Argyresthiinae nouveaux. Entomologica gallica 3(4): 189–192.

——. 1993. Description d'Argyresthia tarmanni, espèce nouvelle pour la science d'Autriche. Entomologica gallica. 4(4): 161–162.

Godart, J.B. & Duponchel, P.A. 1838 [1840]. Histoire naturelle des lépidoptères ou papillons de France. Nocturnes. Tome 11. Méquignon-Marvis, Paris, France. 720 pp.

Goeze, J.A.E. 1783. Entomologische Beyträge zu des Ritter Linne' zwölften Asgabe des Natursystems, Dritten Theiles, Vierter Band. Wiedmann, Leipzig, Germany. 178 pp.

Gómez-Bustillo, M.R. 1979. Reestructuración de los Yponomeutoidea (Fracker, 1915) sensu Heppner, 1977, y puesta al dia de sus categorías familiares en la Península Ibérica I. shilap Revista de Lepidopterologia 7(28): 235–244.

Goodey, B. 2000. Argyresthia cupressella Wals. and A. trifasciata Stgr. (Lep.: Yponomeutidae) in Essex. Entomologist's Record and Journal of Variation 112: 169.

Gozmany, L.A. 1960. Records on Microlepidoptera. Annales Historico-Naturales Musei Nationalis Hungarici 52: 423–427.

Gozmany, L.A. & Vári, L. 1973. The Tineidae of the Ethiopian Region. Transvaal Museum Memoir 18: 1–238.

Grote, A.R. 1865. Notes on the Bombycidae of Cuba. Proceedings of the Entomological Society of Philadelphia 5: 227–255.

——. 1873. Contributions to a knowledge of North American moths. Bulletin of the Buffalo Society of Natural Sciences 1: 73–94.

——. 1874. Notes on American Lepidoptera, with Descriptions of twenty-one new species. Bulletin of the Buffalo Society of Natural Sciences 2: 145–163.

Guenée, M.A. 1845a. Essai sur une nouvelle classification des Microlépidoptères. Annales de la Société entomologique de France 2(3): 297–344.

——. 1845b. Europaeorum Microlepidopterorum index methodicus, sive Pyrales, Tortrices, Tineae et Alucitae Linnaei, secundum novum naturalemque ordinem dispositae, nominibus genuinis restitutis, synonymia accurate elucidata, locis indicatis, novisque speciebus aut larvis brevidescriptis. Pars 1 sistens Tortrices, Phycidas, Crambidas, Tinearumque initium. Roret, Paris. 106 pp.

——. 1879. Étude sur les Yponomeutides. Annales de la Société entomologique de France (Série 5) 9(18–19): 281–290.

Guérin-Méneville, F.E. 1829–1844. Iconographie du règne animal de G. Cuvier, ou représentation d'après nature de l'une des espèces les plus remarquables et souvent non encore figurées de chaque genre d'animaux; avec un texte descriptif mis au courant de la Science. Ouvrage pouvant servir d'atlas a tous les traités de zoologie, Insectes. Paris, France. 576 pp.

Handlirsch, A. 1925. Systematische Übersicht. In: Schröder, C. (ed.), Handbuch der Entomologie, Vol. 3. pp. 377–1143. Gustav Fischer, Jena.

Haworth, A.H. 1802. Prodromus Lepidopterorum Britannicorum: A concise catalogue of British lepidopterous insects, with the times and places of appearance in the winged state. Holt, London, England. 39 + 6 pp.

——. 1811. Lepidoptera Britannica; Sistens digestionem novam insectorum lepidopterorum quæ in magna Britannia reperiuntur, larvarum pabulo, temporeque pascendi; expansione alarum; mensibusque volandi; synonymis atque locis observationibusque variis, Part III. R. Tayor, London. pp. 377–512.

——. 1828. Lepidoptera Britannica; Sistens digestionem novam insectorum lepidopterorum quæ in magna Britannia reperiuntur, larvarum pabulo, temporeque pascendi; expansione alarum; mensibusque volandi; synonymis atque locis observationibusque variis, Part IV. R. Taylor, London. pp. 512–609.

Hayward, K.J. 1969. Datos para el studio de la ontogenia de lepidopteros argentinos. Fundación Miguel Lillo, Miscelánea 31: 1–142.

Heinemann, H.v. 1854. Zehn neue Microlepidoptern. Zeitschrift für Entomologie, Breslau 8: 1–7.

——. 1870. Die Schmetterlinge Deutschlands und der Schweiz. Zweite Abtheilung, Kleinschmetterlinge, Band I. Die Motten und Federmotten, Heft I.C.A. Schwetschke und Sohn, Braunschweig. pp. 1–388.

Heinemann, H.v. & Wocke, M.F. 1876. Die Schmetterlinge Deutschlands und der Schweiz. Zweite Abtheilung, Kleinschmetterlinge, Band II. Die Motten und Federmotten, Heft II. C.A. Schwetschke und Sohn, Braunschweig. pp. 389–825. Note: Kirby, 1878: 187 stated although 1877 is imprinted on the title page, Heft II was printed no later than November, 1876.

Hemming, F. 1937. Hübner: A bibliographical and systematic account of the entomological works of Jacob Hübner, and of the supplements thereto by Carl Geyer, Gotfried Franz von Frölich, and Gottleib August Wilhelm Herrich-Schäffer. London: Royal Entomological Society of London (2 volumes).

Heppner, J.B. 1977. The status of the Glyphipterigidae and a reassessment of relationships in Yponomeutoid families and Ditrysian superfamilies. Journal of the Lepidopterists' Society 31: 124–134.

——. 1982. Synopsis of the Glyphipterigidae (Lepidoptera: Copromorphoidea) of the World. Proceedings of the Entomological Society of Washington 84(1): 38–66.

——. 1984a. Atlas of Neotropical Lepidoptera. Checklist, Part 1 Micropterigoidea – Immoidea. W. Junk Publishers, The Hague, The Netherlands. 112 pp.

——. 1984b. Revision of the Oriental and Nearctic genus Ellabella (Lepidoptera: Copromorphidae). Journal of Research on the Lepidoptera 23(1): 50–73.

——. 1985. Telosphrantis aethiopica Meyrick transferred to Choreutidae. Proceedings of the Entomological Society of Washington 87(4): 859–863.

——. 1992. Yponomeutidae & Argyresthiidae. In: Heppner, J.B. & Inoue, H. (eds.), Lepidoptera of Taiwan 1(2): Checklist. pp. 74–75. Scientific Publishers Inc./Flora & Fauna Books, Gainesville.

——. 1995. Lacturidae, new family. Tropical Lepidoptera 6(2): 146–148.

——. 1998. Classification of Lepidoptera, Part 1. Introduction. Holarctic Lepidoptera 5 (Suppl. 1): 1–148.

——. 2005. Primitive sedge moths from New Zealand and Tasmania: Transfer of Proditrix and relatives to Orthotelinae (Lepidoptera: Glyphipterigidae). Lepidoptera News 2003(1–2): 31–42.

——. 2009a. A new Atteva moth from Peru, with notes on Atteva numeratrix from Brazil. Lepidoptera Novae 2(2): 105–112.

——. 2009b. A new Atteva moth from Ecuador, with notes on Atteva zebra from Panama. Lepidoptera Novae 2(2): 125–128.

Heppner, J.B. & Duckworth, W.D. 1981. Classification of the Superfamily Sesioidea. Smithsonian Contributions to Zoology 314. Smithsonian Institution Press, Washington, d.c., usa. 144 pp.

Hering, E.M. 1951. Biology of the Leaf Miners. W. Junk Publishers, The Hague, The Netherlands. 420 pp.

——. 1957. Bestimmungstabellen der Blasttminen von Europa einschliesslich des Mittelmeerbecken und der Kanarischen Inseln. 3 Bände. Uitgeverij Dr. W. Junk, Gravenhage, The Netherlands. Band 1: 1–648, Band 2: 649–1185, Band 3: 1–221.

Herrebout, W.M., Menken, S.B.J., Povel, G.D.E. & de Water, T.P.M. van 1982. The position of Yponomeuta yanagawanus Matsumura (Lepidoptera, Yponomeutidae). Netherlands Journal of Zoology 32(3): 313–323.

Herrich-Schäffer, G.A.W. 1850–1858. Sammlung neuer, oder wenig bekannter, Aussereuropäischer Schmetterlinge. G.J. Manz, Regensburg, Germany. 84 pp., 120 pls.

——. 1853–1855. Systematische Bearbeitung der Schmetterlinge von Europa zugleich als Text, Revision und Supplement zu Jakob Hübner's Sammlung europäischer Schmetterlinge. 5. Die Schaben und Federmotten. G.J. Manz, Regensburg, Germany. 394 pp., 124 pls.

——. 1854. see Herrich-Schäffer, 1853–1855.

——. 1856–1861. Neue Schmetterlinge aus Europa und den angrenzenden Ländern. Regensburg, Germany. 32 pp., 185 figs.

——. 1866. Correspondez-Blatt des zoologisch-mineralogischen Vereines in Regensburg. Zwanzigster Jahrgang. Neubauer'schen Buchdruckerei, Regensburg, Germany. 216 pp.

Hewitson, W.C. & Moore, F. 1879 [1888]. Descriptions of new Indian lepidopterous insects from the collection of the late Mr. W.S. Atkinson. Asiatic Society of Bengal, Calcutta, India. 299 pp.

Heydenreich, G.H. 1851. Lepidopterorum Europaeorum Catalogus Methodicus. Systematisches Verzeichniss der Europaeischen Schmetterlinge. Dritte Ausgabe. Julius Klinkhardt, Leipzig, Germany. 130 pp.

Hoare, R.J.B. 2005. Hierodoris and overview of Oecophoridae. Fauna of New Zealand 54: 1–100.

Horak, M. & Common, I.F.B. 1996. Heliocosma Group. In: Nielsen, E.S., Edwards, E.D. & Rangsi T.V. (eds.), Checklist of the Lepidoptera of Australia. Monographs on Australian Lepidoptera 4: 137. csiro, Victoria, Australia.

Horsfield, T. & Moore, F. 1858–1859. Catalogue of the lepidopterous insects in the Museum of Natural History at the East-India House, Vol. II. W.H. Allen, London. pp. [279]–440.

Hua, L.-Z. 2005. List of Chinese Insects 3. Sun Yat-sen University Press, Guangzhou, China. 595 pp.

Huang, J. 1982. A new genus and species of Yponomeutidae from China. Entomotaxonomia 4 [numbered 6 in error] (4): 269–271.

Hübner, J. 1786 Beitrage zur Geschichte der Schmetterlinge von Jacob Hübner. Verlag, Augsburg. 1(1): 1–31. Note: Dates and details of publications by Hübner were obtained from Hemming, 1937, A bibliographical and systematic account of the entomological works of Jacob Hübner, and of the supplements thereto by Carl Geyer, Gotfried Franz von Frölich, and Gottlieb August Wilhelm Herrich-Schäffer. 2 volumes.

——. 1786–1789. Beitrage zur Geschichte der Schmetterlinge von Jacob Hübner. Verlag, Augsburg. 128 pp. Note: For details and dates of publication of Hübner's works, see Hemming, 1937.

——. 1793. Sammlung auserlesener Vögel und Schmetterlinge mit ihren Namen herausgegeben auf hundert nach der Natur ausgemalten Kupfern. J. Hübner Verlag, Augsburg. 16 pp. + 100 pls.

——. 1793–1842. Geschichte Europäischer Schmetterlinge. J. Hübner Verlag, Augsburg. 8 pp. + 123 pls. For details and dates of publication of Hübner's works, see Hemming, 1937.

——. 1796–1838. Sammlung europäischer Schmetterlinge. Achte Horde. Tineae Die Schaben; nach der Natur geordnet, beschrieben und vorgestellt. J. Hübner Verlag,

500 pls, 9 volumes (Hordes). Note: Hübner organized Lepidoptera into groups he called hordes; in this publication, we refer to them as volumes. The number of the Horde (volume) follows the word Lepidoptera on each plate. For details and dates of publication of Hübner's works, see Hemming, 1937.

——. 1806 [1873]. Tentamen determinationis digestionis atque denominationis singularum stirpium Lepidopterorum, peritis ad inspiciendum et dijudicandum communicatum. Reprinted in facsimile by S.H. Scudder, 1873, Cambridge, Massachusetts, usa. 2 pp. Note: For details and dates of publication of Hübner's works, see Hemming, 1937.

——. [1809]. Geschichte Europäischer Schmetterlinge. Larvae, Lepidoptera, 8, Tinea 3, Tortriciformes, B, d, figs. 1a–b [see Hübner 1793–1842].

——. [1813]. Sammlung Europäischer Schmetterlinge 3, Tineiformes 2, Pyralidiformes B: pl. 53, fig. 260 (error 360) [see Hübner, 1796–1838].

——. 1816–1826. Verzeichniss bekannter Schmettlinge [sic]. J. Hübner Verlag, Augsburg. 431 pp., 72 pls.

——. 1822. Systematisches-alphabetisches Verzeichniss aller bischer bey den Fürbuldungen zur Sammlung Europäischer Schmetterlinge, ausgegebenen Benennungen; mit Vormerkung auch augsburgscher Gattungen. J. Hübner Verlag, Augsburg. vi + 81 pp.

——. 1825 [1826]. Verzeichniss bekannter Schmettlinge [sic]. J. Hübner Verlag, Augsburg: 412–431 [see Hübner 1816–1826].

——. 1825 [1831]. Zuträge zur Sammlung exotischer Schmetterlinge: bestehend in Bekundigung einzelner Fliegmuster neuer oder rarer nichteuropäischer Gattungen. 1825. 3. 48 pp.

Huemer, P. 2009. Argyresthia friulii sp. n., a new species from the Julian pre-alps, northern Italy. Gortania Botanica, Zoologia 31: 137–142.

Huemer, P. & Tarmann, G. 1991. Westpaläarktische Gespinstmotten der Gattung Kessleria Nowicki: Taxonomie, Ökologie, Verbreitung. Mitteilungen der Münchner Entomologischen Gesellschaft 81: 1–110.

——. 1993. Wissenschaftliche Ergebnisse der lepidopterologischen Forschungen des Tiroler Landesmuseum Ferdinandeum im Alpenraum – 2. Eine neue Kessleria-Art aus den Alpi Orobie. Veröffentlichungen des Tiroler Landesmuseum Ferdinandeum 73: 41–46.

iczn (International Commission on Zoological Nomenclature) 1999. International Code of Zoological Nomenclature, Fourth Edition. The International Trust for Zoological Nomenclature, London, England. 306 pp.

Inoue, H., Sugi, S., Kuroko, H., Moriuti, S. & Kawabe, A. 1982. Moths of Japan, Vol. 1. Text, Vol. 2. Plates and Synonymic Catalogue. Kodansha, Tokyo, Japan, Vol. 1: 966 pp., Vol. 2: 552 pp.

Issiki, S.T. 1957. Lepidoptera. In: Esaki, T. et al. (eds.), Iconographia Insectorum Japonicorum in Coloribus Naturalibus, Part 1. pp. 1–318. Hoikusha, Osaka, Japan.

Janse, A.J.T. 1917. Check-List of the South African Lepidoptera Heterocera. Transvaal Museum, Pretoria, South Africa. 219 pp.

Jin, Q., Fan, X. & Wang, S. 2013. Genus Diaphragmistis Meyrick new to China, with descriptions of two new species. Entomological News 123(3): 237–240.

Jin, Q. & Li, H. 2012. Review of the genus Niphonympha Meyrick, 1914 from China, with description of one new species. Entomotaxonomia 34(2): 253–258.

Jin, Q. & Wang, S. 2008. Taxonomic revision of the genus Metanomeuta Meyrick. Acta Zootaxonomica Sinica 33(1): 49–56.

——. 2009. Euhyponomeuta new to China, with description of a new species. Oriental Insects 43: 271–273.

Jin, Q., Wang, S. & Li, H. 2009. Taxonomic study of the genus Teinoptila Sauber, 1902 from China. Zootaxa 2249: 44–50.

——. 2010. Study of the genus Klausius Moriuti, 1977, with descriptions of two new species. Entomological News 121(1): 1–5.

Junnilainen, J. 2001. Two new species of Swammerdamia from the southern Ural Mountains and southern Siberia. Entomologica Fennica 12(2): 112–118.

Kaçar, G. & Ulusoy, M.R. 2011. Zelleria oleastrella (Milliere) (Lepidoptera: Yponomeutidae), a new record for the fauna of Turkey. Turkish Journal of Zoology 35(6): 891–892.

Kaltenbach, J.H. 1874. Die Pflanzenfeinde aus der Klasse der Insekten. Ein nach Pflanzenfamilien geordnetes Handbuch sämmtlicher auf den einheimischen Pflanzen bisher beobachteten Insekten zum Gebrauch für Entomologen, Insektensammler, Botaniker, Land-und Forstwirthe und Gartenfreunde. Julius Hoffmann, Stuttgart, Germany. 848 pp.

Karisch, T. 2002. Eine neue Argyresthia-Art von Bioko. Lambillionea 102(2): 169.

Karsholt, O. & Nielsen, E.S. 1976. Notes on some Lepidoptera described by Linnaeus, Fabricius, and Ström. Entomologica Scandinavica 7(4): 241–251.

——. 1985. The Lepidoptera described by C.P. Thunberg. Entomologica Scandinavica 16: 433–463.

Karsholt, O & Razowski, J. 1996. The Lepidoptera of Europe. A Distributional Checklist. Apollo Books, Stenstrup, Denmark. 380 pp.

Kearfott, W.D. 1903. Descriptions of new Tineoidea. Journal of the New York Entomological Society 11: 145–165.

——. 1908. New North American Tortricidæ and Tineina. Journal of the New York Entomological Society 16: 167–188.

Keifer, H.H. 1930. Argyresthias found in Golden Gate Park, San Francisco. The Pan-Pacific Entomologist 7(2): 76.

Kirby, W.F. 1870. Lepidoptera. Zoological Record 7: 362–425.

——. 1878. Lepidoptera. Zoological Record 13: 140–191.

——. 1879. Lepidoptera. Zoological Record 16: 122–190.

——. 1892. A synonymic catalogue of Lepidoptera Heterocera, 1. Sphinges and Bombyces. Gurney & Jackson, London, England. 951 pp.

Klimesch, J. 1983. Zelleria wolffi spec. nov. Nota lepidopterologica 6(2–3): 121–126.

——. 1992. Prays friesei sp. nov. Entomofauna zeitschrift für Entomologie 13(4): 113–120.

Kloet, G.S. & Hincks, W.D. 1945. A Check List of British Insects. T. Buncle & Co. Ltd., Stockport, England. 483 pp.

Koçak, A.Ö. 1981. Further notes on the homonym of the specific names of Lepidoptera. Priamus 1(1): 11–17.

——. 1982. Additions and corrections to the names published in "Systematic and synonymic list of the Lepidoptera of France, Belgium, and Corsica" by Leraut, 1980. Priamus 2(3): 97–133.

Krämer, K. 1960. Zur Biologie der Eberschenmotte (Argyresthia conjugella Zell.). Anzeiger für Schädlingskunde 33(7): 102–107

Krulikowski, L. 1909. Petites notices lépidoptérologiques. X. Revue Russe d'Entomologie 8(3/4): 270–276.

Kulfan, M. & Holecová, M. 2010. Potravné nároky húseníc motýľov (Lepidoptera) troficky viazaných na pôvodné druhy borovíc (Pinus spp.) Slovenska. Folia faunistica Slovaca 15(7): 47–54.

Kuo, B.S. 1981. Study on two ermine moths Teinoptila antistatica (Meyrick) Xyrosaris sp. in Yunnan Province. Zoological Research 2(3): 265–271.

Küppers, P.V. 2008. Kleinschmetterlinge, Erkennen und Bestimmen. Fauna Verlag, Nottuln, Germany. 399 pp.

Kutenkova, N.N. 1980. Paraswammerdamia lapponica Pet., new for the ussr. Revue d'entomologie de l'URSS 59: 627–628.

Kuznetzov, V.I. 1958. Two new types of moths (Lepidoptera, Microheterocera) harmful to shrubbery in Armenia. Doklady Akademii Nauk Armyanskoi ssr 27(1): 53–57.

Kuznetzov, V.I. & Stekolnikov, A.A. 1977. Phylogenetic relationships between the superfamilies Psychoidea, Tineoidea and Yponomeutoidea in the light of the functional morphology of the male genitalia. Part 2. Phylogenetic relationships of the families and subfamilies. Entomologicheskoe Obozrenie 56: 19–30.

Kyrki, J. 1984. The Yponomeutoidea: A reassessment of the superfamily and its suprageneric groups (Lepidoptera). Entomologica Scandinavica 15: 71–84.

——. 1988. The systematic position of Wockia Heinemann, 1870, and related genera. Nota lepidopterologica 11(1): 45–69.

——. 1990. Tentative reclassification of holarctic Yponomeutoidea. Nota lepidopterologica 13(1): 28–42.

Landry, B. & Landry, J.-F. 1998. Yponomeutidae of the Galápagos Islands, with description of a new species of Prays. Tropical Lepidoptera 9(1): 31–40.

Landry, J.F., Nazari, V., Dewaard, J.E., Mutanen, M., Lopez-Vaamonde, C., Huemer, P. & Hebert, P.D.N. 2013. Shared but overlooked: 30 species of Holarctic Microlepidoptera revealed by dna barcodes and morphology. Zootaxa 3749(1): 1–93.

Latreille, C. 1796. Précis des caractères génériques des Insectes, disposés dans un ordre naturel. F. Bordeaux, Brive[-la-Gaillarde], France. xiv + 202 pp.

Latreille, P.A. 1802. Histoire naturelle, générale et particulière, des Crustacés et des Insectes, Vol. 3. Dufart, Paris, France. 468 pp.

Lawrence, J.F. & Powell, J.A. 1969. Host relationships in North American fungus-feeding moths. Bulletin of the Museum of Comparative Zoology 138(2): 29–51.

Leather, S.R. 1986. Insects on bird cherry I. The bird cherry ermine moth, Yponomeuta evonymellus (L.) (Lepidoptera: Yponomeutidae). Entomologist's Gazette 37: 209–213.

Leraut, P. 1980. Liste systématique et synonymique des Lépidoptères de France, Belgique et Corse. Supplément A: Alexanor, Revue des Lépidoptéristes français et au Bulletin de la Société entomologique de France. Paris, France. 334 pp.

Lhomme, L. 1946. Catalogue des Lépidoptères de France et de Belgique, Vol. 2 (2). Le Carriol, par Douelle, Lot. pp. 173–487.

Li, H. & Fan, X.-M. 2007. Study on the genus Swammerdamia Hübner from China, with descriptions of two new species. Acta Zootaxonomica Sinica 32(3): 556–560.

Li, H. & Jin, Q. 2012. Analysis of the biogeographical distribution of Yponomeutidae and Ypsolophidae (Lepidoptera: Yponomeutoidea). Sciencepaper Online. http://www.paper.edu.cn/releasepaper/content/201212-1216.

Liebhold, A.M., McManus, M.L., Otvos, I.S. & Fosbroke, S.L.C. (eds.), 2001. Proceedings: Integrated Management and Dynamics of Forest Defoliating Insects. pp. 31–36. Northeastern Research Station, Victoria, Australia.

Lienig, F. & Zeller, P.C. 1846. Lepidopterologische Fauna von Lievland und Curland. Isis von Oken 39(3/4): 202–302.

Linnaeus, C. 1758. Systema Naturae per Regna Tria Naturae, secundum Classes, Ordines, Genera, Species, cum characteribus, differentis, synonymis, locis. Tomus 1. Editio Decima, Reformata. Laurentii Salvii, Holmiæ [=Stockholm, Sweden]. iv + 824 pp.

——. 1761. Fauna Suecica, sistens Animalia Sueciæ Regni: Mammalia, Aves, Amphibia, Pisces, Insecta, Vermes. Distributa per Classes & Ordines, Genera & Species, cum Differentiis specierum, synonymis auctorum, nomnibus incolarum, locis natalium, descriptionibus insectorum. Laurentii Salvii, Stockholmiæ [=Stockholm, Sweden]. 48 + 578 pp.

——. 1767. Systema Naturae per Regna Tria Naturae, secundum Classes, Ordines, Genera, Species, cum characteribus, differentis, synonymis, locis. Tomus 1. Editio Duodecima, Reformata. Laurentii Salvii, Holmiæ [=Stockholm, Sweden]. pp. 553–1327.

Liu, Y. 1980. A study of Chinese Thecobathra Meyrick. Entomotaxonomia 2(1): 33–40.

——. 1983. Two new species of Thecobathra Meyrick. Acta Zootaxonomica Sinica 8: 80–83.

——. 1984. One new species of Thecobathra Meyrick in Yunnan Province. Acta Zootaxonomica Sinica 9: 324–325.

Liu, Y. & Huang, J. 1996. A study of the Chinese Yponomeuta. Forest Pests and Diseases 3: 1–4.

López, S.M. 2007. Sanidad forestal: guía en imágenes de plagas, enfermedades y otros agentes presentes en los bosques. Mundi-Prensa Libros, S.A., Madrid. 575 pp.

Lower, O.B. 1894. On new Australian Heterocera. Transactions of the Royal Society of South Australia 18: 77–113.

——. 1899. Descriptions of new Australian Lepidoptera. Proceedings of the Linnean Society of New South Wales 24: 83–116.

——. 1900. Descriptions of new Australian Lepidoptera. Proceedings of the Linnean Society of New South Wales 25: 403–423.

——. 1908. New Australian Lepidoptera. No. XXV. Transactions and Proceedings of the Royal Society of South Australia 32: 110–120.

Lucas, D. 1955. Nouveaux Lépidoptères nord-africans. Bulletin de la Société des sciences naturelles du Maroc 35: 251–258.

Ma, G. & Sun, Y. 1982. Argyresthia assimilis Moriuti – a new record from China. Entomotaxonomia 4(4): 315–316.

MacKay, M.R. 1972. Larval sketches of some microlepidoptera, chiefly North American. Memoirs of the Entomological Society of Canada 104(S88): 7–83.

Malinoski, M.K. & Clement, D. 2009. Broadleaved shrubs and shade trees: Problems, picture clues, and management options. Natural Resource, Agriculture, and Engineering Service, Ithaca, New York, usa. 180 pp.

Maneval, H. 1934. Notes et remarques sur divers microlépidoptères de la Haute-Loire avec description d'une Zelleria nouvelle. L'Amateur de Papillons. Revue de Lépidoptérologie, Travaux et Observations d'Amateurs 7: 51–57.

Martelli, G. 1915. Intorno a due Specie di Lepidotteri dei Generi Zelleria e Glyphodes viventi sull'olivo. Bollettino del Laboratorio di Zoologia Generale e Agraria della R. Scuola Superiore d'Agricoltura in Portici 10: 89–97.

Matsumura, S. 1931. 6000 Illustrated Insects of the Japan-Empire. Toko Shoin, Tokyo, Japan. 1496 pp.

Menken, S.B.J., Herrebout, W.M. & Wiebes, J.T. 1992. Small ermine moths (Yponomeuta): Their host relations and evolution. Annual Review of Entomology 37: 41–66.

Mey, W. 2007. Microlepidoptera: Smaller families. In: Mey, W. (ed.), The Lepidoptera of the Brandberg Massif in Namibia. Part 2. Esperiana Memoir 4: 9–30.

——. 2011. Basic pattern of Lepidoptera diversity in southwestern Africa. Esperiana Memoir 6: 7–316.

Meyrick, E. 1880. Descriptions of Australian Micro-Lepidoptera. Proceedings of the Linnean Society of New South Wales (1) 5: 204–271.

——. 1887a. Descriptions of new Lepidoptera. Proceedings of the Linnean Society of New South Wales, 2nd Series 1: 1037–1048.

——. 1887b. Descriptions of New Zealand Tineina. Transactions of the New Zealand Institute for 1887 20: 77–106.

——. 1889. Descriptions of New Zealand Micro-Lepidoptera. Transactions of the New Zealand Institute for 1888 21: 155–188.

——. 1892 [1893]. Descriptions of Australian Micro-Lepidoptera. XVI. Tineidae. Proceedings of the Linnean Society of New South Wales for the year 1892. Second Series. 7: 477–611. Note: the cover of the Proceedings states it was for sale by the society in 1893, although, the heading at the top of the of the title page for the article, in brackets, reports it was printed Dec. 28, 1892.

——. 1894. On a collection of Lepidoptera from Upper Burma. Transactions of the Entomological Society of London for the year 1894(1): 1–29.

——. 1901. Descriptions of new Lepidoptera from New Zealand. Transactions of the Entomological Society of London 1901(4): 565–579.

——. 1905. Descriptions of Indian Micro-Lepidoptera I. Journal of the Bombay Natural History Society 16(4): 580–619.

——. 1906. Descriptions of Indian Micro-Lepidoptera II. Journal of the Bombay Natural History Society 17(2): 403–417.

——. 1907a. Descriptions of Indian Micro-Lepidoptera III. Journal of the Bombay Natural History Society 17(3): 730–754.

——. 1907b. Descriptions of Indian Micro-Lepidoptera IV. Journal of the Bombay Natural History Society 17(4): 976–994.

——. 1907c. Descriptions of Indian Micro-Lepidoptera V. Journal of the Bombay Natural History Society 18(1): 137–160.

——. 1907d. Descriptions of Australasian Micro-Lepidoptera. XIX. Plutellidae. Proceedings of the Linnean Society of New South Wales 32(1): 47–150.

——. 1908a. New Micro-Lepidoptera from India and Burma. Records of the Indian Museum 2, Part 4(39): 395–400.

——. 1908b. Descriptions of Indian Micro-Lepidoptera VII. Journal of the Bombay Natural History Society 18(3): 613–638.

——. 1908c. Descriptions of African Micro-Lepidoptera. Proceedings of the Zoological Society of London 1908(46–48): 716–755.

——. 1909a. Descriptions of Indian Micro-Lepidoptera IX. Journal of the Bombay Natural History Society 19(2): 410–437.

——. 1909b. Descriptions of Transvaal Micro-Lepidoptera. Annals of the Transvaal Museum 2(1): 1–28.

——. 1909c. New South African Micro-Lepidoptera. Annals of the South African Museum 5(7): 349–379.

——. 1910a. Notes and descriptions of Indian Micro-Lepidoptera. Records of the Indian Museum 5, Part 4(22): 217–232.

——. 1910b. Descriptions of Malayan Micro-Lepidoptera. Transactions of the Entomological Society of London 1910(4): 430–478.

——. 1911a. Reports of the Percy Sladen Trust Expedition to the Indian Ocean in 1905, under the leadership of Mr. J. Stanley Gardiner, m.a. III, part 2, XII. Tortricina and Tineina. Transactions of the Linneaen Society of London, Zoology (2) 14(2): 263–307.

——. 1911b. Descriptions of Indian Micro-Lepidoptera XIV. Journal of the Bombay Natural History Society 21(1): 104–131.

——. 1911c. Descriptions of South African Microlepidoptera. III. Annals of the Transvaal Museum 3(1): 63–83.

——. 1912. New South African Microlepidoptera. III. Annals of the South African Museum 10(3): 53–74.

——. 1913a. Exotic Microlepidoptera (Marlborough) 1(5): 129–160.

——. 1913b. Descriptions of South African Micro-Lepidoptera. Annals of the Transvaal Museum 3(4): 267–336.

——. 1913c Descriptions of South American Lepidoptera. Transactions of the Entomological Society of London for the year 1913. Part 1: 170–200.

——. 1914a. Lepidoptera Heterocera Fam. Glyphipterigidae. Fascicule 164. pp. 1–39. In: Wytsman, P. (ed.), Genera Insectorum, Fascicules. pp. 164–169. V. Verteneuil & L. Desmet, Brussels, Belgium. 556 pp. + pls (each Fascicule numbered beginning with page 1).

——. 1914b. Descriptions of South African Micro-Lepidoptera. V. Annals of the Transvaal Museum 4(4): 187–205.

——. 1914c. Lepidopterorum Catalogus, Pars 19: Hyponomeutidae, Plutellidae, Amphitheridae. W. Junk, Berlin, Germany. 64 pp.

——. 1914d. Exotic Microlepidoptera (Marlborough) 1(6): 161–192.

——. 1914e. Exotic Microlepidoptera (Marlborough) 1(7): 193–224.

——. 1914f. Descriptions of Indian Micro-Lepidoptera. XVIII. Journal of the Bombay Natural History Society 23(1): 118–130.

——. 1915a. Descriptions of South American Micro-Lepidoptera. Transactions of the Entomological Society of London 1915: 201–256.

——. 1915b. Revision of New Zealand Tineina. Transactions and Proceedings of the New Zealand Institute 47: 205–241.

——. 1917. Descriptions of South African Micro-Lepidoptera. Annals of the South African Museum 17(1): 1–21.

——. 1918a. Exotic Microlepidoptera (Marlborough) 2(6): 161–192.
——. 1918b. Descriptions of South African Micro-Lepidoptera. Annals of the Transvaal Museum 6: 1–59.
——. 1919. Exotic Microlepidoptera (Marlborough) 2(8): 225–256.
——. 1921a. Descriptions of South African Micro-Lepidoptera. Annals Mededelingen of the Transvaal Museum 8(2): 49–148.
——. 1921b. Exotic Microlepidoptera 2(14): 417–448.
——. 1921c. Notes and descriptions of New Zealand Lepidoptera. Transactions and Proceedings of the New Zealand Institute 53: 334–336.
——. 1921d. New Micro-Lepidoptera. Zoologische Mededelingen Leiden 6: 145–202.
——. 1922a. Exotic Microlepidoptera (Marlborough) 2(18): 545–576.
——. 1922b. New Microlepidoptera of the German Entomological Institute. Entomologische Mitteilungen 11(2): 44–47.
——. 1922c. New Micro-Lepidoptera. Zoologische Mededeelingen uitgegeven vanwege 'sRijks Museum van Natuurlijke Historie te Leiden 7(1–2): 80–89.
——. 1923a. Micro-Lepidoptera of Rodriguez. Transactions of the Entomological Society of London 1923(3–4): 544–557.
——. 1923b. Voyage du Compte Jacques de Rohan Chabot dans l'Angola. Descriptions d'espèces nouvelles de Microlépidoptères. Bulletin du Muséum national d'histoire naturelle de Paris 1923: 563–566.
——. 1924. Exotic Microlepidoptera (Marlborough) 3(4): 97–128.
——. 1925a. Exotic Microlepidoptera (Marlborough) 3(5): 129–160.
——. 1925b. Lepidoptera Heterocera Fam. Gelechiadae. Fascicule 184. pp. 1–300. In: Wytsman, P. (ed.), Genera Insectorum, Fascicules 184–185. V. Verteneuil & L. Desmet, Brussels, Belgium. 634 pp + pls. (each Fascicule numbered beginning with page 1)
——. 1925c. New Malayan Micro-Lepidoptera. Treubia 6(3–4): 428–433.
——. 1925d. Descriptions of twenty-four new Egyptian Microlepidoptera. Bulletin de la Société royale d'entomologie d'Egypte 1925: 206–219.
——. 1926a. New South African Micro-Lepidoptera. Annals of the South African Museum 23(2): 325–351.
——. 1926b. On Micro-Lepidoptera from the Galapagos Islands and Rapa. Transactions of the Entomological Society of London 74(2): 269–278.
——. 1926c. XI. Microlepidoptera from Northern Sarawak. The Sarawak Museum Journal 3(2)9: 147–169.
——. 1927a. Exotic Microlepidoptera (Marlborough) 3(12): 353–384.
——. 1927b. Descriptions of New Zealand Lepidoptera. Transactions and Proceedings of the New Zealand Institute 58(3): 313–316.
——. 1927c. Insects of Samoa and other Samoan Terrestrial Arthropoda. Part 3. Lepidoptera. Fascicle 2. Micro-Lepidoptera. British Museum of Natural History, London. pp. 65–116.

——. 1928a. Lepidoptera collected during a zoological mission to the Great Atlas of Morocco, 1927 by F. Le Cerf and G. Talbot. Bulletin of the Hill Museum 2(3): 232–240.

——. 1928b. Exotic Microlepidoptera (Marlborough) 3(13): 385–416.

——. 1928c. Exotic Microlepidoptera (Marlborough) 3(14): 417–448.

——. 1928d. A revised handbook of British Lepidoptera. Watkins and Doncaster, London, England. 914 pp.

——. 1928e. The Micro-Lepidoptera of the "St. George" expedition. Transactions of the Entomological Society of London 76: 489–521.

——. 1930a. Ergebnisse einer zoologischen Sammelreise nach Brasilien, insbesondere in das Amazonasgebiet, ausgeführt von Dr. H. Zerny. V. Tiel. Micro-Lepidoptera. Annalen des Naturhistorischen Museums in Wien 44: 223–271.

——. 1930b. Exotic Microlepidoptera (Marlborough) 3(18): 545–576.

——. 1930c. Exotic Microlepidoptera (Marlborough) 3(19): 577–608.

——. 1931a. Exotic Microlepidoptera (Marlborough) 4(3): 65–96.

——. 1931b. Exotic Microlepidoptera (Marlborough) 4(6): 161–192.

——. 1931c. Micro-Lepidoptera from south Chile and Argentina. Anales del Museo Nacional de Historia Natural Bernardino Rivadavia, Buenos Aires (Entomologia) 36: 377–415.

——. 1931d. Die Lepidopteren der Deutschen Gran-Chaco-Expedition 1925/1926. III. Pterophoridae – Tineidae. Mitteilungen der Münchner Entomologischen Gesellschaft 21: 37–46.

——. 1931e. Reports of an expedition to Brazil and Paraguay in 1926–7, supported by the Trustees of the Percy Sladen Memorial Fund and the Executive Committee of the Carnegie Trust for Scotland. Microlepidoptera. Journal of the Linnean Society of London, Zoology 37(252): 277–284.

——. 1932a. Exotic Microlepidoptera (Marlborough) 4(8): 225–256.

——. 1932b. Exotic Microlepidoptera (Marlborough) 4(9): 257–288.

——. 1932c. Exotic Microlepidoptera (Marlborough) 4(11): 321–352.

——. 1932d. Entomological Expedition to Abyssinia 1926–7. Transactions of the Entomological Society of London 80(1): 107–120.

——. 1933. Exotic Microlepidoptera (Marlborough) 4(12): 353–384.

——. 1934. Exotic Microlepidoptera (Marlborough) 4(15): 449–480.

——. 1935a. List of Microlepidoptera of Chekiang, Kiangsu and Hunan. In: Caradja, A. & Meyrick, E. (eds.), Materialien zu einer Microlepidopteren Fauna der Chinesischen Provinzen Kiangsu, Chekiang und Hunan. pp. 44–96. R. Friedländer & Sohn, Berlin.

——. 1935b. Exotic Microlepidoptera (Marlborough) 4(18): 545–576.

——. 1935c. Exotic Microlepidoptera (Marlborough) 4(19): 577–608.

——. 1936a. New species of Pyrales and Microlepidoptera from the Deutsches Entomologisches Institut. Arbeiten über morphologische und taxonomische Entomologie aus Berlin-Dahlem 3(2): 94–109.

——. 1936b. Exotic Microlepidoptera (Marlborough) 4(20): 609–642.
——. 1936c. Exotic Microlepidoptera (Marlborough) 5(2): 33–64.
——. 1937. Exotic Microlepidoptera (Marlborough) 5(5): 129–160.
——. 1938a. Microlepidoptera excl. Pyralidae. In: Caradja, A. & Meyrick, E. (eds.), Materialien zu einer Mikrolepidopterenfauna des Yülingshanmassivs (Provinz Yunnan). Deutsche Entomologische Zeitschrift "Iris" 52(1): 1–29.
——. 1938b. New Javanese Lepidoptera. Deutsche Entomologische Zeitschrift "Iris" 52(6): 73–88.
——. 1938c. Papuan Microlepidoptera. Transactions of the Royal Entomological Society of London 87(21): 503–528.
Mille, C. & Cazeres, S. 2007. Un nouveau ravageur en Nouvelle-Calédonie: La Teigne des fleurs du Citronnier, Prays nephelomima Meyrick (Lepidoptera Yponomeutidae). L'Entomologiste 63: 135–136.
Miller, J.C. & Hammond, P. 2003. Lepidoptera of the Pacific Northwest: Caterpillars and Adults. H.J. Andrews Publication No. 3739. usda Forest Service, Morgantown, West Virginia, usa. 324 pp.
Miller, S.E. & Hodges, R.W. 1990. Primary types of Microlepidoptera in the museum of comparative zoology (with a discursion on V.T. Chambers' work). Bulletin of the Museum of Comparative Zoology 152(2): 45–87.
Millière, P. 1854. Description de nouvelles espèces de microlépidoptères. Annales de la Société entomologique de France 1854: 59–68.
——. 1864. Iconographie et description de chenilles et lépidoptères inédits 2(12). Annales de la Société Linnéenne de Lyon, 1864: 37–67.
——. 1867. Iconographie et description de chenilles et lépidoptères inédits 2(18). Annales de la Société Linnéenne de Lyon, 1867: 281–328.
——. 1869. Iconographie et description de chenilles et lépidoptères inédits 3(24). Annales de la Société Linnéenne de Lyon, 1869: 42–80.
——. 1873. Lépidoptères nouveaux de France. Petites Nouvelles Entomologiques 5(77): 310.
——. 1880 [1882]. Lépidoptérologie. 8. H. Vidal, Cannes, France. 27 pp., 4 pls.
——. 1883. Lépidoptérologie. 8. Annales de la Société Linnéenne de Lyon Nouvelle série 29: 153–179, 4 pls.
Minelli, A., Ruffo, S. & La Posta, S. 1996. Checklist delle Specie della Fauna Italiana. Commission of the European Communities, Edizioni Calderini, Bologna. pp. 4–6.
Minet, J. 1986. Ebauche d'une classification moderne de l'ordre des Lépidoptères. Alexanor 14: 291–313.
Mirzoyan, S.A. & Grigoryan, A.D. 1987. Apple fruit moth, Argyresthia conjugella Zeller (Lepidoptera, Hyponomeutidae): A dangerous pest of apple trees. Biologicheskii Zhumal Armenii 40(10): 848–853.

Mishra, S.C. & Pandey, N.D. 1966. Observations on the life history of Atteva fabriciella Swed. (Yponomeutidae: Lepidoptera) a pest of Ailanthus excelsa at Kanpur. Indian Forester 92: 461–464.

Mitterberger, K. 1912. Die Arten der Gattung Argyresthia Hb. (Microlepidoptera) um Steyr in Oberösterreich und im angrenzenden Teile von Steiermark. Entomologische Zeitschrift 26: 142–143.

Mohanadas, K. & Verma, R.V. 1984. A new host record for Atteva fabriciella (Lepidoptera: Yponomeutidae) a pest of Ailanthus. Journal of Tree Sciences 3: 128.

Moore, F. 1867. On the Lepidopterous Insects of Bengal. Proceedings of the Scientific Meetings of the Zoological Society of London for the year 1867: 612–687.

——. 1887. The Lepidoptera of Ceylon, Vol. 3. L. Reeve & Co., London. 587 pp.

Moriuti, S. 1961. Japanese species of Saridoscelis Meyrick. Transactions of the Lepidopterological Society of Japan (Tyō To Ga) 11(4): 64–69.

——. 1963a. Studies on the Yponomeutoidea (II). Two yponomeutid genera, Niphonympha and Pseudocalantica, of Japan and Formosa. Kontyû 31(3): 215–223.

——. 1963b. Studies on the Yponomeutoidea (VI). Genus Lycophantis Meyrick. Kontyû 31(4): 261–266.

——. 1964. Yponomeutoiden-studien (IX). Eine neue Argyresthia-art aus Japan. Transactions of the Lepidopterological Society of Japan (Tyō To Ga) 15(1): 20–21.

——. 1965a. Studies on the Yponomeutoidea (XI). Two new species of the Pseudocalantica Friese from Nepal. Kontyû 33(1): 7–10.

——. 1965b. Studies on the Yponomeutoidea (XII). Argyresthia-species attacking Coniferous plants in Japan. Bulletin of the University of Osaka Prefecture, Series B. 16: 65–80.

——. 1968. Studies on Yponomeutoidea (IV). A new Argyresthia from Formosa. Transactions of the Lepidopterological Society of Japan (Tyō To Ga) 19(3–4): 85–86.

——. 1969a. Argyresthiidae of Japan. Bulletin of the University of Osaka Prefecture, Series B. 21: 1–50.

——. 1969b. Argyresthiidae. In: Issiki, S.T. et al. (eds.), Early Stages of Japanese Moths in Color, Vol. II. pp. 125–127. Hoikusha Publishing Co., Ltd., Osaka.

——. 1971a. A revision of the world species of Thecobathra. Kontyû 39(3): 230–251.

——. 1971b. Two new genera and a new species of the Indian Yponomeutidae. Kontyû 39(3): 251–255.

——. 1972. Taxonomic notes on Yponomeuta polystigmellus C. et R. Felder and allied species of temperate East Asia, with the description of a new species. Kontyû 40(3): 139–159.

——. 1973. Toxonomic [sic] notes on the Yponomeutid genera Psychromnestra and Macrophanta. Transactions of the Lepidopterological Society of Japan (Tyō To Ga) 24(2–3): 49–54.

——. 1974a. A new generic synonym and a new species of the genus Comocritis Meyrick. Transactions of the Lepidopterological Society of Japan (Tyō To Ga) 25(4): 105–108.

——. 1974b. A new Formosan Argyresthiid. Transactions of the Lepidopterological Society of Japan (Tyō To Ga) 25(4): 113–114.

——. 1977a. Fauna Japonica: Yponomeutidae s. lat. Keigaku Pub. Co., Tokyo, Japan. 327 pp.

——. 1977b. A review of the new monograph of Japanese Yponomeutidae s. lat.: Japan Heterocerists' Journal 94: 566–569.

——. 1978. Amphitheridae: Four new species from Asia, Telethera blepharacma Meyrick New to Japan and Formosa, and Sphenograptis Meyrick Transferred to the Family. Bulletin of the Osaka Prefecture Series B 30: 1–17.

——. 1981. A new Kessleria from New Guinea. Transactions of the Lepidopterological Society of Japan (Tyō To Ga) 32(1/2): 83–84.

——. 1982a. A new and an unrecorded species of Thecobathra Meyrick from Thailand. Transactions of the Lepidopterological Society of Japan (Tyō To Ga) 33(1/2): 61–64.

——. 1982b. A redescription of Callithrinca Meyrick, with a new species from Thailand. Transactions of the Lepidopterological Society of Japan (Tyō To Ga) 33(1/2): 65–69.

——. 1983. Angoonopteryx siamensis, a new genus and species of Yponomeutidae from Thailand. Bulletin of the University of Osaka Prefecture, Series B 35: 11–15.

Möschler, H.B. 1890. Dia Lepidopteren der insel Portorico. Abhandlungen Senckenbergischen Naturforschenden Gesellschaft 16: 330–346.

Mosher, E. 1916. A classification of the Lepidoptera based on characters of the pupa. Bulletin of the Illinois State Laboratory of Natural History 12: 17–159.

Müller-Rutz, J. 1920. Aus der welt der kleinschmetterlinge mit beschreibungen neuer arten und formen. Mitteilungen der Entomologia Zürich und Umgebung [1920] 5: 334–349.

——. 1934. Ueber Microlepidopteren. Beschreibung neuer Arten und Formen aus der Schweiz, sowie ergänzende Angaben über ältere, weniger bekannte Arten. Mitteilungen der Schweizerischen Entomologischen Gesellschaft 16: 118–128.

Mutanen, M., Wahlberg, N. & Kaila, L. 2010. Comprehensive gene and taxon coverage elucidates radiation patterns in moths and butterflies. Proceedings of the Royal Society B 277: 2839–2848.

Neumoegen, B. 1891. New Rhopalocera and Heterocera. The Canadian Entomologist 23: 122–126.

Nielsen, E.S. 1996. Roeslerstammiidae & Lyonetiidae. In: Nielsen, E.S., Edwards, E.D. & Rangsi, T.V. (eds.), Checklist of the Lepidoptera of Australia. Monographs on Australian Lepidoptera 4: 44 & 57. csiro, Victoria, Australia.

Nielsen, E.S., Edwards, E.D. & Rangsi, T.V. (eds.) 1996. Checklist of the Lepidoptera of Australia. Monographs on Australian Lepidoptera 4. csiro, Victoria, Australia. 529 pp.

Nieukerken, E.J. van Kaila, L., Kitching, I.J., Kristensen, N.P., Lees, D.C., Minet, J., Mitter, C., Mutanen, M., Regier, J.C., Simonsen, T.J., Wahlberg, N., Yen, S.-H., Zahiri, R., Adamski, D., Baixeras, J., Bartsch, D., Bengtsson, B.A., Brown, J.W., Bucheli, S.R., Davis, D.R., De Prins, J., De Prins, W., Epstein, M.E., Gentili-Poole, P., Gielis, C., Hattenschwiler, P., Hausmann, A., Holloway, J.D., Kallies, A., Karsholt, O., Kawahara, A., Koster, J.C., Kozlov, M., Lafontaine, J.D., Lamas, G., Landry, J.-F., Lee, S., Nuss, M., Park, K.-T., Penz, C., Rota, J., Schmidt, B.C., Schintlmeister, A., Sohn, J.-C., Solis, M.A., Tarmann, G.M., Warren, A.D., Weller, S., Yakovlev, R.V., Zolotuhin, V.V. & Zwick, A. 2011. Order Lepidoptera Linnaeus, 1758. In: Zhang, Z.-Q. (ed.), Animal Biodiversity: An Outline of Higher-Level Classification and Survey of Taxonomic Richness. Zootaxa 3148: 212–221.

Nordman, A.F. 1926. Über die Biologie und Verbreitung von Zelleria ribesiella de Joann. Notulae Entomologicae 6(2): 51–55.

Nowicki, M.S. 1860. Enumeratio Lepidopterorum Haliciae Orientalis. Sumptibus Auctoris, Leopoli [=Lviv], Ukraine. 269 pp.

——. 1864. Microlepidopterorum Species Novae – Cracoviae, typis Universitatis Jagellonicae: 1–31(22–24), 1 pl.

Nye, I.W.B. & Fletcher, D.S. 1991. The Generic Names of Moths of the World, Vol. 6. Natural History Museum Publications, London, England. 368 pp.

Ochsenheimer, F. & Treitschke, F. 1830. Die Schmetterlinge von Europa, Vol. 8. G. Fleischer, Leipzig, Germany. 330 pp.

Oku, T. 2003. Microlepidoptera of the Iwate Prefecture. Transactions of the Iwate Entomological Society, Supplement 2: 1–157.

Osthelder, L., 1951. Die Schmetterlinge Südbayerns und der andgrenzenden nördlischen Kalkalpen, II, Die Kleinschmetterlinge, 2. Mitteilungen der Münchner Entomologische Gesellschaft 41: 1–112.

osu (Oregon State University) 2010. Pacific Northwest Nursery ipm. http://oregonstate.edu/dept/nurspest/cherry_ermine_moth_closeup1.htm [accessed: Apr. 2012].

Packard, A.S. Jr. 1871. First Annual Report on the Injurious and Beneficial Insects of Massachusetts. Wright & Potter, Boston, Massachusetts, usa. 31 pp.

Paclt, J. 1951. A propos de la nomenclature de deux genres de Microlépidoptères nuisibles aux essences forestières. Revue française de lépidoptérologie 13(7/8): 127.

Pagenstecher, A. 1884. Beiträge zur Lepidopteren-Fauna von Amboina. (Mit zwei Tafeln). Jahrbücher des Nassauischen Vereins für Naturkunde 37: 150–326.

Parenti, U. 2000. A Guide to the Microlepidoptera of Europe. Museo Regionale di Scienza Naturali, Torino, Italy. 426 pp.

Pathania, P.C. & H.S. Rose, 2003. Comments on the Yponomeuta complex from western Himalaya. Journal of Nature Conservation 15(2): 529–536.

Pelekassis, C.E.D. 1962. A contribution to the study of nomenclature, taxonomy, biology, ecology and the natural parasitization of the Olive Kernel Borer [Prays oleae

(Bernard) Lesne] basic recommendations for a rational and economic control. Annales de l' Institut Phytopathologique Benaki, Nouvelle Série 4(3): 185–308.

Peterson, W. 1932. Die Arten der Gattung Swammerdamia Hb. mit Bemerkungen zur Mutationslehre. Archiv für Naturgeschichte Zeitschrift für Systematische Zoologie (Neue Folge) 1(2): 197–224.

Peyerimhoff, H. de 1877. Diagnoses de Microlépidoptères nouveaux ou peu connus. Petites Nouvelles Entomologiques 2(164): 101–102.

Philpott, A. 1918. Descriptions of new species of Lepidoptera. Transactions and Proceedings of the New Zealand Institute 50(5): 125–132.

——. 1930. New species of Lepidoptera in the collection of the Auckland museum. Records of the Auckland Institute Museum 1(1): 1–16.

——. 1931. Notes and descriptions of New Zealand Lepidoptera. Transactions of the New Zealand Institute 62: 26–36.

Pierce, F.N. & Metcalfe, J.W. 1935. The Genitalia of the Tineid Families of Lepidoptera of the British Islands. An Account of the Morphology of the Male Clasping Organs and the Corresponding Organs of the Female. Oundle, Northants. Warmington, England. 116 + 22 pp., 68 pls.

Pitkin, B., Ellis, W., Plant, C. & Edmunds, R. 2012. The leaf and stem mines of British flies and other insects. http://www.ukflymines.co.uk/index.php [accessed: Dec. 2012].

Pitkin, B. & Jenkins, P. 2004. The Butterflies and Moths of the World, Generic Names and Their Type-Species. http://www.nhm.ac.uk/research-curation/research/projects/butmoth/ [accessed: Feb. 2014].

Poole, R.W. 1989. Lepidopterorum Catalogus (New Series) Fascicle 118, Noctuidae Part 1. E.J. Brill/Flora & Fauna Publications, New York, usa. 500 pp.

Popescu-Gorj, A. 1992. Le catalogue des types de Lépidoptères gardés dans les collections du Muséum d'Histoire Naturelle "Grigore Antipa" (Bucarest) (Fam. Micropterigidae-Pterophoridae). Travaux du Muséum National d'Histoire Naturelle "Grigore Antipa" 32: 131–184.

Povel, G.D.E. 1985. Eumonopyta mintenna spec. nov. and Tienoptila antistatica (Meyrick) from Java. Proceedings of the Koninklijke Nederlandse Akademie van Wetenschappen. Series C. Biological and medical Sciences 88(4): 441–447.

Povolný, D. & Zacha, V. 1990. Blastotere thujella Packard – a new arborvitae pest in Czechoslovakia. Ochrana Rostlin 26(1): 67–71.

Powell, J.A. & Opler, P.A. 2009. Moths of Western North America. University of California Press, Berkeley, Los Angeles and London. 383 pp.

Pozmogova, Z.N. 1969. Buryat miner Friese – Ocnerostoma friesei transbaicalensis Pozmogova, ssp. n. (Lepidoptera, Hyponomeutidae). Trudy Buryatskogo Instituta Estestevennykh Nauk 7: 53–147.

Prentice, R.M. 1965. Forest Lepidoptera of Canada recorded by the Forest Insect Survey, Vol. 4. Microlepidoptera. Department of Forestry Canada Publication 1142. Roger Duhamel, Ottawa, Ontario, Canada. pp. 545–840.

Ratzeburg, J.T.C. 1840. Die Forst.-Insecten oder Abbildung und Beschreibung der in den Wäldern Preufsens und der Nachbarstaaten als schädlich oder nützlich bekannt gewordenen Insecten; in systematischer Folge und mit besonderer Rücksicht auf die Vertilgung der Schädlichen. 2. Nicholaische Buchhandlung, Berlin, Germany. 252 pp. + pls.

Réal, R. 1990. Argyresthia albistria Haw. subsp. tarnis ssp. nova. Bulletin de la Société Entomologique de Mulhouse (1–3): 1–6.

Rebel, H. 1892. Beitrag zur Microlepidopterenfauna des canarischen Archipels. Annalen des K.K. Naturhistorischen Hofmuseums. Wien 7(3): 241–285.

——. 1899. Zweiter Beitrag zur Lepidopteren-Fauna Südtirols. Verhandlungen der k.k. Zoologisch-Botanischen Gesellschaft in Wien 49(3): 158–184.

——. 1901a. Neue Lepidopteren aus Bosnien und der Hercegovina. Verhandlungen der k.k. Zoologisch-Botanischen Gesellschaft in Wien 51(10): 798–804.

——. 1901b. Catalog der Lepidopteren des Palaearctischen Faunengebietes 2 Famil. Pyralidae-Micropterygidae. Friedlander & Sohn, Berlin, Germany. 368 pp.

——. 1910a. Beitrag zur Microlepidopterenfauna der Ionischen Inseln. Verhandlungen der k.k. Zoologisch-Botanischen Gesellschaft in Wien 60: 418–431.

——. 1910b. Neue palaearktische Pyraliden, Tortriciden und Tineen. Deutsche Entomologische Zeitschrift "Iris" herausgegeben vom Entomologischen Verein Iris zu Dresden. 27: 1–14.

——. 1940. Die Lepidopterenfauna des Azorischen Archipels. Societas Scientiarum Fennica. Commentationes Biologicae 8(1): 35–60.

Retzius, A.J. 1783. In: De Geer, Genera et Species Insectorum e generosissimi auctoris scriptis extraxit, digessit, latine quoad partem reddidit, et terminologiam insectorum Linneanam addidit. S.L. Crusium, Lipsiae [=Leipzig, Germany]. 220 pp.

Reutti, C. 1898. Übersicht der Lepidopteren-Fauna des grossherzogtums Baden [und der Anstossenden Länder]. Verhandlungen des Naturwissenschaftlichen Vereins in Karlsruhe 12. Gebrüder Borntraeger, Berlin, Germany. 361 pp.

Riley, C.V. 1891. Tineina. In: Smith, J.B. (ed.), List of the Lepidoptera of Boreal America. pp. 94–115. American Entomological Society, Philadelphia.

Ritsema, C. 1875. Description de deux espèces nouvelles de Microlépidoptères. Petites Nouvelles Entomologiques 7(120): 479.

Roberti, D. 1952. Contributo alla conoscenza dell'Hyponomeuta rorellus [sic] Hb. Bollettino del Laboratorio di Entomologia Agraria "Filippo Silvestri" Portici 11: 220–233.

Robinson, G.S., 2002. Global Taxonomic Database of Tineidae (Lepidoptera). http://www.nhm.ac.uk/research-curation/research/projects/tineidae/ [accessed: Oct. 2012].

——. 2009. Biology, Distribution and Diversity of Tineid Moths. Natural History Museum, London, England. 143 pp.

Robinson, G.S. & Nielsen, E.S. 1983. The Microlepidoptera described by Linnaeus and Clerck. Systematic Entomology 8: 191–242.

——. 1993. Tineid Genera of Australia (Lepidoptera). Monographs on Australian Lepidoptera 2. csiro, Victoria. xvi + 344 pp.

Robinson, G.S., Ackery, P.R., Kitching, I.J., Beccaloni, G.W. & Hernández, L.M. 2010. hosts – A Database of the World's Lepidopteran Hostplants. Natural History Museum, London, England. http://www.nhm.ac.uk/hosts [accessed: 18 Aug. 2010].

Robinson, G.S. & Tuck, K.R. 1996. A revisionary checklist of the Tineidae (Lepidoptera) of the Oriental Region. Occasional Papers on Systematic Entomology, bm 9: 1–29.

Robinson, G.S., Tuck, K.R. & Shaffer, M. 1994. A Field Guide to the Smaller Moths of South-East Asia. Malaysian Nature Society, Kuala Lumpur, Malaysia. 309 pp.

Rosenthal, S.S., Frankie, G.W. & Koehler, C.S. 1969. Biological studies of Argyresthia franciscella and A. cupressella on ornamental Cupressaceae. Annals of the Entomological Society of America 62(1): 109–112.

Salmon, J.T. & Bradley, J.D. 1956. Lepidoptera from the Cape Expedition and Antipodes Islands. Records of the Dominion Museum 3(1): 61–81.

Sarwar, M. 2006. Occurrence of insect pests on Guava (Psidium guajava) tree. Pakistan Journal of Zoology 38(3): 197–200.

Sattler, K. 1967. Microlepidoptera Palearctica 2. Ethmiidae. George Fromme & Co., Vienna, Austria. 185 pp., 9 pls.

——. 1973. A catalogue of the family-group and genus-group names of the Gelechiidae, Holcopogonidae, Lecithoceridae and Symmocidae. Bulletin of the British Museum of Natural History (Entomology) 28(4): 155–282.

Sattler, K. & Tremewan, W.G. 1973. The Entomological Publications of Pierre Millière (1811–1887). Bulletin of the British Museum (Natural History), Historical series 4(4): 221–280.

——. 2009. The authorship of the so-called 'Wiener Verzeichnis'. Nota lepidopterologica 32(1): 3–10.

Sauber, C.J.A. 1902. Familie Tineina. In: Semper, G. (ed.), Die Schmetterlinge der Philippinschen Inseln. Beitrag zur Indo-Malayischen Lepidopteren-Fauna. Die Nachtfelter. Heterocera. 2(6): 696–712.

Savigny, C.-J. 1816. Mémoires sur les animaux sans vertèbres 1: 1–37.

Schrank, F.v.P. 1796 [1802]. Sammlung Naturhistorischer und Physikalischer Aufsätze. Nürnberg, Raspe. 456 pp.

Schreiter, R. 1943. Notas entomo-biológicas y otras. Acta Zoológica Lilloana 1: 7–44 + 26 pls.

Schultze, W. 1908. New and little-known Lepidoptera of the Philippine Islands. Philippine Journal of Science. A. General Science 3: 27–37.

Schütze, K.T. 1931. Die Biologie der Kleinschmetterlinge unter besonderer Berücksichtigung ihrer Nährpflanzen und Erscheinnungszeiten. Verlag des Internationalen Entomologischen Vereins e.V., Frankfurt/Main. 235 pp.

Scopoli, I.A. 1763. Entomologia Carniolica exhibens insecta Carnioliæ indigena et distributa in ordines, genera, species, varietates, Methodo Linnæana. Trattner, Vienna, Austria. 552 pp.

Semper, G. 1902. Die Schmetterlinge der Philippinschen Inseln. Beitrag zur Indo-Malayischen Lepidopteren-Fauna. Die Nachtfalter. Heterocera. 2(6): 625–728.

Servadei, A. 1930. Contributo alla conoscenza delle Hyponomeuta padellus L., cognatellus Hbn. e vigintipunctatus Retz. Bollettino del Laboratorio di Entomologia del R. Istituto Superiore Agrario di Bologna 8: 254–299, pls. 4–8.

Seven, S. & Ozdikmen, H. 2005. Turneriella nom. nov., a replacement name for the genus Amblyzancla Turner, 1939. New Zealand Journal of Zoology 32: 235.

Sharma, J.P., Khajuria, D.R. & Dogra, G.S. 1988. Chemical control of apple-fruit moth, Argyresthia conjugella Zeller (Lepidoptera, Yponomeutidae). Indian Journal of Agricultural Science 58(5): 378–381.

Sharp, D. 1893. Arthropoda. Zoological Record 30. Zoological Society of London. 371 pp.

——. 1914. Arthropoda. Zoological Record (7) 51. Zoological Society of London. 342 pp.

Silva, L., Markin, G. & Tavares, J. 1995. Argyresthia atlanticella Rebel (Insecta: Lepidoptera) an excluded agent for Myrica faya Aiton (Myricaceae) biocontrol. Arquipélago, Life and Marine Sciences 13A: 105–113.

Silver, G.T. 1951. The Arborvitae leaf-miners. Bi-monthly Program Report, Division of Forest Biology, Ottawa 7(5): 1.

——. 1957. Separation of the species of Arborvitae leaf miners in New Brunswick. The Canadian Entomologist 89(3): 97–107

Silvestri, F. 1915. Contributo alla conoscenza degli insetti dell'olivo dell' Eritrea e dell' Africa meridionale. Bollettino del Laboratoria di Zoologia Generale e Agraria della R. Scuola Superiore d'Agricoltura in Portici 9: 240–267.

Sinev, S.Y. 2008. Catalogue of the Lepidoptera of Russia. kmk Scientific Press Ltd., St. Petersburg & Moscow, Russia. 424 pp.

Smith, J.B. 1891. List of the Lepidoptera of Boreal America. American Entomological Society, Philadelphia. Pennsylvania, usa. 136 pp.

Snellen, P.C.T. 1877. Heterocera op Java verzameld door Mr. M.C. Piepers, met Aanteekeningen en Beschrijvingen der nieuwe soorten. Tijdschrift voor Entomologie 20: 1–50.

——. 1878. Nieuwe exotische Tineinen van 'S Rijks Museum van Natuurlijke Historie te Leiden. Tijdschrift voor Entomologie 21: 129–147.

——. 1901. Beschrijvingen van nieuwe exotische Tortricinen, Tineinen en Pterophorinen benevens aanteekeningen over reeds bekend gemaakte soorten. Tijdschrift voor Entomologie 44: 67–98.

Sodoffsky, W. 1837. Etymologische Untersuchungen über die Gattungsnamen der Schmetterlinge. Bulletin de la Société Impériale des naturalistes de Moscou 10(6): 76–97.

Sohn, J.-C. 2007. Faunistic contribution to the Korean Microlepidoptera and Pyralids (1) – 40 species new to Korea. Tinea 20(1): 12–27.

——. 2012. A new genus and species of Praydidae from Vietnam. Tinea 22(2): 120–124.

Sohn, J.-C. & Lewis, J.A. 2012. Catalogue of the type specimens of Yponomeutoidea in the collection of the United States National Museum of Natural History. Zootaxa 3573: 1–17.

Sohn, J.-C. & Nishida, K. 2011. A taxonomic review of Eucalantica Busck with descriptions of six new species. Zookeys 118: 75–96.

Sohn, J.-C. & Park, K.-T. 2010. Three new species of Lycophantis Meyrick from Vietnam, with a key to the world species. Oriental Insects 44: 225–233.

Sohn, J.-C. & Peralta, M.P. 2014. A new species of Atemelia feeding on the ornamental shrub Mahonia (Ranunculales: Berberidaceae) in Chile. Annals of the Entomological Society of America 107(2): 339–346.

Sohn, J.-C., Regier, J.C., Mitter, C., Davis, D., Landry, J.-F., Zwick, A. & Cummings, M.P. 2013. A molecular phylogeny for Yponomeutoidea (Insecta, Lepidoptera, Ditrysia) and its implications for classification, biogeography and the evolution of host plant use. PLoS ONE [Public Library of Science] 8(1): e55066. DOI: 10.1371/journal.pone.0055066 (in press).

Sohn, J.-C. & Wu, C.-S. 2011. A taxonomic review of Prays Hübner, 1825 from China with descriptions of two new species. Tijdschrift voor Entomologie 154: 25–32.

——. 2013. A taxonomic review of Attevidae from China with descriptions of two new species and a revised identity of the Ailanthus webworm moth, Atteva fabriciella, from the Asian tropics. Journal of Insect Science 13(66): 1–16.

Sohn, J.-C., Wu, C.-S. & Han, H.-L. 2010. Three new species of Yponomeutinae from China with faunistic supplements and an updated list of the Chinese species of the subfamily. Journal of Natural History 44(45/46): 2803–2816.

Spuler, A. 1910. Die Schmetterlinge Europas 2. E. Schweizerbart, Stuttgart, Germany. 523 pp.

Stainton, H.T. 1849a. A Monograph on the European Species of the Genus Argyresthia. The Zoologist: a popular miscellany of Natural History 7. Appendix to the Zoologist for 1849: I–XXIV.

——. 1849b. Description of a new British Species of Argyresthia. The Zoologist: A popular miscellany of Natural History. Appendix to the Zoologist for 1849: XL.

——. 1849c. An attempt at a systematic catalogue of the British Tineidae & Pterophoridae. Systematic Catalogue of the British Tineidae & Pterophoridae 8: 1–32.

——. 1851. A Supplementary Catalogue of the British Tineidæ & Pterophoridæ. [Including Appendix: A Catalogue of the Tineidæ obtained from Herr Joseph Mann, of Vienna, in 1849.]. John van Voorst, London, England. 28 pp.

——. 1854. Insecta Britannica. Lepidoptera: Tineina, Vol. 3. Reeve and Benham, London, England. 331 pp.

——. 1858. Communications – Lepidoptera. The Entomologist's Weekly Intelligencer 4(92): 115–119.

——. 1859. Lepidoptera. New British species in 1858. The Entomologist's Annual for 1859: 145–157.

——. 1860. Descriptions of South African Tineina collected by R. Trimen, Esq., in 1858–9. The Transactions of the Entomological Society of London New Series 5(6): 220–223.

——. 1868. A few observations on the synonymy of Tinea (?) alpicella, and Zelleria saxifragæ (n. sp.). The Transactions of the Entomological Society of London 1868(1): 137–139.

——. 1869. Entomological Society of London, 7th June, 1869. The Entomologist's Monthly Magazine 6: 42.

——. 1870. General observations on the genus Swammerdamia. The Natural History of the Tineina 11: 40–87.

——. 1871. Lepidoptera. New British Tineina in 1870. The Entomologist's Annual 1871: 96–100.

——. 1876. On a new species of the genus Zelleria. The Entomologist's Monthly Magazine 13: 1–2.

Standfuss, M. 1851. Lepidopterologische Beiträge zur Kenntniß der Seefelder und ihrer Umgebung bei Reinerz (Schluß). Zeitschrift für Entomologie Herausgegeben von dem Verein für schlesische Insekten-kunde zu Breslau 17(1): 59–60.

Staudinger, O. 1870 [1871]. Beschreibung neuer Lepidopteren des europäischen Faunengebiets. Berliner Entomologische Zeitschrift. Herausgegeben von dem Entomologischen Vereine in Berlin 14(3/4): 273–330.

——. 1877. Neue Lepidopteren des europäischen Faunengebiets aus meiner Sammlung. Entomologische Zeitung. Herausgegeben von dem entomologischen Vereine zu Stettin 38: 175–208.

——. 1880. Lepidopteren-Fauna Kleinasien's. Horae Societatis Entomologicae Rossicae 15: 159–435.

Staudinger, O. & Rebel, H. 1901. Catalog der Lepidopteren des Palararctischen Faunengebietes. Vol. 3. R. Friedländer & Sohn, Berlin, 368 pp.

Staudinger, O. & Wocke, M. 1871. Catalog der Lepidopteren des Europæischen Faunengebietes. Hermann Burdach, Dresden, Germany. 473 pp.

Stephens, J.F. 1829. A Systematic Catalogue of British Insects: Being an attempt to arrange all the hitherto discovered indigenous insects in accordance with their natural affinities, containing also the references to every English writer on Entomology, and to the principal foreign authors. Insecta Haustellata. Ordo I: (8). Lepidoptera. Baldwin & Cradock, London, England. 388 pp.

——. 1834. Illustrations of British Entomology; or, a Synopsis of Indigenous Insects: containing their generic and specific distinctions; with an account of their metamorphoses, times of appearance, localities, food, and economy, as far as practicable. Haustellata 4. Baldwin & Craddock, London, England. 433 pp.

Steuer, H. 1993. Paraswammerdamia ruthiella nov. spec., eine neue Yponomeutide aus dem Schwarzatal bei Bad Blankenburg (Thüringen). Rudolstädter Naturhistorische Schriften 5: 15–16.

Stoll, C. 1782. In: Cramer, P. (ed.), De Uitlandsche Kapellen voorkomende in de drie Waereld-deelen Asia, Africa en America 4: Papillons exotiques des trois parties du monde l'Asie, l'Afrique et l'Amérique. 4. S.J. Baalde, Amsterdam, The Netherlands. 252 pp.

Strand, E. 1901. Beitrag zur Schmetterlingsfauna Norwegens. Nyt Magazin for Naturvidenskaberne 39(1): 25–72.

——. 1920. Beiträge zur Lepidopterenfauna Norwegens und Deutschlands. Archiv für Naturgeschichte (Abteilung A) 85(4): 1–82.

Stretch, R.H. 1873. Illustrations of the Zygaenidae & Bombycidae of North America. San Francisco, California, usa. 242 pp.

Stringer, H. 1930. New species of Microlepidoptera in the collection of the British museum. Annals and Magazine of Natural History, Series 10(6): 415–422.

Sugi, S. & Tominaga, S. 2002. Larva of Teinoptilia guttella Moriuti feeding on Maytenus diversifolia (Celastraceae) in Okinawa Island. Japan Heterocerists' Journal 220: 375–376.

Sullivan, J.J., Burrows, C.J. & Dugdale, J.S. 1995. Insect predation of seeds of native New Zealand woody plants in some central South Island localities. New Zealand Journal of Botany 33: 355–364.

Surhone, L.M., Tennoe, M.T. & Henssonow, S.F. 2010. Prays Citri. Betascript Publishing, Forlag. 144 pp.

Svensson, I. 1966. New and confused species of Microlepidoptera. Opuscula Entomologica 31(3): 183–202.

Swainson, W. 1851. Review. Australian Lepidoptera and their Transformations, drawn from the Life, by Harriet and Helena Scott; with Descriptions General and Systematic, by A.W. Scott, m.a. The Sydney Morning Herald, 30 August 1851 31(4459): 2.

Swederus, N.S. 1787. Fortsattning af Beskrisningen på 50 nya Species af Insecter. Kongliga Vetenskaps Academiens Nya Handlingar 8: 276–290.

Swinhoe, C., Walsingham, T.L. & Durrant, J.H. 1900. Catalogue of Eastern and Australian Lepidoptera Heterocera in the collection of the Oxford University Museum 2. Clarendon Press, Oxford, England. 630 pp.

Tams, W.H.T. 1935. Résultats Scientifiques du Voyage aux Indes Orientales Néerlandaises de LL. AA. RR. le Prince et la Princesse Léopold de Belgique. Lepidoptera II 2. Heterocera. Mémoires du Musée Royal d'Histoire naturelle de Belgique 4(12): 31–64.

Tengström, J.M.J. 1848. Bidrag till Finlands Fjäril-Fauna. Notiser ur Sällskapets Pro Fauna et Flora Fennica Förhandlingar 1: 69–164.

——. 1869. Catalogus Lepidopterorum Faunæ Fennicæ præcursorius. Notiser ur Sällskapets Pro Fauna et Flora Fennica Foerhandlingar 10: 287–370.

Tenow, O., Nilssen, A.C., Holmgren, B. & Elverum, F. 1999. An insect (Argyresthia retinella, Lep., Yponomeutidae) outbreak in northern birch forests, released by climatic changes? The Journal of Applied Ecology 36(1): 111–122.

Thunberg, C.P. 1788. Præside C.P. Thunberg, Publico examini subjicit, Carolus Gustavus Schalén. D. [Docente] D. [Deo] Museum Naturalium Academiæ Upsaliensis 6: 69–84.

——. 1794. D.D. Dissertatio entomologica sistens Insecta Suecica. Præside C.P. Thunberg, Publico examini subjicit Gustavus Magnus Wenner. D. [Docente] D. [Deo] Museum Naturalium Academiæ Upsaliensis 7: 83–98 + pl.

Toll, S.G. 1941. Die Genitalien der europäischen "Hyponomeuta." Zeitschrift des Weiner Entomologen-Vereines 26: 170–176.

——. 1947(1948). Beitrag zur Microlepidopterenfauna von Nordost-Persien. Zeitschrift des Wiener Entomologischen Gesellschaft 32: 107–116.

Traugott-Olsen, E. & Nielsen, E.S. 1977. The Elachistidae of Fennoscandia and Denmark. Scandinavian Science Press Ltd., Klampenborg, Denmark. 299 pp.

Treitschke, F. 1830. Die Schmetterlinge von Europa 8. G. Fleischer, Leipzig. 312 pp.

——. 1832. Die Schmetterlinge von Europa 9(1). G. Fleischer, Leipzig. 272 pp.

——. 1833. Die Schmetterlinge von Europa 9(2). G. Fleischer, Leipzig. 284 pp.

Tremewan, W.G. 1977. The publications on Lepidoptera by O.G. and A. Costa and the nominal taxa described therein. Bulletin of the British Museum (Natural History) Historical Series 5(3): 213–232.

Turati, E. 1924. Spedizione Lepidoptterologica in Cirenaica 1921–1922. Atti della Società Italiana di Scienze Naturali e del Museo Civico di storia naturale in Milano 63: 21–191.

——. 1930. Novità di Lepidotterologia in Cirenaica III. Atti della Società Italiana di Scienze Naturali e del Museo Civico di storia naturale in Milano 69: 46–92.

Turner, A.J. 1898. Descriptions of new Micro-Lepidoptera from Queensland. Transactions of the Royal Society of South Australia 22: 200–214.

——. 1903. Revision of Australian Lepidoptera. Proceedings of the Linnean Society of New South Wales 28: 42–92.

——. 1913. Studies in Australian Microlepidoptera. Proceedings of the Linnean Society of New South Wales 38: 174–228.

——. 1917. Lepidopterological Gleanings. Proceedings of the Royal Society of Queensland 29: 70–106.

——. 1923. New Australian Micro-Lepidoptera. Transactions of the Royal Society of South Australia 47: 165–194.

——. 1926a. New and little-known Tasmanian Lepidoptera. Papers and Proceedings of the Royal Society of Tasmania 1925: 81–117.

——. 1926b. Studies in Australian Lepidoptera. Transactions of the Royal Society of South Australia 50: 120–155.

——. 1927a. New and little-known Tasmanian Lepidoptera II. Papers and Proceedings of the Royal Society of Tasmania 1926: 119–164.

——. 1927b. A revision of the Lepidoptera of Tasmania. II. Papers and Proceedings of the Royal Society of Tasmania 1927: 29–65.

——. 1939. A Second Revision of the Lepidoptera of Tasmania. Papers and Proceedings of the Royal Society of Tasmania 1938: 57–116.

——. 1942. Fragmenta Lepidopterologica. Proceedings of the Royal Society of Queensland 53: 61–96.

Turner, H., Lieshout, N., van Ginkel, W.E. & Menken, S.B.J. 2010. Molecular phylogeny of the Small Ermine Moth genus Yponomeuta (Lepidoptera, Yponomeutidae) in the Palaearctic. PLoS ONE 5(3): e9933.

Tutt, J.W. 1896. Scientific notes and observations. Entomologist's Record and Journal of Variation 7(8): 197–201.

Uffeln, K. 1930. Die sogenannten "Kleinschmetterlinge" Westfalens. Sonderabdruck aus den "Abhandlungen aus dem Westfälischen Provinzialmuseum für Naturkunde" 1: 1–98.

Ulenberg, S.A. 2009. Phylogeny of the Yponomeuta species (Lepidoptera, Yponomeutidae) and the history of their host plant associations. Tijdschrift voor Entomologie 152: 187–207.

Unruh, T.R., Congdon, B.D. & Lagasa, E. 1993. Yponomeuta malinellus Zeller (Lepidoptera: Yponomeutidae), a new immigrant pest of apples in the Northwest: Phenology and distribution expansion, with notes on efficacy of natural enemies. Pan-Pacific Entomologist 69(1): 57–70.

van Deurs, W. 1965. Nye Sommerfugle (Lepidoptera) 1964. Entomologische Meddelelser 34: 154.

Vári, L. & Kroon, D.M. 1986. Southern African Lepidoptera. A series of cross-referenced indices. Lepidopterist's Society of Southern Africa & The Transvaal Museum, Pretoria, South Africa. 198 pp.

Vári, L., Kroon, D.M. & Krüger, M. 2002. Classification and Checklist of the Species of Lepidoptera Recorded in Southern Africa. Simple Solutions Australia (Pty) Ltd. Chatswood, Australia. 385 pp.

Verma, R.V. 1992. Impact of Atteva fabriciella (Lepidoptera: Yponomeutidae) feeding on seed production in Ailanthus triphysa. Indian Journal of Forestry 15: 326–328.

Viette, P.E.L. 1949. Catalogue of the Heterocerous Lepidoptera from French Oceania. Pacific Science 3(4): 315–334.

——. 1954. Nouveaux Tineoidea (s.l.) de Madagascar. Annales de la Société Entomologique de France 123: 75–114.

——. 1957. Lepidopteres (Excepte les Tordeuses et les Geometrides). Mémoires de l'Institut scientifique de Madagascar. Série E. 8: 139–226.

——. 1967. Microlépidoptères de Madagascar inédits ou peu connus. Bulletin de la Société entomologique de France 72: 291–296.

Vives Moreno, A. 1992. Catalogo Sistematico y Sinonimico de los Lepidopteros de la Peninsula Iberica y Baleares. Ministerio de Agricultura, Pesca y Alimentación, Madrid, Spain. 378 pp.

——. 2001. Contribución al conocimiento de los microlepidópteros de España, con la descriptción de ocho nuevas especies para la Ciencia. shilap Revista de Lepidopterologia 29(114): 165–178.

Vollenhoven, S.C.S. van 1863. Description de quelques nouvelles espèces de Lépidoptères des Indes Orientales. Tijdschrift voor Entomologie 6: 133–144.

Wakely, S. 1959. Notes on Hyponomeuta irrorella Hb. Entomologist's Record and Journal of Variation 71(2): 61–63.

Walker, F. 1854. List of the Specimens of Lepidopterous Insects in the Collection of the British Museum. Part II – Lepidoptera Heterocera. The Trustees, London. pp. 279–581.

——. 1856. List of the Specimens of Lepidopterous Insects in the Collection of the British Museum. Part VII – Lepidoptera Heterocera. The Trustees, London. pp. 1508–1808.

——. 1863. List of the Specimens of Lepidopterous Insects in the Collection of the British Museum. Part XXVIII. Tortricites & Tineites. The Trustees, London. pp. 287–561.

——. 1864a. List of the Specimens of Lepidopterous Insects in the Collection of the British Museum. Part XXIX. Tineites. The Trustees, London. pp. 563–835.

——. 1864b. List of the Specimens of Lepidopterous Insects in the Collection of the British Museum. Part XXX. Tineites. The Trustees, London. pp. 837–1096.

——. 1864c [1865]. List of the Specimens of Lepidopterous Insects in the Collection of the British Museum. Part XXXI. Supplement. The Trustees, London, England. 321 pp.

——. 1865. List of the Specimens of Lepidopterous Insects in the Collection of the British Museum. Part XXXII. Supplement, Part 2. The Trustees, London. pp. 324–706.

——. 1866a. List of the Specimens of Lepidopterous Insects in the Collection of the British Museum. Part XXXIV. Supplement Part 4. The Trustees, London. pp. 1121–1533.

——. 1866b. List of the Specimens of Lepidopterous Insects in the Collection of the British Museum. Part XXXV. Supplement Part 5. The Trustees, London. pp. 1535–2040.

——. 1869. Characters of undescribed Lepidoptera Heterocera. E.W. Janson. London, England. 112 pp.

Wallengren H.D.J. [1861]. Lepidoptera. In: Virgin, C.A. (ed.), Kongliga Svenska Fregatten Eugenies Resa Omkring Jorden. Zoologie 1, Insecta, pp. 357–388, pls 6–7. P.A. Nordtedt & Söner, Stockholm, Sweden.

Walsingham, T.L. 1880. On some new and little known Species of Tineidae. Proceedings of the Scientific Meetings of the Zoological Society of London for the year 1880: 77–93.

——. 1881a. On the Tortricidæ, Tineidæ, and Pterophoridæ of South Africa. Transactions of the Entomological Society of London for the year 1881(2): 219–288.

——. 1881b. On some North-American Tineidæ. Proceedings of the Scientific Meetings of the Zoological Society of London for the year 1881(2): 301–324.

——. 1882. Notes on Tineidae of North America. Transactions of the American Entomological Society, and Proceedings of the Entomological Section of the Academy of Natural Sciences 10(3): 165–204.

——. 1887. Family Tineidae. In: Moore, F. (ed.), The Lepidoptera of Ceylon, Vol. 3. pp. 498–507. Reeve & Co., London.

——. 1891a. Notes on the genus Argyresthia Hb., with descriptions of new species. Insect Life 3(3): 117–120.

——. 1891b. III. African Microlepidoptera. Transactions of the Entomological Society of London for the year 1891: 63–132.

——. 1891c [1892]. On the Micro-Lepidoptera of the West Indies. Proceedings of the general meetings for scientific business of the Zoological Society of London 1891(33–37): 492–548.

——. 1894a. Micro-Lepidoptera from Norfolk & Scotland: Including an addition to the British list. The Entomologist's Monthly Magazine. Second Series 5(30): 50–52.

——. 1894b. Catalogue of the Pterophoridæ, Tortricidæ, and Tineidæ of the Madeira Islands, with notes and descriptions of new species. Transactions of the Entomological Society of London 1894(4): 535–555.

——. 1897a. Revision of the West-Indian Micro-Lepidoptera, with descriptions of new species. Proceedings of the general meetings for scientific business of the Zoological Society of London 1897(4–12): 54–182.

——. 1897b. Western Equatorial African Microlepidoptera. Transactions of the Entomological Society of London 1897(1): 33–67.

——. 1900. Walsingham. In: Swinhoe, C., Walsingham, T.L. & Durrant, J.H. 1900.

——. 1907a. Descriptions of new North American Tineid moths, with a generic table of the family Blastobasidæ. Proceedings of the United States National Museum 33(1567): 197–228.

——. 1907b. Microlepidoptera of Tenerife. Proceedings of the general meetings for scientific business of the Zoological Society of London 1907(31–68): 911–1034.

——. 1907c. Microlepidoptera. In: Sharp, D. (ed.), Fauna Hawaiiensis or the Zoology of the Sandwich (Hawaiian) Isles. Fauna Hawaiiensis 1(5): 469–759, pls. 10–25.

——. 1911. Algerian Microlepidoptera. Entomologist's Monthly Magazine 47: 12–15.

——. 1914. Biologia Centrali-Americana. Insecta. Lepidoptera-Heterocera. 4. Tineina, Pterophorina, Orneodina, and Pyralidina and Hepialina (part.). R.H. Porter. London, England. 482 pp.

Walsingham, T.L. & Durrant, J.H. 1909. Revision of the Nomenclature of Microlepidoptera. The Entomologist's Monthly Magazine. Second Series 20(45): 155–157.

Wang, S. 2013. In: Jin, Q., Fan, X. & Wang, S. 2013.

Warren, W. 1888. On Lepidoptera collected by Major Yerbury in Western India in 1886 and 1887. Proceedings of the general meetings for scientific business of the Zoological Society of London 1888(21–23): 292–339.

Weber, P. 1945. Die Schmetterlinge der Schweiz 7. Nachtrag, Microlepidopteren. Mitteilungen der Schweizerischen Entomologischen Gesellschaft 19: 347–406.

Werner, K. 1958. Die Larvalsystematik einiger Kleinschmetterlingsfamilien (Hyponomeutidae, Orthoteliidae, Acrolepiidae, Tineidae, Incurvariidae und Adelidae). Abhandlungen zur Larvalsystematik der Insekten, Nr. 2. Akademie-Verlag, Berlin, Germany. 145 pp.

Westwood, J.O. 1840. Synopsis of the genera of British insects; in An introduction to the modern classification of insects; founded on the natural habits and corresponding organisation of the different families 2. Longman, Orme, Brown, Green, and Longmans. London, England. 154 pp.

White, M.J. 2000. Some moths new to Monmouthshire. Entomologist's Record and Journal of Variation 112: 41–42.

Wilson, J.J., Landry, J.-F., Janzen, D.H., Hallwachs, W., Nazari, V., Hajibabaei, M. & Hebert, P.D.N. 2010. Identity of the ailanthus webworm moth, a complex of two species: evidence from dna barcoding, morphology and ecology. ZooKeys 46: 41–60.

Wytsman, P. 1914. Genera Insectorum, Fascicules 164–169. V. Verteneuil & L. Desmet, Bruxelles. 556 pp. + pls. (each Fascicule beginning with page numbered 1).

——. 1925. Genera Insectorum, Fascicules 184–185. V. Verteneuil & L. Desmet, Bruxelles. 634 pp. + pls. (each Fascicule beginning with page numbered 1).

Yang, J.K. 1979. Moths of North China. 1. Harbin (China): Northeast Agricultural University Press. 299 pp.

Yano, K. 1964. Insects of Campbell Island. Lepidoptera: Tineidae, Elachistidae, Hyponomeutidae, Oecophoridae, Tortricidae, Pterophoridae, Noctuidae. Pacific Insects Monograph 7: 255–259.

Yu, H.L. & Li, H.H. 2001. A taxonomic study on the genus Thecobathra Meyrick, 1922 from China. shilap Revista de Lepidopterologia 29(114): 115–120.

——. 2002. Two new species of Niphonympha Meyrick from China. Reichenbachia 34(45): 375–380.

——. 2004. Three new species of Prays Hübner, 1825 from China. shilap Revista de Lepidopterologia 32(125): 13–18.

Zagulajev, A.K. 1969. New species of the genus Yponomeuta Latr. from the Far East. Entomologicheskoye Obozrenie 48(1): 193–198.

Zaki, F.A. & Masoodi, M.A. 1997. Biology of Willow Small Ermine moth, Yponomeuta rorellus Hubner (Lepidoptera) on willows in Ladakh. Journal of Insect Science 10(2): 146–149.

Zeller, P.C. 1838. Kritische Bestimmung der in Reaumur's Mémoires pour servir à l'histoire des insectes vortommenden Lepidopteren. Isis von Oken 31(9/10): 626–736.

——. 1839. Versuch einer naturgemässen Eintheilung der Schaben. Isis von Oken 32(3): 167–220.

——. 1844. Monographie des Genus Hyponomeuta. Isis von Oken 37(3): 198–238.

——. 1846. In: Schlaeger, F. 1842–1847. Berichte des Lepidopterologischen Tauschvereins in Jena über die jahre 1842–1847. p. 165. Jena.

——. 1847a. Bemerkungen über die aufeiner Reise nach Italien und Sicilien gesammelten Schmetterlingsarten. Isis von Oken 40(11): 801–859.

——. 1847b. Die Argyresthien. Linnaea Entomologica 2: 234–302.

——. 1852 [1854]. Microptera, quæ J.A. Wahlberg in Caffrorum terra collegit, descripsit P.C. Zeller. Kongliga Vetenskapsakademiens Handlingar 1852: 1–120.

——. 1863. Zwölf amerikanische Nachtfalter. Entomologische Zeitung. Herausgegeben von dem entomologischen Vereine zu Stettin 24(4–6): 136–155.

——. 1868. Beiträge zur Kenntniss der Lepidopteren-Fauna der Umgegend von Raibl in Oberkärnthen und Preth im angrenzenden Küstengebiet. Verhandlungen der kaiserlich-königlichen Zoologisch-Botanischen Gesellschaft in Wien 18: 563–628.

——. 1872. Beiträge zur Kenntniss der nordamerikanischen Nachtfalter, besonders der Microlepidopteren. Erste Abtheilung. Verhandlungen der kaiserlich-königlichen Zoologisch-Botanischen Gesellschaft in Wien 22: 447–566.

——. 1873. Beiträge zur Kenntniss der nordamericanischen Nachtfalter, besonders der Microlepidopteren. Verhandlungen der kaiserlich-königlichen Zoologisch-Botanischen Gesellschaft in Wien 23(3): 201–334.

——. 1877. Exotische Microlepidoptera. Horae Societatis Entomologicae Rossicae 13: 3–493.

Zetterstedt, J.W. 1838–1840. Insecta Lapponica descripta. Voss, Leipzig. vi + 1139 pp.

Zimmerman, E.C. 1978. Insects of Hawaii, Vol. 9. Microlepidoptera. University of Hawaii Press, Honolulu. xx + 1903 pp.

Animal Genus and Species Names Index

www.ingramcontent.com/pod-product-compliance
Lightning Source LLC
Chambersburg PA
CBHW071332230926
51849CB00002B/376
* 9 7 8 9 0 0 4 2 5 4 7 4 9 *